AAPS Advances in the Pharmaceutical Sciences Series

Series Editors

Daan J.A. Crommelin
Robert A. Lipper

For further volumes:
http://www.springer.com/series/8825

Markus Müller
Editor

Microdialysis in Drug Development

aapspress

Springer

Editor
Markus Müller
Department of Clinical Pharmacology
Allgemeines Krankenhaus
Medical University Vienna
Vienna
Austria

ISSN 2210-7371 ISSN 2210-738X (electronic)
ISBN 978-1-4614-4814-3 ISBN 978-1-4614-4815-0 (eBook)
DOI 10.1007/978-1-4614-4815-0
Springer New York Heidelberg Dordrecht London

Library of Congress Control Number: 2012947069

Printed on acid-free paper

Springer is part of Springer Science+Business Media (www.springer.com)

Preface

Following a number of international scientific meetings on the appropriate use of microdialysis, a joint FDA-AAPS workshop on microdialysis took place in Nashville, TN in 2005 where principles, application, and regulatory perspectives were discussed and reported. The present book, which was conceived during a follow-up workshop at the AAPS annual meeting 2010 in New Orleans, contains a number of up-to-date and state-of-the-art chapters on current microdialysis technology by international experts. I sincerely thank all the authors for their efforts and hope that this guide will provide useful information for colleagues in academia and industry who are interested in the technique of microdialysis.

Vienna, May 2012

Contents

Contributors

Aline Barth Department of Pharmaceutics, University of Florida, Gainesville, FL, USA

Edward D. Bashaw Office of Clinical Pharmacology, Center for Drug Evaluations and Research, United States Food and Drug Administration, Silver Spring, MD, USA

Eva Benfeldt Department of Dermatology, Faculty of Health Sciences, Roskilde Hospital, University of Copenhagen, Roskilde, Denmark

Thomas Birngruber Institute for Biomedicine and Health Sciences, Johanneum Research, Graz, Austria

Manfred Bodenlenz Institute for Biomedicine and Health Sciences, Johanneum Research, Graz, Austria

Martin Brunner Department of Clinical Pharmacology, Medical University of Vienna, Austria

Martin K. Church Charité Universitätsmedizin, Berlin, Germany

John R. Cirrito Department of Neurology, Hope Center for Neurological Disorders and the Knight Alzheimer's Disease Research Center, Washington University School of Medicine, St. Louis, MS, USA

Geraldine F. Clough Southampton General Hospital, Institute of Developmental Sciences, University of Southampton, Southampton, UK

Justin C. Cooley Ralph N. Adams Institute for Bioanalytical Chemistry and Department of Chemistry, University of Kansas, Lawrence, KS, USA

William Couet Laboratoire de Pharmacocinétique, School of Medicine and Pharmacy, Pôle Biologie Santé (PBS), Médecine-Sud, Poitiers Cedex, France

Hartmut Derendorf Department of Pharmaceutics, University of Florida, Gainesville, FL, USA

Michael W. Ducey Ralph N. Adams Institute for Bioanalytical Chemistry and Department of Chemistry, Missouri Western State University, St. Joseph, MO, USA

Margareta Hammarlund-Udenaes Division of Pharmacokinetics and Drug Therapy, Faculty of Pharmacy, Department of Pharmaceutical Biosciences, Uppsala University, Uppsala, Sweden

Christian Höfferer Institute for Biomedicine and Health Sciences, Johanneum Research, Graz, Austria

Per-Anders Jansson The Wallenberg Laboratory, Department of Molecular and Clinical Medicine, Sahlgrenska University Hospital, Institute of Medicine, Sahlgrenska Academy at University of Gothenburg, Gothenburg, Sweden

Jing Li Office of Generic Drugs, Center for Drug Evaluations and Research, United States Food and Drug Administration, Rockville, MD, USA

Robert Lionberger Office of Generic Drugs, Center for Drug Evaluations and Research, United States Food and Drug Administration, Rockville, MD, USA

Craig E. Lunte Southampton General Hospital, Institute of Developmental Sciences, University of Southampton, Southampton, UK

Susan M. Lunte Department of Pharmaceutical Chemistry and Ralph N. Adams Institute for Bioanalytical Chemistry and Department of Chemistry, University of Kansas, Lawrence, KS, USA

Nanda K. Mandava Division of Pharmaceutical Sciences, University of Missouri-Kansas City School of Pharmacy, Kansas, MO, USA

Sandrine Marchand Laboratoire de Pharmacocinétique, School of Medicine and Pharmacy, Pôle Biologie Santé (PBS), Médecine-Sud, Poitiers Cedex, France

Selma Mautner Institute for Biomedicine and Health Sciences, Johanneum Research, Graz, Austria

Ashim K. Mitra Division of Pharmaceutical Sciences, University of Missouri-Kansas City School of Pharmacy, Kansas, MO, USA

Markus Müller Department of Clinical Pharmacology, Medical University of Vienna, Austria

Pradyot Nandi Department of Pharmaceutical Chemistry and Ralph N. Adams Institute for Bioanalytical Chemistry and Department of Chemistry, University of Kansas, Lawrence, KS, USA

Mitesh R. Patel Division of Pharmaceutical Sciences, University of Missouri-Kansas City School of Pharmacy, Kansas, MO, USA

Thomas Pieber Institute for Biomedicine and Health Sciences, Johanneum Research, Graz, Austria

Anne R. Regel Ralph N. Adams Institute for Bioanalytical Chemistry and Department of Chemistry, University of Kansas, Lawrence, KS, USA

Robert Sauermann Department of Clinical Pharmacology, University of Vienna, Vienna, Austria

Chinmay Shukla Office of Clinical Pharmacology, Center for Drug Evaluations and Research, United States Food and Drug Administration, Silver Spring, MD, USA

Frank Sinner Institute for Biomedicine and Health Sciences, Johanneum Research, Graz, Austria

Julie A. Stenken Department of Chemistry and Biochemistry, University of Arkansas, Fayetteville, AR, USA

Katrin Tiffner Institute for Biomedicine and Health Sciences, Johanneum Research, Graz, Austria

Markus Zeitlinger Department of Clinical Pharmacology, Medical University of Vienna, Austria

Part I
General Aspects

Chapter 1
Introduction to the Microdialysis Technology

Markus Müller

Abstract Microdialysis (μD) has been established as a reliable in vivo tool to measure interstitial drug concentrations quasi online in virtually every given tissue and organ in animals and humans. Following its first description in the 1970s μD has been further developed and employed as a research and drug development tool in >14,000 PubMed publications. The rationale for the application of μD in Pharma R&D is based on the presence of blood to tissue barriers which hamper drug penetration to target sites. A clear understanding of target site pharmacokinetics (PK), as provided by μD and complementary techniques is a prerequisite for understanding drug action in vivo.

1.1 Access to the Tissue Compartment

It has long been a key objective of PK to link the drug load at the site of drug action to drug effects (Eichler and Müller 1998). Traditionally, this has been accomplished by modeling blood-derived data, mostly because a direct quantification of tissue drug concentrations was precluded for technical and ethical reasons. In the last decade a number of technologies have matured which today offer a solution for this problem by providing direct access to the tissue compartment (Müller et al. 2004). These comprise non-invasive imaging technologies (Enblad et al. 1996; Fischman et al. 2002; Brunner et al. 2004; Müller et al. 2004; Langer et al. 2005; Shingaki et al. 2012, see also Chap. 14) and invasive technologies

M. Müller (✉)
Department of Clinical Pharmacology, Medical University of Vienna, Vienna, Austria
e-mail: markus.mueller@meduniwien.ac.at

M. Müller (ed.), *Microdialysis in Drug Development*, AAPS Advances in the Pharmaceutical Sciences Series, DOI: 10.1007/978-1-4614-4815-0_1,

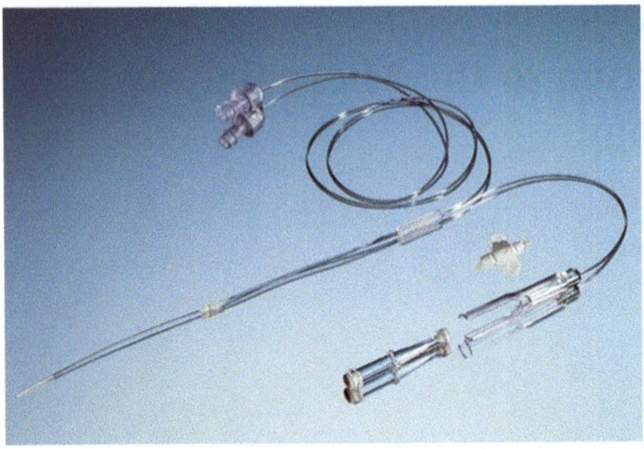

Fig. 1.1 Typical μD catheter: the catheter consists of an inlet and an outlet tubing, a shaft which is introduced into the tissue and a tip, equipped with a semipermeable membrane (*white*) where the exchange between the free drug fraction in the interstitium and the probe fluid takes place. Courtesy and with permission from Mdialysis AB, Solna, SE (http://www.mdialysis.se/)

which are based on direct sampling of tissue, cells, or tissue fluids (Müller et al. 2004).

For all these methodological approaches two major limitations have emerged: (1) limited applicability for select analytes only and (2) an imprecise topical resolution. Positron emission tomography (PET) imaging, for example, constitutes a tool to study drug concentrations in virtually all organs simultaneously. However, it is only applicable to analytes which can be labeled with 11C or 18F and a clear discrimination of tissue compartments, i.e., intra- and extracellular (EC) spaces is not possible.

A methodology which overcomes both limitations is μD (for the principle of μD see Figs. 1.1, 1.2, 1.3). μD, which is a minimal invasive technology, provides continuous access to the unbound drug fraction in a clearly defined compartment, i.e., the EC space fluid, and is applicable to a magnitude of diffusible analytes. The EC site is a particularly important space for drug research because unbound EC concentrations determine the drug load available to cell surface receptors. For this reason μD has become a standard tool for the measurement of EC target site concentrations in fields where the attainment of suitable EC concentrations is rate limiting, e.g., anti-infective research and development. Intuitively, it could be assumed that blood measurements would suffice for many scenarios, but it has been convincingly shown that concentrations in the blood may differ substantially from target sites (Müller et al. 2004). Besides its obvious advantages, μD only provides data from a single site within a tissue, which is a clear disadvantage versus imaging techniques which offer simultaneous multi-organ and multi-site information.

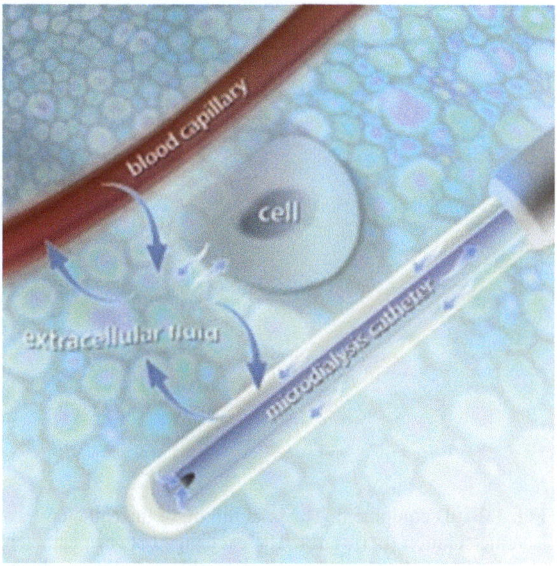

Fig. 1.2 Principle of μD: the catheter tip is located in the tissue and the catheter fluid equilibrates with interstitial space fluid. The catheter, thus, resembles a closed, non-fenestrated capillary. Courtesy and with permission from Mdialysis AB, Solna, SE (http://www.mdialysis.se/)

1.2 Development of μD

μD has gone through a number of development cycles. The conceptual beginnings can be traced back to the 1960s, when various types of in vivo sampling techniques were first tested for characterization of tissue concentrations of drugs, transmitters, mediators, and metabolites. The credit for inventing μD in a strict sense is traditionally given to Ungerstedt and Pycock, two researchers who first reported on the use of "hollow fibers" in 1974 (Ungerstedt and Pycock 1974). Another conceptual milestone was the invention of μD-probe calibration by Lönnroth et al. in 1987 (Lönnroth et al. 1987), which allowed an exact, quantitative determination of EC concentrations. In the last two decades, the extensive use of μD has been catalyzed by (1) the increasing availability of highly sensitive chemical–analytical methods (see also Chap. 4), (2) refined catheter manufacturing, and (3) CE- and FDA-certification of various catheters which has also led to the implementation of μD in clinical pharmacology.

To date μD has been used as a research and drug development tool in virtually all organs in animals and humans, including human lung tissue (Herkner et al. 2002), myocardium (Kennergren et al. 1999), and brain (Hillered and Persson 1999; Hillman et al. 2005) and a number of recommendable review articles have been published for μD use in various scenarios, including PK and pharmacodynamics (PD) in drug development (Ungerstedt 1991; Fettweis and Borlak 1996; Elmquist and Sawchuk 1997; Chaurasia 1999; Hansen et al. 1999; Müller 2000, 2002, 2009; Stahl et al. 2002; Höcht et al. 2006; Schmidt et al. 2008, see also Chap. 5). μD has proven to be a useful tool in preclinical and clinical drug development, which is also underlined by the large number of μD-related

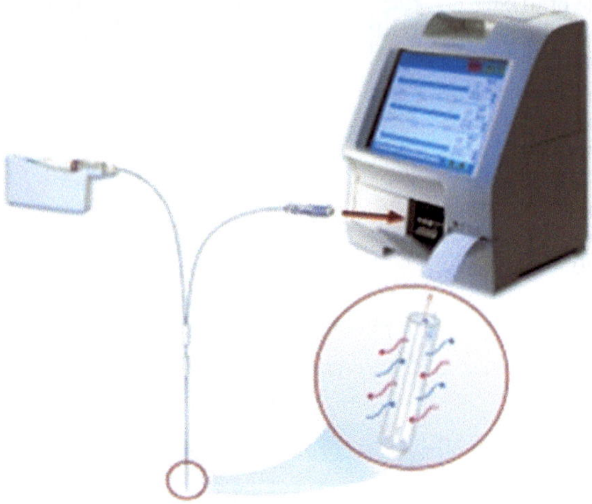

Fig. 1.3 μD equipment. The inlet tubing of the catheter is connected to a pump and flushed with isotonic solution (perfusate). The exchange between catheter perfusate and interstitial fluid takes place at the probe tip with semipermeable membrane which is localized in the tissue of interest (*red circle*). The equilibrated perfusate ("dialystae") leaves the catheter via the outlet tubing and is subjected to chemical analysis. Courtesy and with permission from Mdialysis AB, Solna, SE (http://www.mdialysis.se/)

publications: as of March 16, 2012, 14,132 PubMed hits about μD could be retrieved, 10,123 of which relate to "pharmacology" and 3,155 to studies in humans. In addition to μD, a number of derivative techniques such as ultrafiltration, and open flow microperfusion (Holmgaard et al. 2012, see also Chap. 15) have been described in the literature. A comprehensive handbook on μD, edited by Westerink and Cremers, was published in 2007, including chapters about clinical drug development (Brunner and Müller 2007).

1.3 μD Methodology

Before its broad application as a research tool took place, a number of technical details were addressed in the literature (de Lange et al. 2000). First, the question about a potential tissue trauma was investigated in a number of tissues under various conditions. Following μD probe implantation, various tissue reactions to implantation were described. These range from transient changes in blood flow (Anderson et al. 1994) or increases in trauma metabolites immediately after implantation to gliosis in the nervous system or foreign body-like reactions after chronic implantation. Altogether it is now recommended to allow the tissue to adjust for 30–120 min after probe implantation before actual measurement takes place and limits long-term measurements to a couple of days. Second, quantitative μD was enabled by the development of several calibration techniques which allow calculating "true" tissue concentrations from "perfusate" concentrations (for details see Chap. 2, de Lange et al. 2000; Lönnroth et al. 1987). Third, quantification of low analyte concentrations has traditionally been a large obstacle for μD but was catalyzed by the availability of highly sensitive analytical methods which allow for the measurement in the nM range in the dialysate samples which typically have volumes between 1 and 20 μl (Lunte and Lunte 1996; Davies et al. 2000; Nandi and Lunte 2009; Van Eeckhaut et al. 2011; see also Chap. 4). Despite these developments, limitations of μD are still related to the measurement of highly lipophilic compounds or compounds which interact with probe materials (Carneheim and Ståhle 1991; Kurosaki et al. 1998; Whitaker and Lunte 2010). For lipophilic and high molecular weight compounds approaches with the use of different perfusate solutions, e.g., antibody-spiked perfusates, were proposed (Duo et al. 2006; Herbaugh and Stenken 2011). Therefore, meticulous in vitro experiments should be carried out with each new potential analyte before moving to in vivo experiments. Given the above considerations, μD has turned out to be a very reliable and reproducible technique, in particular for diffusible analytes with a molecular weight <3 kD (Ståhle et al. 1991).

1.4 Rationale for μD in Drug Research and Development

The ultimate rationale for employing μD in drug research and development is based on the presence of blood-to-tissue barriers, which hinder drug penetration to the tissue target site. These have been described for a number of tissues, most notably for the brain in form of ABC transporters at the blood–brain barrier (BBB). In the presence of ABC transporters, even under steady-state conditions, no complete equilibration between blood and tissue is achieved. Besides the BBB, however, numerous other barriers have been described for a number of organs and pathological disease states, e.g., tumors (Jain 1994; Müller et al. 1997a, b; Eichler et al. 2011). In some of these cases, e.g., in inflammatory states (Müller 2004), the barriers present as a reduction in the capillary surface area which is available for diffusion processes. Thus, ultimately, equilibration between tissue and blood is attained; however, the rate of drug exchange between blood and tissue is decreased.

If barriers are present, a clear understanding of the target site concentration is crucial for drug development purposes. Relying on blood concentrations might, thus, be misleading as concentrations at the target and also PD, i.e., overall drug effects are typically underestimated. One important factor determining PD and target penetration is protein binding (Zeitlinger et al. 2011). These considerations have also been addressed repeatedly by regulatory authorities, which view μD as a potentially useful tool in drug regulation (Chaurasia et al. 2007; Shah 2004, see also Chap. 3).

1.5 Examples of μD Use in Drug Research and Development

μD has been initially conceived as a methodology to measure neurotransmitter concentrations. As of today there is an abundance of experimental settings where μD was employed to study biogenic amines and other transmitters in health, disease, and under various experimental conditions. The research in this area has even led to experiments where it was shown that dopamine release increases in the human amygdala during performance of cognitive tasks (Fried et al. 2001). Important insights have also been gathered about EC amyloid beta dynamics and neuronal vulnerability (Brody et al. 2008; Bero et al. 2011, see also Chap. 9). Besides neurochemistry, endocrinology and neurointensive care have been fields where μD has been exploited as a biosensor for continuous measurement of glucose and insulin (Jansson et al. 1993; Bolinder et al. 1993; Müller et al. 1996b; Petersen et al. 2004; Cline et al. 1999; see also Chap. 12) and metabolites predictive of brain damage and neuronal pathophysiology (During and Spencer 1993; Kendrick et al. 1997; Bellander et al. 2004). Relevant progress to study the effect of tissue barriers on PK and PD has also been enabled by μD in preclinical and clinical settings with regard to BBB function (see also Chap. 8), pain research (see Chap. 10) antiepileptic drug efficacy (Clinckers et al. 2009), tissue penetration of anti-infectives (Müller et al. 1996a, 2004; Delacher et al. 2000, see also Chap. 6) and antivirals (Yang et al. 1997), anti-tumor agents (Müller et al. 1997a, b; Zhou and Gallo 2005, see also Chap. 16), ocular—(see Chap. 11), and transdermal penetration (Anderson et al. 1991; Müller et al. 1995; Cross et al. 1997; Benfeldt and Groth 1998; Kreilgaard 2002; Holmgaard et al. 2010 see also Chap. 7). In addition, μD was successfully used to measure target tissue concentrations of biomarkers as PD outcome parameters (Müller et al. 1997a, b; Li et al. 2006).

1.6 Challenges in μD

The traditional μD membrane pore size of 20 kD easily allows for diffusion of analytes <3 kD. However, many important research questions relate to high molecular weight compounds. One of the biggest challenges in μD research is the measurement of "large" molecules, i.e., drugs with molecular weights >3 kD (Clough 2005; Ao and Stenken 2006, see also Chap. 13) and lipophilic compounds (Carneheim and Ståhle 1991; Kurosaki et al. 1998; Benfeldt and Groth 1998; Whitaker and Lunte 2010). Although membranes with a cutoff of >100 kD are available, and a number of convincing and elegant approaches for larger molecules have been described (Clough 2005; Ao and Stenken 2006; Maurer et al. 2008), it is noticed that technical difficulties increase with increasing pore size and increasing lipophilicity. Therefore, complementary techniques might also be considered under special circumstances such as open flow microperfusion (Holmgaard et al. 2012, see also Chap. 15) or molecular imaging technologies such as PET

(Fischman et al. 2002; Brunner et al. 2004; Langer et al. 2005; Matthews et al. 2012; Shingaki et al. 2012, see also Chap. 14).

References

Anderson C, Andersson T, Molander M (1991) Ethanol absorption across human skin measured by in vivo microdialysis technique. Acta Derm Venereol 71:389–393

Anderson C, Andersson T, Wårdell K (1994) Changes in skin circulation after insertion of a microdialysis probe visualized by laser Doppler perfusion imaging. J Invest Dermatol 102:807–811

Ao X, Stenken JA (2006) Microdialysis sampling of cytokines. Methods 38:331–341

Bellander BM, Cantais E, Enblad P, Hutchinson P, Nordström CH, Robertson C, Sahuquillo J, Smith M, Stocchetti N, Ungerstedt U, Unterberg A, Olsen NV (2004) Consensus meeting on microdialysis in neurointensive care. Intensive Care Med 30:2166–2169

Benfeldt E, Groth L (1998) Feasibility of measuring lipophilic or protein-bound drugs in the dermis by in vivo microdialysis after topical or systemic drug administration. Acta Derm Venereol 78:274–278

Bero AW, Yan P, Roh JH, Cirrito JR, Stewart FR, Raichle ME, Lee JM, Holtzman DM (2011) Neuronal activity regulates the regional vulnerability to amyloid-β deposition. Nat Neurosci 14:750–756

Bolinder J, Ungerstedt U, Arner P (1993) Long term continuous glucose monitoring with microdialysis in ambulatory insulin-dependent diabetic patients. Lancet 342:1080–1085

Brody DL, Magnoni S, Schwetye KE, Spinner ML, Esparza TJ, Stocchetti N, Zipfel GJ, Holtzman DM (2008) Amyloid-beta dynamics correlate with neurological status in the injured human brain. Science 321:1221–1224

Brunner M, Langer O, Dobrozemsky G, Müller U, Zeitlinger M, Mitterhauser M, Wadsak W, Dudczak R, Kletter K, Müller M (2004) [^{18}F]Ciprofloxacin, a new PET tracer for non-invasive assessment of ciprofloxacin tissue pharmacokinetics in humans. Antimicrobial Agents Chemother 48:3850–3857

Brunner M, Müller M (2007) Microdialysis in clinical drug delivery studies. In: Westerink BH, Cremers TI (eds) Handbook of microdialysis. Elsevier, pp 625–644

Carneheim C, Ståhle L (1991) Microdialysis of lipophilic compounds: a methodological study. Pharmacol Toxicol 69:378–380

Chaurasia CS (1999) In vivo microdialysis sampling: theory and applications. Biomed Chromatogr 13:317–332

Chaurasia CS, Müller M, Benfeldt E, Sawchuk RJ, Cheung WY, Elmquist WF, Bungay PM, Derendorf H, DeLange ECM, Lunte CE, Hammarlund-Udenaes M, Bolinder J, Bullock R, Kellogg DL, Joukhadar C, Henrik Nordstrom C, Rollema H, Welty DF, Yeo H, Stahle L, Bashaw ED, Shah VP, Ungerstedt U (2007) AAPS-FDA Workshop white paper: microdialysis principles, application and regulatory perspectives. J Clin Pharmacol 47:589–603

Clinckers R, Smolders I, Vermoesen K, Michotte Y, Danhof M, Voskuyl R, Della Pasqua O (2009) Prediction of antiepileptic drug efficacy: the use of intracerebral microdialysis to monitor biophase concentrations. Expert Opin Drug Metab Toxicol 5:1267–1277

Cline GW, Petersen KF, Krssak M, Shen J, Hundal RS, Trajanoski Z, Inzucchi S, Dresner A, Rothman DL, Shulman GI (1999) Impaired glucose transport as a cause of decreased insulin-stimulated muscle glycogen synthesis in type 2 diabetes. N Engl J Med 341:240–246

Clough GF (2005) Microdialysis of large molecules. AAPS J 7:E686–E692

Cross SE, Anderson C, Thompson MJ, Roberts MS (1997) Is there tissue penetration after application of topical salicylate formulations? Lancet 350:636

Davies MI, Cooper JD, Desmond SS, Lunte CE, Lunte SM (2000) Analytical considerations for microdialysis sampling. Adv Drug Deliv Rev 45:169–188
Delacher S, Derendorf H, Hollenstein U, Brunner M, Joukhadar C, Hofmann S, Georgopoulos A, Eichler HG, Müller M (2000) An in vivo PK—in vitro PD model to simulate antibiotic activity at the target site. J Antimicrob Agents Chemother 46:733–739
de Lange EC, de Boer AG, Breimer DD (2000) Methodological issues in microdialysis sampling for pharmacokinetic studies. Adv Drug Deliv Rev 45:125–148
Duo J, Fletcher H, Stenken JA (2006) Natural and synthetic affinity agents as microdialysis sampling mass transport enhancers: current progress and future perspectives. Biosens Bioelectron 22:449–457
During MJ, Spencer DD (1993) Extracellular hippocampal glutamate and spontaneous seizure in the conscious human brain. Lancet 341:1607–1610
Enblad P, Valtysson J, Andersson J, Lilja A, Valind S, Antoni G, Långström B, Hillered L, Persson L (1996) Simultaneous intracerebral microdialysis and positron emission tomography in the detection of ischemia in patients with subarachnoid hemorrhage. J Cereb Blood Flow Metab 16:637–644
Eichler AF, Chung E, Kodack DP, Loeffler JS, Fukumura D, Jain RK (2011) The biology of brain metastases-translation to new therapies. Nat Rev Clin Oncol 8:344–356
Eichler HG, Müller M (1998) Drug distribution—the forgotten relative of clinical pharmacokinetics. Clin Pharmacokinet 34:95–99
Elmquist WF, Sawchuk RJ (1997) Application of microdialysis in pharmacokinetic studies. Pharm Res 14:267–288
Fettweis G, Borlak J (1996) Topics in xenobiochemistry—application of microdialysis techniques in pharmacokinetic studies. Xenobiotica 26:473–485
Fischman AJ, Alpert NM, Rubin RH (2002) Pharmacokinetic imaging: a noninvasive method for determining drug distribution and action. Clin Pharmacokinet 41:581–602
Fried I, Wilson CL, Morrow JW, Cameron KA, Behnke ED, Ackerson LC, Maidment NT (2001) Increased dopamine release in the human amygdala during performance of cognitive tasks. Nat Neurosci 4:201–206
Hansen DK, Davies MI, Lunte SM, Lunte CE (1999) Pharmacokinetic and metabolism studies using microdialysis sampling. J Pharm Sci 88:14–27
Herbaugh AW, Stenken JA (2011) Antibody-enhanced microdialysis collection of CCL2 from rat brain. J Neurosci Methods 202:124–127
Herkner H, Müller MR, Kreischitz N, Mayer BX, Frossard M, Joukhadar C, Klein N, Lackner E, Müller M (2002) Closed chest microdialysis to measure antibiotic penetration into human lung tissue. Am J Resp Crit Care Med 165:273–276
Hillered L, Persson L (1999) Neurochemical monitoring of the acutely injured human brain. Scand J Clin Lab Invest Suppl 229:9–18
Hillman J, Aneman O, Anderson C, Sjögren F, Säberg C, Mellergård P (2005) A microdialysis technique for routine measurement of macromolecules in the injured human brain. Neurosurgery 56:1264–1268
Höcht C, Opezzo JA, Bramuglia GF, Taira CA (2006) Application of microdialysis in clinical pharmacology. Curr Clin Pharmacol 1:163–183
Holmgaard R, Nielsen JB, Benfeldt E (2010) Microdialysis sampling for investigations of bioavailability and bioequivalence of topically administered drugs: current state and future perspectives. Skin Pharmacol Physiol 23:225–243
Holmgaard R, Benfeldt E, Nielsen JB, Gatschelhofer C, Sorensen JA, Höfferer C, Bodenlenz M, Pieber TR, Sinner F (2012) Comparison of open-flow microperfusion and microdialysis methodologies when sampling topically applied fentanyl and benzoic acid in human dermis ex vivo. Pharm Res 29:1808–1820
Jain RK (1994) Barriers to drug delivery in solid tumors. Sci Am 271:58–65
Jansson PA, Fowelin JP, von Schenck HP, Smith UP, Lönnroth PN (1993) Measurement by microdialysis of the insulin concentration in subcutaneous interstitial fluid. Importance of the endothelial barrier for insulin. Diabetes 42:1469–1473

Kendrick KM, Guevara-Guzman R, Zorrilla J, Hinton MR, Broad KD, Mimmack M, Ohkura S (1997) Formation of olfactory memories mediated by nitric oxide. Nature 388:670–674

Kennergren C, Mantovani V, Lönnroth P, Nystrom B, Berglin E, Hamberger A (1999) Monitoring of extracellular aspartate aminotransferase and troponin T by microdialysis during and after cardioplegic heart arrest. Cardiology 92:162–170

Kreilgaard M (2002) Assessment of cutaneous drug delivery using microdialysis. Adv Drug Deliv Rev 54(Suppl 1):S99–S121

Kurosaki Y, Nakamura S, Shiojiri Y, Kawasaki H (1998) Lipo-microdialysis: a new microdialysis method for studying the pharmacokinetics of lipophilic substances. Biol Pharm Bull 21:194–196

Langer O, Brunner M, Karch R, Müller U, Dobrozemsky G, Abrahim A, Paul M, Zeitlinger M, Joukhadar C, Dudczak R, Kletter K, Müller M (2005) Combined positron emission tomography and microdialysis for in vivo assessment of intracellular drug pharmacokinetics in humans. J Nucl Med 46:1835–1841

Li Y, Peris J, Zhong L, Derendorf H (2006) Microdialysis as a tool in local pharmacodynamics. AAPS J 8:E222–E235

Lönnroth P, Jansson PA, Smith U (1987) A microdialysis method allowing characterization of intercellular water space in humans. Am J Physiol 253:E228–E231

Lunte SM, Lunte CE (1996) Microdialysis sampling for pharmacological studies: HPLC and CE analysis. Adv Chromatogr 36:383–432

Matthews PM, Rabiner EA, Passchier J, Gunn RN (2012) Positron emission tomography molecular imaging for drug development. Br J Clin Pharmacol 73:175–186

Maurer MH, Haux D, Unterberg AW, Sakowitz OW (2008) Proteomics of human cerebral microdialysate: from detection of biomarkers to clinical application. Proteomics Clin Appl 2:437–443

Müller M, Schmid R, Wagner O, Osten BV, Shayganfar H, Eichler HG (1995) In vivo characterization of transdermal drug transport by microdialysis. J Controlled Release 37: 49–57

Müller M, Haag O, Burgdorff T, Georgopoulos A, Weninger W, Jansen B, Stanek G, Pehamberger H, Agneter E, Eichler HG (1996a) Characterization of peripheral compartment kinetics of antibiotics by in vivo microdialysis in humans. Antimicrob Agents Chemother 40:2703–2709

Müller M, Holmäng A, Anderson OK, Eichler HG, Lönnroth P (1996b) Measurement of interstitial muscle glucose and lactate concentrations during an oral glucose tolerance test. Am J Physiol 271:E1003–E1007

Müller M, Mader RM, Steiner B, Steger GG, Jansen B, Gnant M, Helbich T, Jakesz R, Eichler HG, Blöchl-Daum B (1997a) 5-Fluorouracil kinetics in the interstitial tumor space and clinical response in breast cancer patients. Cancer Res 57:2598–2601

Müller M, Burgdorff T, Jansen B, Singer EA, Agneter E, Dorner G, Brunner M, Eichler HG (1997b) In vivo drug-response measurements in target tissues by microdialysis. Clin Pharmacol Ther 62:165–170

Müller M (2000) Microdialysis in clinical drug delivery studies. Adv Drug Deliv Rev 45:255–269

Müller M (2002) Science, medicine and the future: microdialysis. BMJ 324:588–591

Müller M, delaPena A, Derendorf H (2004) Issues in PK-PD of antibiotics: tissue penetration. Antimicrobial Agents Chemother 48:1441–1453

Müller M (2009) Monitoring tissue drug levels by microdialysis. Altern Lab Anim 37:57–59

Nandi P, Lunte SM (2009) Recent trends in microdialysis sampling integrated with conventional and microanalytical systems for monitoring biological events: a review. Anal Chim Acta 651:1–14

Petersen KF, Dufour S, Befroy D, Garcia R, Shulman GI (2004) Impaired mitochondrial activity in the insulin-resistant offspring of patients with type 2 diabetes. N Engl J Med 350:664–671

Schmidt S, Banks R, Kumar V, Rand KH, Derendorf H (2008) Clinical microdialysis in skin and soft tissues: an update. J Clin Pharmacol 48:351–364

Shingaki T, Takashima T, Wada Y, Tanaka M, Kataoka M, Ishii A, Shigihara Y, Sugiyama Y, Yamashita S, Watanabe Y (2012) Imaging of gastrointestinal absorption and biodistribution of an orally administered probe using positron emission tomography in humans. Clin Pharmacol Ther 91:653–659

Shah VP (2004) Topical drug products—microdialysis: regulatory perspectives. Int J Clin Pharmacol Ther 42:379–381

Stahl M, Bouw R, Jackson A, Pay V (2002) Human microdialysis. Curr Pharm Biotechnol 3:165–178

Ståhle L, Arner P, Ungerstedt U (1991) Drug distribution studies with microdialysis. III: Extracellular concentration of caffeine in adipose tissue in man. Life Sci 49:1853–1858

Ungerstedt U, Pycock C (1974) Functional correlates of dopamine neurotransmission. Bull Schweiz Akad Med Wiss 30:44–55

Ungerstedt U (1991) Microdialysis—principles and applications for studies in animals and man. J Intern Med 230:365–373

Van Eeckhaut A, Maes K, Aourz N, Smolders I, Michotte Y (2011) The absolute quantification of endogenous levels of brain neuropeptides in vivo using LC-MS/MS. Bioanalysis 3:1271–1285

Whitaker G, Lunte CE (2010) Investigation of microdialysis sampling calibration approaches for lipophilic analytes: doxorubicin. J Pharm Biomed Anal 53:490–496

Yang Z, Brundage RC, Barbhaiya RH, Sawchuk RJ (1997) Microdialysis studies of the distribution of stavudine into the central nervous system in the freely-moving rat. Pharm Res 14:865–872

Zeitlinger MA, Derendorf H, Mouton JW, Cars O, Craig WA, Andes D, Theuretzbacher U (2011) Protein binding: do we ever learn? Antimicrob Agents Chemother 55:3067–3074

Zhou Q, Gallo JM (2005) In vivo microdialysis for PK and PD studies of anticancer drugs. AAPS J 7:E659–E667

Chapter 2
Recovery and Calibration Techniques: Toward Quantitative Microdialysis

Elizabeth C. M. de Lange

Abstract One of the most important questions in microdialysis is how to relate the concentrations in the microdialysate to the true concentrations outside the probe as in vitro recovery was not equal per se to the recovery in vivo. Many calibration techniques and approaches have been proposed, with different levels of practicality and usefulness, to obtain quantitative information on extracellular concentrations as a function of time. Today, we have build up enough experience to know how to obtain quantitative data using the microdialysis technique. With that, it is the only technique that provides data on the unbound concentration of compounds in extracellular fluid spaces in the body, being highly important information in drug development as the unbound concentrations are the drivers for the effect.

2.1 Introduction

Microdialysis technique was developed about 30 years ago, actually as improvement of the push–pull technique to obtain information on the extracellular fluid in the brain (brainECF), mostly focused on neurotransmitter functioning. By the use of a microdialysis, probe with a tip consisting of a semipermeable membrane and a continuous flush of the microdialysis probe with perfusion fluid, small enough molecular weight molecules diffuse from the outside to the inside of the probe

E. C. M. de Lange (✉)
Division of Pharmacology, Leiden/Amsterdam Center for Drug Research,
Leiden University, Leiden, Netherlands
e-mail: lange_l@lacdr.leidenuniv.nl

M. Müller (ed.), *Microdialysis in Drug Development*, AAPS Advances in the Pharmaceutical Sciences Series, DOI: 10.1007/978-1-4614-4815-0_2,

according to their concentration gradient, providing the dialysate fluid to be collected in intervals and analyzed on its content.

Then, questions arose about the possible impact of the microdialysis probe on its surrounding tissue. Quite a few studies have dedicated to investigate the changes of the periprobe tissue and its functionality. It was concluded that provided that surgery is carefully performed and experimental conditions are well chosen, the microdialysis probe had only minimal influence on the periprobe tissue.

After the application of the microdialysis technique was broadened to the pharmacokinetic research field, being quantitative by nature, from this field the question rose about the relationship between the concentrations measured in the dialysate in relation to the concentrations around the microdialysis tip. This ratio was termed recovery and was first estimated in vitro, by measuring the recovery of the compound of interest in the dialysate from a bulk solution with known concentration. This value was used to correct in vivo microdialysis concentrations to obtain what was supposed to be the free concentrations around the microdialysis probe. With time, it became clear that in vitro recovery was not equal per se to the recovery in vivo.

Thus, one of the most important questions in microdialysis is how to relate the concentrations in the microdialysate to the true concentrations outside the probe. In other words: to which extent is the drug recovered by the dialysate (or lost if a certain drug is added to the perfusate) and how fast? This chapter deals with the concept of recovery, and factors that will influence recovery, followed by theoretical and experimental approaches to determine the extracellular concentrations around the microdialysis probe, and pharmacokinetic data analysis. Finally, some emerging techniques and applications will be discussed.

2.2 Recovery

2.2.1 The Concept of Recovery

The term recovery describes the relation between concentrations of the drug in the periprobe fluid and those in the dialysate. These concentrations will differ from each other because a constant flow of the perfusate prevents the establishment of true concentration equilibrium between the periprobe fluid and the dialysate. Two types of recoveries can be distinguished: (1) absolute recovery (or mass recovery), which is defined as the amount of drug that is extracted by the dialysate as a function of time, and (2) relative recovery (or concentration recovery) being defined as the concentration of the drug in the dialysate divided by its uniform concentration in the periprobe fluid. In general, the term "recovery" is used to refer to concentration recovery. The recovery is defined as

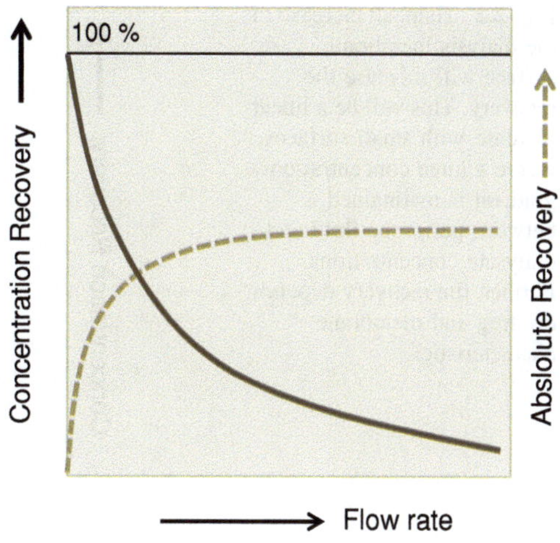

Fig. 2.1 At least in vitro the relative or concentration recovery is inversely dependent on the flow rate and often linearly dependent on the concentration in the periprobe fluid. The absolute (mass) recovery increases with flow rate up to a maximum when the mass flux of molecules become rate limiting

$$R = (C_{\text{in}} - C_{\text{dial}})/C_{\text{in}}$$

In this equation, C_{in} and C_{dial} are the concentrations of a compound in the perfusate and in the dialysate, respectively. This equation should be adapted if concentrations are added to the perfusate, as in some of the calibration methods discussed later in this chapter, i.e., the retrodialysis method, the no-net-flux, and the dynamic-no-net-flux method.

2.2.2 In Vitro Recovery

A number of parameters influence absolute and concentration recovery (Hsiao et al. 1990). These parameters include perfusate flow rate, temperature, characteristics of the semipermeable membrane, probe geometry, surface of the semipermeable membrane, perfusate composition, tubing characteristics, and length, as well as the characteristics of the drug. These factors can be investigated in vitro, knowing the concentration of the drug in the bulk solution.

Flow rate. The in vitro recovery is inversely dependent on the *flow rate* and often linearly dependent on the concentration in the periprobe fluid (Fig. 2.1). Most investigators now use a flow rate that ranges from 0.1 to 5 μl/min, with a typical value being 2 ul/min.

Membrane surface. Then, an increase of the *dialysis membrane surface* will increase the recovery (Fig. 2.2). This will be a linear increase with small surfaces, and dependent on drug characteristics. Note that, at larger surfaces the increase in recovery will start to lag behind the increase in surface because the concentration difference between periprobe fluid and dialysate traversing along the semipermeable membrane will gradually diminish with the longer length of the membrane

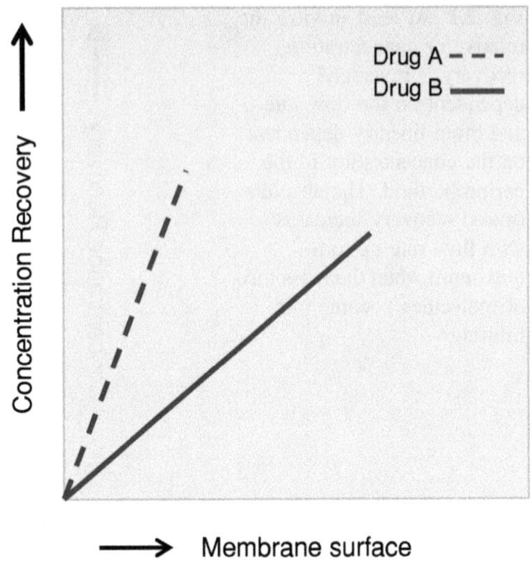

Fig. 2.2 Then, an increase of the dialysis membrane surface will increase the recovery. This will be a linear increase with small surfaces, where a huge concentration-gradient is maintained between periprobe fluid and dialysate concentrations. Further, the recovery depends on drug and membrane characteristics

Temperature. As diffusion increases with higher *temperature*, a higher temperature will increase recovery. As probes in vivo are exposed to body temperature, it is important that testing of the probe characteristic in vitro also occur at this temperature.

Membrane material. Different membrane materials can be used. For self-made probes, the choice among the different membranes is wide, e.g., those used for artificial kidneys. Commercially available probes have limited choice of membrane material (De Lange 1997). It is important to investigate the rate toward and extent of equilibrium of concentration recovery of the drug for a selected membrane (Tao and Hjorth 1992). If the drug has interaction with the membrane, it may lead to delayed response of the dialysate concentration to changes in the external medium (Fig. 2.3). The faster the changes occur in the external medium (read here in vivo periprobe concentrations), the more important this aspect is. The problem of interaction of a drug with the membrane material is more often encountered for lipophilic drugs (Carneheim and Stahle 1991).

Tubing material and dimensions. Tubing material from the microdialysis probe to the sample collector should ideally have no interaction with the drug as this may have a profound effect on the relation between the concentration of the drug found in the dialysate and the periprobe fluid concentrations, similar to effects of interaction with the probe membrane (Lindberger et al. 2002). The inner diameters of the tubing connections, however, may be of importance with respect to build up of fluid pressure and thereby fluid loss over the semipermeable membrane. This may be prevented by using inlet tubing (from perfusion pump to the probe) with an inner diameter smaller than that of the probe itself, and an outlet tubing (from probe to collection site) with an inner diameter being larger than that of the probe.

Perfusion medium composition. Many different perfusion medium composites in microdialysis studies have been used in the past (De Lange et al. 1997).

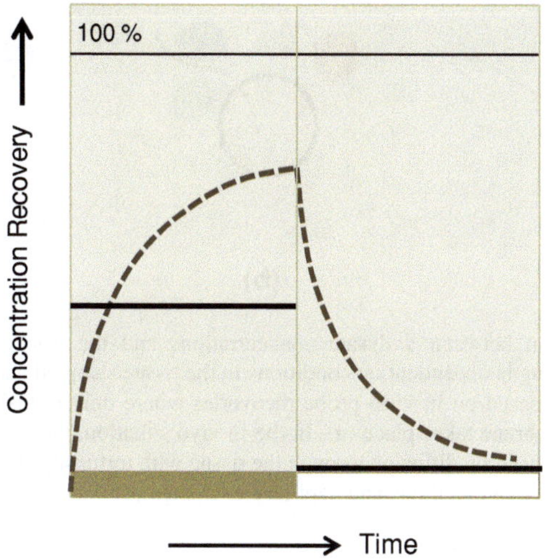

Fig. 2.3 It is important to investigate the rate toward and extent of equilibrium of concentration recovery of the drug for a selected membrane as the drug may interaction with the membrane (as well as tubing). This may lead to a delayed ("blunted") response of the dialysate concentration to changes in the external medium. The faster the changes occur in the external medium (read here in vivo periprobe concentrations), the more important this aspect is. Drugs that interact with probe membrane and/or tubing are called "sticky" drugs. Care should be taken to avoid a blunted response

The most commonly used today is the minimal perfusion medium, consisting of 140 mM sodium, 3.0 mM potassium, 1.2 mM calcium, 1.0 mM magnesium, and 147 mM chloride.

To prevent or reduce the interaction of a drug with the probe membrane and/or tubing (or improve the rate and/or extent of concentration recovery), several additions to the perfusion medium have been proposed (Carneheim and Stahle 1991). The addition of bovine serum albumin has often been proven useful to that end, and 0.2–0.5 % appears to be enough in many cases. Higher concentrations will increase the osmotic value of the perfusion medium, which may lead water movement from the external medium into the dialysate, a condition that should be avoided. To enhance recovery of lipophilic compounds, the addition of intralipid and encapsin has been suggested based on in vitro studies (Ward et al. 2003).

Another approach was the use of trapping agents such as native beta-cyclodextrin and a water-soluble beta-cyclodextrin polymer to study the temporal in vitro concentration response for the lipophilic drug carbamazepine to concentration changes outside of the probe (Ao et al. 2006). The in vitro concentration response thereby was no longer delayed.

Then, with time the microdialysis technique has been expanded from small molecules toward larger ones. In the light of having pharmacokinetic information on biological the work reviewed by Duo et al. (2006) is of interest. They reported on the inclusion of affinity-based trapping agents into the microdialysis perfusion fluid to improve the recovery via the binding reaction of low molecular weight

(a) (b) (c)

Fig. 2.4 The relation between dialysate concentrations and the true extracellular (in vivo) concentration of a drug is dependent on conditions in the tissue sampled and cannot be calculated from "calibrations" based on in vitro probe recoveries where only diffusion through the buffer and through the membrane takes place (**a**). In the in vivo situation, the recovery of the drug into the dialysate is influenced by diffusion through the tissue with tortuosity (**b**), and in virtually all of the cases also by extra-intracellular and capillary exchange (**c**)

hydrophobic analytes and larger analytes such as peptides and proteins. Antibody-immobilized bead microdialysis sampling enhancement was investigated for various endocrine hormones (amylin, GLP-1, glucagon, insulin, and leptin). The antibody-bead enhancement approach allowed for recovery enhancements that ranged between 3- and 20-fold for these peptides. Using the enhanced recovery approach, endocrine peptides at pM concentrations could be quantified. Then, non-antibody based enhancement agents using bovine serum albumin-heparin conjugates covalently bound to polystyrene microspheres were presented for the cytokine, tumor necrosis factor-alpha (TNF-alpha). Unlike antibodies, heparin provides the advantage of being reusable as an enhancement agent and served to improve the relative recovery of TNF-alpha by threefold.

2.2.3 In Vivo Recovery

It has been common practice in microdialysis studies for probes to be "calibrated" in bulk solution and to use these in vitro recoveries values to estimate in vivo extracellular concentrations. However, the relation between dialysate concentrations and the true extracellular concentration of a drug is dependent on conditions in the tissue sampled and cannot be calculated from in vitro probe recoveries (Fig. 2.4). In vivo, effective diffusion of the drug through the extracellular fluid of a tissue, uptake into cells, metabolic conversion rate, active transport across membranes, extent of tissue vascularization and blood flow all may play a role. Scheyer et al. (1994) investigated recovery of carbamazepine and carbamazepine epoxide, and found differences between the in vitro and human in vivo recovery, due to tissue factors. Another example is provided by the study of Ross et al. (2006) on factors that determine the in vivo recovery of salicylate and norepinephrine.

2.2.4 Mathematical Framework

The mathematical framework for in vivo concentration recovery from tissue, developed by Bungay et al. (1990), provided lots of insight into the contribution of different transport processes to the in vivo recovery. The mathematical model was developed given steady-state concentrations of the compound of interest. They approached the recovery process by division into three components, by means of diffusion in (1) probe lumen, (2) dialysis membrane, and (3) the periprobe environment Diffusion in probe lumen is limiting only with the use of very low flow rates. Diffusion through the dialysis membrane is limiting only when transport through the periprobe environment is rapid.

For the mathematical framework, it was assumed that the probe is inserted in tissue with intimate contact between tissue and outer surface of the probe, for which diffusion through tissue for hydrophilic drugs takes place through the extracellular fluid, and for which transport processes are linear in the studied concentration range. Other assumptions may lead to other expressions, which according to the authors are nevertheless likely to fit within the general mass transport laws. For in vivo concentration recovery the following equation can be used, viewing the transport of a molecule from the tissue to the dialysate being determined by a series of mass transfer resistances:

$$E_d = (C_{\mathrm{out}} - C_{\mathrm{in}})/(\mathrm{brain}_{\mathrm{ECF}} - C_{\mathrm{in}}) = 1 - \exp\{-1/[F\,(R_d + R_m + R_e)]\}.$$

Here E_d = in vivo recovery (or dialysate extraction fraction); C_{out} = dialysate concentration; C_{in} = perfusate concentration; F = flow rate of the perfusate; R = mass transfer resistance with subscripts d = dialysate, m = membrane and e = external medium. For a probe in tissue it generally holds that $R_e \gg R_m \gg R_d$, leaving only the R_e to be further considered:

$$R_e = \{(K_0/K_1)/(2\,r_o\,LD_e\,\varphi_e)\}\,\Gamma$$

Here K_0 and K_1 are modified Bessel functions, r_0 is the radius of the probe, L is the length of the semipermeable part of the membrane, D_e is the diffusion coefficient for the extracellular phase, φ_e is the accessible volume fraction of the extracellular phase, and Γ is the profile depth parameter. The profile depth parameter is defined as:

$$\Gamma = \sqrt{\{D_e/\left(k_{\mathrm{ep}} + k_{\mathrm{em}} + k_{e\to im}\right)\}}$$

in which k_{ep}, k_{em}, and $k_{\mathrm{e\to im}}$ represent first-order rate constants for efflux to the microvasculature, irreversible extracellular metabolism, and the composite of irreversible intracellular metabolism and extracellular–intracellular exchange respectively. Thus, it can be understood that under steady-state conditions all processes that contribute to elimination of a drug will affect in vivo concentration recovery. In contrast, influx of the drug does not affect the in vivo concentration recovery, but plays a role in determining the actual concentrations of the drug in the extracellular fluid.

As a steady-state concentration of endogenous compounds and drugs in vivo is more exception than a rule, Morrison et al. (1991) developed time-dependent equations for the dialysate concentration and for concentration profiles about the probe. The model predictions were compared with in vivo observations. It was found that microvasculature transport and metabolism play major roles in determining microdialysate transient responses. The higher rates of metabolism and/or capillary transport the shorter was the time required to approach steady-state recovery. This indicates that for substances characterized by low permeabilities and negligible metabolism, experimental situations exist that are predicted to have very slow approaches to microdialysis steady state.

2.3 In Vivo Methods to Obtain Quantitative Microdialysis Data

As in vitro recovery will not be equal to in vivo recovery, special quantification methods are needed to determine the actual relation between dialysate concentrations and those in the external medium which in vivo is the extracellular fluid.

2.3.1 Zero and Ultraslow Flow Rate Method

The lower the flow rate, the more the dialysate concentrations will approach the external medium concentration (Jacobson et al. 1985). In vivo, if steady-state concentrations are present in the tissue, flow rates can be sequential changed from high to low. The resulting dialysate concentrations can be related to the flow rate, and by regression will provide the true extracellular concentrations at zero flow rate. The ultraslow microdialysis was proposed by Kaptein et al. (1998), for endogenous glucose and lactate. It is however, very difficult to properly handle very low volumes of dialysate, while also the analytical methodology may be too insensitive to measure the very low amounts of drugs in the microdialysis samples.

2.3.2 No-Net-Flux or Difference Method

The principle of the "no-net-flux" (NNF) or "difference"method was developed by Lonnroth et al. (1987) for endogenous glucose, but was not yet named as such at that time, but later by Menacherry and Hubert (1992) in their study on cocaine. The method involves consecutive perfusion of the microdialysis probe with different concentrations, if steady-state conditions of the drug are present in the tissue. The resulting dialysate concentrations are measured and the difference between the perfusion medium concentration and the dialysate concentration is calculated and plotted as a function of the perfusion medium concentration. Regression analysis of the results yields the perfusate concentration, which is in equilibrium with the surrounding tissue.

With this method, no assumptions on periprobe behavior of the drug have to be made due to the fact that at NNF conditions no mass-transfer of the drug from further positions to the probe is taking place. Stahle (1991) used this approach to compare in vitro and in vivo recoveries of theophylline and caffeine, two highly identical compounds. It was found that the recovery in vitro was virtually identical for caffeine and theophylline. The in vivo recovery of theophylline, however, was significantly smaller than the recovery of caffeine in brain, liver, muscle, and adipose tissue. The difference in recovery was significantly larger in the brain than in other tissues.

2.3.3 Extended or Dynamic No-Net-Flux Method

Quantitative determination of extracellular drug concentration under transient conditions can be obtained using the extended version of the NNF approach proposed by Olson and justice (1993). The method provides the extracellular concentration and the in vivo probe recovery as a function of time. The technique is based on the NNF method for steady-state conditions, but differs in the use of a between-group rather than a within-group design. Instead of serial perfusion of individual animals with different concentrations via the probe, a group of animals are continuously perfused with one selected perfusion concentration. Different groups receive different concentrations and the results are combined at each time point. Regression of the mean data points of the different groups at a particular point in time will give the actual extracellular concentration with the associated in vivo concentration recovery value at that time. An important observation of Olson and Justice (1993), using this approach, was that the administration of a neuro-active drug may alter the in vivo recovery of neurotransmitters. Although this is a powerful experimental setup, more experimental animals are needed, which in part reduces the advantage of minimizing the use of living experimental animals by the microdialysis technique.

2.3.4 Internal Standard Technique

Another approach to determine in vivo recovery for every dialysate sample during the experiment is the use of an internal standard, which is added to the perfusate during the course of the experiment (Larsson 1991; Yokel et al. 1992; Scheller and Kolb 1991).

The internal standard should match the characteristics of the drug as close as possible, so that the concentration loss of the internal standard will predict the concentration recovery of the drug. In vitro, the recoveries of both drugs can be measured, and the ratio between the values for internal standard and drug can be determined. With the assumption that the obtained ratio in vitro would remain the same in vivo, it can be used to calculate in vivo recovery of the drug as a function of time. Recovery of the drug of interest (R_{drug}) is defined as:

$$\text{In vitro } R_{drug} = \left[C_{in,drug} - C_{dial,drug}\right] / C_{in,drug}$$

The recovery of the internal standard is defined as:

$$\text{In vitro } R_{IS} = \left[C_{in,IS} - C_{dial,IS}\right]/C_{in,IS}$$

Then, loss of the internal standard (L_{IS}) is defined as:

$$\text{In vivo } L_{IS} = \left[C_{inIS} - C_{dial,IS}\right]/C_{in,IS}$$

Assuming that the ratio of recovery of the drug and internal standard in vitro will be the same in vivo, and that in vivo loss and recovery of the internal standard is equal to the in vivo gain of the internal standard, the concentration of the drug in the extracellular fluid ($C_{ECF,drug}$) can be calculated by:

$$C_{ECF,drug=}C_{dial,drug} \times 1/[\text{In vivo } L_{IS}] \times [\text{In vitro } R_{drug}/\text{In vitro } R_{IS}]$$

This method would be suited to determine changes in recovery if brought about by factors that decrease probe efficiency, such as the formation of air bubbles on the inside of the semipermeable membrane or occlusion of membrane pores by cells or sticky drugs. However, in vivo, effective diffusion of the internal standard and drug is assumed to be equal. This may not be realistic as illustrated by Stahle et al. (1991) for theophylline and caffeine, for which a difference in in vivo recovery was found using the NNF method. Moreover, the difference found was dependent on the tissue, with the highest deviation in brain. Also interaction of the internal standard with the drug should be ruled out. This stresses prudence to be exercised in the use of internal standards. Other investigators also used and validated this method against the no-net-flux method.

2.3.5 Retrodialysis or Reverse Dialysis of the Drug

In reverse dialysis or retrodialysis, the drug itself may be added to the perfusate and its in vivo loss may be used as a measure for in vivo recovery (Wang et al. 1991, 1993). It is based on the calculation of in vivo recovery of a drug on the basis of its loss from the perfusate during so-called retrodialysis, assuming that in vivo recovery equals in vivo loss, and was used to estimate the concentration of drug in the extracellular fluid (Fig. 2.5).

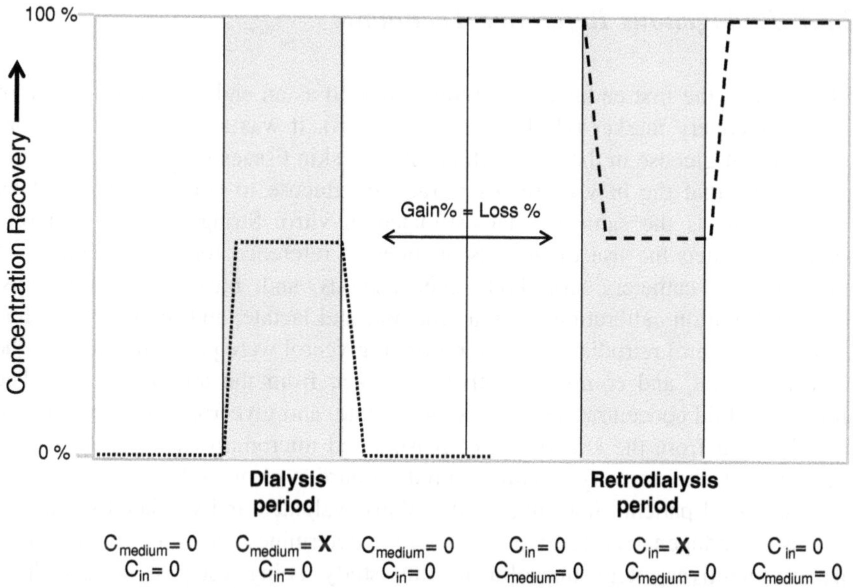

Fig. 2.5 Microdialysis probes should first be investigated in vitro by measuring the loss of drug using drug-containing perfusate or the gain of drug using drug-containing sample solutions. This will reveal drug recovery as being fully governed by simple diffusion and probe performance (membrane characteristics, potential interaction of the drug with the membrane of the probe, and/or tubing). When significant interaction of the drug and microdialysis probe and/or tubing is observed, one has to search for other probe and/or tubing materials without significant interaction(s). Furthermore, as the recovery determined in vitro may differ from the recovery in humans. Its actual value therefore needs to be determined in every in vivo experiment

This measurement has to be conducted before administration of the drug to the body. It is a relatively easy and useful approach, and is therefore most often used, but should be validated, as has been shown by a tissue specific asymptotic profile in recovery that results under increasing concentration gradient (LeQuellec 1995).

2.3.6 Combined Retrodialysis by Drug and Calibrator

A stronger approach, a combination of internal standard (calibrator) and retrodialysis, has been developed by Bouw and Hammarlund-Udenaes (1998). In vivo recovery of morphine was determined by morphine itself, retrodialysis by drug, and by the calibrator nalorphine, retrodialysis by calibrator. The calibrator is valuable as a quality control during the whole experiment. For morphine, this method has been compared with the DNNF method in a study on blood–brain barrier transport of morphine as discussed later in this chapter.

2.3.7 Endogenous Reference Technique

Glucose was the first endogenous compound used as an endogenous reference or internal recovery marker (Hashimoto et al. 1998). It was assumed that the concentration of glucose in the extracellular fluid of skin tissues is the same as that in plasma and that the in vivo recovery ratio of glucose to heir drug tranilast by microdialysis is the same as that estimated in vitro. Stringberg and Lonnroth (2000) validated the use of urea as an internal reference for the calibration of microdialysis catheters, for both subcutaneous and muscle measurements. The equilibration calibration of urea, glucose, and lactate, and the internal reference (3H-glycerol retrodialysis) calibration of glycerol were performed in both the muscle and fat, and compared with those taken from the arterial plasma. The interstitial fluid concentrations of glucose, lactate, and glycerol can were found to be calculated from the knowledge of plasma and microdialysate content of urea together with the recovery in vitro. To reduce burden of microdialysis procedures in critically ill patient, Schwalbe et al. (2006) evaluated and validated the use of urea as an endogenous reference compound to determine relative in vivo recovery of anti-infectives, with linezolid in this study being used as model drug. Concentration- and flow rate-dependence were evaluated in vitro to determine the recovery of urea. The in vivo recovery of urea was correlated with in vivo recovery of linezolid obtained by the traditional retrodialysis method, with reasonable correlation.

2.3.8 Some Method Comparisons

Le Quellec et al. (1995), investigated the NNF and the reverse dialysis methods. First, the accuracy of the No-Net-Flux method to estimate in vivo recovery was compared in two situations: diffusion from the probe into the dialysis medium and diffusion from the outer medium into the probe.

The point of no net transport was used to estimate the concentration surrounding the probe. Neither differences between extracellular concentrations (intercept values) nor differences between recoveries were observed. Then, the reverse dialysis method was tested to estimate the relative loss of drug from the perfusate when the probe was placed in a drug-free medium. Finally, comparisons of the behavior of the drug diffusion across the membrane under increasing gradient conditions have shown an asymptotic profile, specific of the tissue; blood, muscle, and adipose tissue. It was found that (in line with predictions from the mathematical model of Bungay) that the faster a drug was removed by microvascular transport (blood > muscle > adipocytes), the higher was the recovery, until the perfusate concentration reached a threshold value where the transport process became gradient limited and no more tissue limited.

The urea reference technique was compared with the NNF and the retrodialysis technique for glucose and paracetamol as model compounds healthy volunteers (Brunner et al. 2000). For glucose, recovery values, calculated by the urea reference technique differed significantly from those values, which were assessed by the no-net-flux technique whereas for paracetamol recovery values did not differ significantly, albeit high intramethod variability was observed. Thus, the authors could not confirm the hypothesis that recovery values calculated by the urea reference technique provide equivalent results compared with standard calibration techniques.

For microdialysis studies in the skin, laboratory-made linear probes are often used, which requires careful assessment of the linear probes' performance to ensure validity of the data obtained using this technique. Zheng et al. (2007). For sinomenine, in vitro and vivo recovery was determined using different lengths of the dialysis membrane and different perfusion flow rates. In vitro recovery of sinomenine from the microdialysis probe was independent of concentration, stable over an 8-h period. Comparable in vitro recoveries were obtained by different established approaches including recovery estimation by gain, loss, and the NNF method. Recovery by loss was used to study the in vivo recovery of sinomenine from rat subcutaneous tissue. The performance of the microdialysis system was stable over an 8-h study, resulting identical in vivo recovery values for the retrodialysis and the NNF method.

2.3.9 Modified Ultraslow Microdialysis

The microdialysis technique is to date still of enormous value in obtaining information on unbound (therewith highly relevant) drug concentrations in the tissue of choice. However, obtaining quantitative data by calibration of the microdialysis probe recovery by the earlier described calibration methods includes assumptions, is time-consuming and is not without limitations. To improve robustness and practicality of quantitative microdialysis sampling, Cremers et al. (2009) modified the ultraslow microdialysis approach. Ultraslow microdialysis uses very low microdialysis flow rates (<200 nl/min) which increase recovery (both in vivo and in vitro) to over 90 %. However, new practical issues arise when attempting to work with these flow rates. The resulting very low volumes and long lag times make this method very impractical for general application. In the modified version, addition of a carrier flow after the dialysis process has been completed which negates the problems of long lag times and low volumes. The resulting dilution of the dialysis sample concentration can simply be mathematically corrected. To show the potential of this approach, Cremers et al. (2009) measured the free brain levels of two CNS compounds using the classic DNNF and the new modified ultraslow dialysis method. Modified ultraslow microdialysis was shown to generate robust data with the use of only small numbers of rats. Currently lots of experience is built on quantitatively measuring blood–brain barrier penetration of compounds into the brain. Although researchers needs to become skilled in using technique, it is very promising as it circumvents the assumptions and limitations of calibration methods, and may be used for a much more straightforward screening of drugs.

2.4 Considerations in Experimental Conditions

Appropriate experimental conditions are a prerequisite for obtaining valuable quantitative data (De Lange et al. 1997). The relationship between dialysate concentrations and those in the extracellular fluid of the periprobe tissue, the recovery of the drug, depends on periprobe processes governing the actual concentration of the drug at that site. These include pressure gradients, extracellular–microvascular exchange, metabolism, and tissue diffusion of the drug. Experimental conditions, also between a control and treated group of subjects, or even disease conditions may influence these processes and may therefore have an impact on the actual in vivo recovery.

2.4.1 Static and Pulsatile Pressure

As already described for in vitro recovery, it has been considered that a higher flow rate will increase the fluid pressure inside the probe, which *may* result in net transport of fluid across the dialysis membrane and thereby counteract the diffusion of the drug into the dialysis fluid). In a study of Siaghy et al. (1999), this is exemplified by a microdialysis study in the heart, in which it was shown that mass transfer across a microdialysis membrane is dependent on the direction of the transmembrane pressure gradient and increases with heart rate, and must be taken into account in calculation of extracellular fluid concentrations.

2.4.2 Anesthesia

Pan et al. (1995) evaluated the effects of chloral hydrate and pentobarbital sodium on the basal extracellular concentrations of dopamine (DA) and norepinephrine (NA) as well as their in vivo extraction fraction (relative recovery) by the NNF method with conscious rats as controls. There were no significant effects of anesthesia on the basal extracellular concentrations of DA and NA. In vivo recovery of DA was not influenced by anesthesia, while for NA a higher recovery was observed for the conscious group.

2.4.3 Blood Flow

Kurosawa et al. (1991) found that changes in cerebral blood flow do not directly affect in vivo recovery of extracellular lactate from the rat brain through microdialysis probe. Relative radioactivity loss of (14C)-lactate from perfusate medium through a microdialysis probe was continuously measured in vivo as an indicator of relative recovery of extracellular lactate through the probe because both the relative recovery of lactate and the relative hot loss of (14C)-lactate through a microdialysis probe were similar to each other in vitro. The relative hot loss of (14C)-lactate decreased in parallel with decreases in the striatal blood flow, while it did not significantly change in response to increases in the blood flow up to 200 % of control. These results demonstrate that recovery of extracellular lactate in vivo through the microdialysis probe was not directly influenced by changes in the cerebral blood flow.

Clough et al. (2002) investigated the impact of changes in local blood flow on the recovery of a small, diffusible molecule (sodium fluorescein) from the extravascular tissue space of the human skin, by microdialysis in vivo, under conditions of basal, reduced (noradrenaline), and increased (glyceryl trinitrate) blood flow. Whereas loss of tracer from the delivery probe appeared unaffected by changes in local blood flow, retrieval of fluorescein by the second probe was directly related to blood flux, measured using scanning laser Doppler imaging. They concluded that clearance of solute by the blood will have a significant impact on microdialysis probe recovery and that, in the skin, the magnitude of this clearance is directly related to blood flow.

2.4.4 Transporter Functionality and Metabolism

Intracerebral microdialysis probe recovery (extraction fraction) may be influenced by several mass transport processes, including efflux and uptake exchange between brain and blood. The mdr1a-encoded P-glycoprotein is an active pump being expressed among other tissue linings, on the luminal face of the cells constituting the blood–brain barrier. Thereby, P-glycoprotein will counteract brain penetration of its widely diverse substrates as implicated by large differences in brain distribution of a number of P-glycoprotein substrates observed between mdr1a (−/−) and wild-type mice. Intracerebral microdialysis has been applied in mdr1a (−/−) and wild-type mice and the importance of quantification of microdialysate data will be exemplified beneath for two drugs for which in vivo recovery values were influenced by differences in active elimination of these drugs by absence versus presence of mdr1a-encoded P-glycoprotein. Using the NNF and DNNF techniques this was shown for Rhodamine-123 (De Lange et al. 1998), morphine (Xie et al. 1999) and sparfloxacin (De Lange et al. 2000). In all cases, the in vivo recovery values were lower for the mdr1a (−/−) mice. This is in line with the theoretical mathematical framework on in vivo concentration recovery developed by Bungay et al. (1990), which indicates lower in vivo recovery values in case of absence of an eliminating process (absence of Pgp) compared with presence of this elimination process.

Sun et al. (2001) investigated further the effect of capillary efflux transport inhibition on the determination of probe recovery during in vivo microdialysis in the brain. The effect of inhibiting transport on probe recovery was investigated for two capillary efflux transporters, Pgp and an organic anion transporter, with potentially different membrane localization and transport mechanisms. Fluorescein/probenecid and quinidine/LY-335979 were chosen as the substrate/inhibitor combinations for organic anion transport and P-glycoprotein-medicated transport, respectively. Probenecid decreased the probe recovery of fluorescein in the frontal cortex, in line with a reduction of the total brain elimination rate constant of fluorescein. In contrast, the microdialysis recovery of quinidine, delivered locally to the brain via the probe perfusate, was not sensitive to P-glycoprotein inhibition by systemically administered LY-335979, a potent and specific inhibitor of P-glycoprotein. Recovery of difluorofluorescein, an analog of fluorescein, was also decreased by probenecid in the frontal cortex but not in the ventricle cerebrospinal fluid. These studies suggest that only in certain circumstances will efflux inhibition at the blood–brain barrier and blood–cerebrospinal fluid barrier influence the microdialysis probe recovery, and this may depend upon the substrate and inhibitor examined and their routes of administration, the localization and mechanism of the membrane transporter, as well as the microenvironment surrounding the probe.

Stenken et al. (1997) investigated the impact of metabolism on microdialysis recovery of phenacetin and antipyrine in the liver, by local infusion of these drugs with/without intravenous administration of an inhibitor of metabolism. In the rat, the results indicated that microdialysis recoveries measured after inhibition of a concentration-dependent kinetic process via pharmacological challenge will change only when the kinetic process that is being challenged is large compared to the contribution of all concentration-dependent kinetic processes, including other metabolism routes, capillary exchange, or uptake that remove the analyte from the tissue space. It is concluded that the microdialysis recovery of a substance from the liver is not generally affected by liver metabolism.

2.4.5 Composition and Temperature of the Perfusate

Perfusion media used in microdialysis experiments vary widely in composition and pH. Ideally the composition, ion strength, osmotic value, and pH of the perfusion solution should be as close as possible to those of the extracellular fluid of the dialyzed tissue. Extracellular fluids mostly contain only very small concentrations of proteins. But, in some cases, proteins have been added to the perfusion medium to prevent sticking of the drug to the microdialysis probe and tubing connections Also, the composition of the perfusate can be changed with the intention to study the effects on the system under investigation. Deviations in ion composition have been shown to affect brain dialysate levels of neurotransmitters and drugs. De lange et al. (1994) have shown that a non-physiological (hypotonic) perfusion medium used in daily repeated experiments to measure blood–brain barrier permeability resulted in a substantial increase of the dialysate levels of the hydrophilic drug atenolol with days, presumably reflecting increased blood–brain barrier permeability.

Most investigators use a perfusion medium at room temperature before entering the probe. As a result, a temperature gradient exists between the probe and its environment. This may have an effect on tissue processes and consequently on the results. De Lange et al. (1995) used a subcutaneous cannula by which the perfusion fluid could equilibrate to body temperature before entering the cortical brain probe. The results on area under the concentration–time curve values for acetaminophen following intravenous administration obtained with "prewarmed" isotonic and hypotonic perfusate were compared to those obtained with perfusate at room temperature. A temperature effect was observed only for the use of the hypotonic perfusate, with a twofold higher dialysate area under the concentration–time curve value obtained with the room temperature perfusion medium). It was hypothesized that the periprobe tissue, already "stressed" by the hypotonic condition, loses its capability to compensate temperature effects. This indicates that perfusate temperature may be especially important in pathological circumstances. However, it is recommended to perform all microdialysis experiments with perfusion fluids at body temperature.

2.4.6 Interactions of Probe Material with Drug or Tissue

Jensen et al. (2007) investigated the application of microdialysis for sampling of free fatty acids (FFA). In vitro and in vivo studies were performed. The recovery of oleic acid was found to be dependent on the concentration of analyte in the medium surrounding the membrane, but this addition of BSA to the perfusion fluid resulted, however, in a concentration-independent recovery. This indicated an interaction of the FFA with the microdialysis membrane and or tubing. However, in vitro, it was observed that not for glycerol but for FFA's the recovery of FFA's declined indicating clogging of the membrane pores by these compounds.

Other examples of changes in transport of compounds over microdialysis membrane exist, and may also point to tissue reactions at the implantation site (Wisniewski et al. 2001), upon long-term use (Mou et al. 2010; Georgieva et al. 1993). Changes in periprobe tissues (Georgieva et al. 1993) while also tissue compatibility has been reported (De Lange et al. 1995, 1997).

2.5 Considerations in Data Analysis

Finally, as concentration–time profiles of drugs can be obtained by the microdialysis method, pharmacokinetic parameters can be derived. To that end, it should be realized that microdialysate data represent "average" concentrations obtained within the sampling interval, typically during a period between 5 and 20 min. So, this should be considered in the description of the concentration–time profiles (Ståhle 1992). For pharmacokinetic data, in most cases, the use of the interval midpoint to relate the dialysate concentrations to is valid. However, adjustments are required when the half-life of a process, such as absorption or elimination, is short in comparison with the sampling interval. Calculations of half-life and slopes are similar to standard methods for equal sample intervals. Calculation of area-under-the-curve and clearance values may be even more accurate for microdialysis data than for normal sampling at discrete time-points. This is because of the time-integral character of the microdialysis technique as well as the typical more frequent data points. Ståhle (1992) used a general method to deal with multicompartment models. Then, adjusted time points are not equidistant although sample intervals are equal. This is because the slope will be different in different portions of the natural logarithmic of the concentrations depicted versus time. Then, the time point T, measured from the initial time-point t_i of the sampling interval at which the extracellular concentration coincides with the dialysate concentration should be found. With the assumption that the data can be locally approximated by a mono-exponential function the following relation can be derived:

$$T/\text{ª}t = [\ln(q) - \ln(1 - e^{-q})]/q$$

Here $q = K\text{ª}t$, with k being the slope of the log(conc) versus time profile. When $\text{ª}t \ll t2$, $T/\text{ª}t$ will be about 0.5, the interval midpoint. For longer sampling intervals, $T/\text{ª}t$ will shift to values lower than 0.5. Thus, a drug with a first phase elimination half-life of 10 min, a second phase elimination half-life of 30 min and a sampling interval of 10 min the values of the sampling midpoint $T/\Delta t$ will be 0.42 and 0.47 in the first and second elimination phase respectively

2.6 Conclusions

Microdialysis is a very powerful technique to obtain quantitative data on concentration–time profiles of unbound drugs in virtually any tissue in animals and human. One issue is the need for calibration when using microdialysis probe perfusion flow rates by which equal concentrations between periprobe extracellular fluids will not be reached. In such conditions, many calibration techniques and approaches have been proposed, with different levels of practicality, usefulness, and assumptions, to obtain quantitative information on extracellular tissue concentrations of the drug as a function of time. The recently developed modified ultraslow microdialysis approach seems to set aside many limitations of the calibration techniques as concentrations obtained in the dialysate only have to be corrected for dilution by the carrier flow.

Thus, today we have build up enough experience to know how to obtain quantitative data using the microdialysis technique. With that, it is the only technique that provides data on the unbound concentration of compounds in extracellular fluid spaces in the body, being highly important information in drug development as the unbound concentrations in target tissues are the drivers for the effect (De Lange et al. 2005).

References

Ao X, Stenken JA (2006) No delayed temporal response to sample concentration changes during enhanced microdialysis sampling using cyclodextrins and antibody-immobilized microspheres. Analyst 131(1):62–67. Epub 2005 Oct 27

Bouw MR, Hammarlund-Udenaes M (1998) Methodological aspects of the use of a calibrator in in vivo microdialysis- further improvement of the retrodialysis method. Pharm Res 15:1673–1679

Brunner M, Joukhadar C, Schmid R, Erovic B, Eichler HG, Müller M (2000) Validation of urea as an endogenous reference compound for the in vivo calibration of microdialysis probes. Life Sci 67(8):977–984

Bungay PM, Morrison PF, Dedrick RL (1990) Steady-state theory for quantitative microdialysis of solutes and water in vivo and in vitro. Life Sci 46:105–119

Carneheim C, Stahle L (1991) Microdialysis of lipophilic compounds a methodological study. Pharmacol Toxicol 69:378–380

Clough GF, Boutsiouki P, Church MK, Michel CC (2002) Effects of blood flow on the in vivo recovery of a small diffusible molecule by microdialysis in human skin. J Pharmacol Exp Ther 302(2):681–686

Cremers TI, de Vries MG, Huinink KD, van Loon JP, v d Hart M, Ebert B, Westerink BH, De Lange EC (2009) Quantitative microdialysis using modified ultraslow microdialysis: direct rapid and reliable determination of free brain concentrations with the MetaQuant technique. J Neurosci Methods 178(2):249–254 Epub 2008

De Lange EC, Danhof M, de Boer AG, Breimer DD (1997) Methodological considerations of intracerebral microdialysis in pharmacokinetic studies on drug transport across the blood-brain barrier. Brain Res Brain Res Rev 25(1):27–49

De Lange EC, Marchand S, van den Berg D, van der Sandt IC, de Boer AG, Delon A, Bouquet S, Couet W (2000) In vitro and in vivo investigations on fluoroquinolones; effects of the P-glycoprotein efflux transporter on brain distribution of sparfloxacin. Eur J Pharm Sci 12(2):85–93

De Lange EC, Ravenstijn PG, Groenendaal D, van Steeg TJ (2005) Toward the prediction of CNS drug-effect profiles in physiological and pathological conditions using microdialysis and mechanism-based pharmacokinetic-pharmacodynamic modeling. AAPS J 7(3):E532–E543

De Lange ECM, Danhof M, De Boer AG, Breimer DD (1994) Critical factors of intracerebral microdialysis as a technique to determine the pharmacokinetics of drugs in rat brain. Brain Res 666:1–8

De Lange ECM, De Bock G, Schinkel AH, De Boer AG, Breimer DD (1998) BBB transport of rhodamine-123 in mdr1a(-/-) and wild-type mice. Total brain versus microdialysis concentration profiles. Pharm Res 15:1657–1665

De Lange ECM, Zurcher C, Danhof M, De Boer AG, Breimer DD (1995) Repeated microdialysis perfusions: periprobe tissue reactions and BBB permeability. Brain Res 702:261–265

Duo J, Fletcher H, Stenken JA (2006) Natural and synthetic affinity agents as microdialysis sampling mass transport enhancers: current progress and future perspectives. Biosens Bioelectron 22(3):449–457 Epub

Georgieva J, Luthman J, Mohringe B, Magnusson O (1993) Tissue and microdialysate changes after repeated and permanent probe implantation in the striatum of freely moving rats. Brain Res Bull 31(5):463–470

Hashimoto Y, Murakami T, Kumasa C, Higashi Y, Yata N, Takano M (1998) In vivo calibration of microdialysis probe by use of endogenous glucose as an internal recovery marker: measurement of skin distribution of tranilast in rats. J Pharm Pharmacol 50(6):621–626

Hsiao JK, Ball BA, Morrison PF, Mefford IN, Bungay PM (1990) Effects of different semipermeable membranes on in vitro and in vivo performance of microdialysis probes. J Neurochem 54:1449–1452

Jensen SM, Hansen HS, Johansen T, Malmlöf K (2007) In vivo and in vitro microdialysis sampling of free fatty acids. J Pharm Biomed Anal 43(5):1751–1756. Epub 2006

Jacobson I, Sandberg M, Hamberger A (1985) Mass transfer in brain dialysis devices- a new method for the estimation of extracellular amino acids concentration. J Neurosci Meth 15:263–268

Kaptein WA, Zwaagstra JJ, Venema K, Korf J (1998) Continuous ultraslow microdialysis and ultrafiltration for subcutaneous sampling as demonstrated by glucose and lactate measurements in rats. Anal Chem 70(22):4696–4700

Kurosawa M, Hallström A, Ungerstedt U (1991) Changes in cerebral blood flow do not directly affect in vivo recovery of extracellular lactate through microdialysis probe. Neurosci Lett 126(2):123–126

Larsson CI (1991) The use of an "internal standard" for control of the recovery in microdialysis. Life Sci 49:73–78

Le Quellec A, Dupin S, Genissel P, Saivin S, Marchand B, Houin G (1995) Microdialysis probes calibration: gradient and tissue dependent changes in no net flux and reverse dialysis methods. J Pharmacol Toxicol Methods 33(1):11–16

Lindberger M, Tomson T, Lars S (2002) Microdialysis sampling of carbamazepine, phenytoin and phenobarbital in subcutaneous extracellular fluid and subdural cerebrospinal fluid in humans: an in vitro and in vivo study of adsorption to the sampling device. Pharmacol Toxicol 91(4):158–165

Lonnroth P, Jansson PA, Smith U (1987) A microdialysis method allowing characterization of intercellular water space in humans. Am J Physiol 253:E228–E231 Endocrinol Metab 16

Menacherry S, Hubert W Jr (1992) In vivo calibration of microdialysis probes for exogenous compounds. Anal Chem 64(6):577–583

Morrison PF, Bungay PM, Hsiao JK, Ball BA, Mefford IN, Dedrick RL (1991) Quantitative microdialysis: analysis of transients and application to pharmacokinetics in brain. J Neurochem 57:103–119

Mou X, Lennartz MR, Loegering DJ, Stenken JA (2010) Long-term calibration considerations during subcutaneous microdialysis sampling in mobile rats. Biomaterials 31(16):4530–4539. Epub 2010

Olson RJ, Justice JB (1993) Quantitative microdialysis under transient conditions. Anal Chem 65:1017–1022

Pan WH, Lai YJ (1995) Anesthetics decreased the microdialysis extraction fraction of norepinephrine but not dopamine in the medial prefrontal cortex. Synapse 21(1):85–92

Ross HA, van Gurp PJ, Willemsen JJ, Lenders JW, Tack CJ, Sweep FC (2006) Transport within the interstitial space, rather than membrane permeability, determines norepinephrine recovery in microdialysis. J Pharmacol Exp Ther 319(2):840–846. Epub 2006

Scheller D, Kolb J (1991) The internal reference technique in microdialysis: a practical approach to monitoring dialysis efficiency and to calculating tissue concentration from dialysate samples. J. Neurosci. Meth 40:31–38

Scheyer RD, During MJ, Spencer DD, Cramer JA, Mattson RM (1994) Measurement of carbamazepine and carbamazepine epoxide in the human brain using in vivo microdialysis. Neurology 44:1469–1472

Schwalbe O, Buerger C, Plock N, Joukhadar C, Kloft C (2006) Urea as an endogenous surrogate in human microdialysis to determine relative recovery of drugs: analytics and applications. J Pharm Biomed Anal 41(1):233–239. Epub 2005

Siaghy EM, Oesterlé B, Kheiri A, Halejcio-Delophont P, Ungureanu-Longrois D, Villemot JP, Mertes PM (1999) Consequences of static and pulsatile pressure on transmembrane exchanges during in vitro microdialysis: implication for studies in cardiac physiology. Med Biol Eng Comput 37(2):196–201

Ståhle L (1991) Drug distribution studies with microdialysis: I. Tissue dependent difference in recovery between caffein and theophylline. Life Sci 49:1835–1842

Ståhle L (1992) Pharmacokinetic estimations from microdialysis data. Eur J Clin Pharmacol 43:289–294

Stenken JA, Lunte CE, Southard MZ, Ståhle L (1997) Factors that influence microdialysis recovery. Comparison of experimental and theoretical microdialysis recoveries in rat liver. J Pharm Sci 86(8):958–966

Strindberg L, Lönnroth P (2000) Validation of an endogenous reference technique for the calibration of microdialysis catheters. Scand J Clin Lab Invest 60(3):205–211

Sun H, Bungay PM, Elmquist WF (2001) Effect of capillary efflux transport inhibition on the determination of probe recovery during in vivo microdialysis in the brain. J Pharmacol Exp Ther 297(3):991–1000

Tao R, Hjorth S (1992) Differences in the in vitro and in vivo 5-hydroxytryptamine extraction performance among three common microdialysis membranes. J Neurochem 59:1778–1785

Wang Y, Wong SL, Sawchuck RJ (1991) Comparison of in vitro and in vivo calibration of microdialysis probes using retrodialysis. Curr Sep 10:87

Wang Y, Wong SL, Sawchuck RJ (1993) Microdialysis calibration using retrodialysis and zero-net-flux: application to a study of the distribution of zidovudine to rabbit cerebrospinal fluid and thalamus. Pharm Res 10:1411–1419

Ward KW, Medina SJ, Portelli ST, Mahar Doan KM, Spengler MD, Ben MM, Lundberg D, Levy MA, Chen EP (2003) Enhancement of in vitro and in vivo microdialysis recovery of SB-265123 using Intralipid and Encapsin as perfusates. Biopharm Drug Dispos 24(1):17–25

Wisniewski N, Klitzman B, Miller B, Reichert WM (2001) Decreased analyte transport through implanted membranes: differentiation of biofouling from tissue effects. J Biomed Mater Res 57(4):513–521

Xie R, Hammarlund-Udenaes M, de Boer AG, De Lange ECM. (1999) The role of P-glycoprotein in blood-brain barrier transport of morphine: transcortical microdialysis studies in mdr1a (−/−) and mdr1a (+/+) mice. Br J Pharmacol (in press)

Yokel RA, Allen DD, Burgio DE, McNamara PJ (1992) Antipyrine as a dialyzable reference to correct differences in efficiency among and within sampling devices during in vivo microdialysis. J. Pharm. Meth 27:135–142

Zheng H, Shi LF, Hu JH (2007) Assessment of in vitro and in vivo recovery of sinomenine using microdialysis. Skin Res Technol 13(3):323–329

Chapter 3
Analytical Considerations for Microdialysis Sampling

Justin C. Cooley, Michael W. Ducey, Anne R. Regel, Pradyot Nandi, Susan M. Lunte and Craig E. Lunte

Abstract In this chapter, a wide variety of analytical techniques that can be used in combination with microdialysis sampling will be described. Most of these methods are also commonly used for other biological samples, such as blood, urine, and tissues. However, when using microdialysis, the analysis and sampling must be viewed as a unified experiment where a parameter directly affecting one aspect will indirectly affect all the other components of the experiment. It is therefore almost never appropriate to simply apply an analytical method developed for a different sampling system to microdialysis sampling. The analyst must develop a method with the recognition of the small sample volumes available, the difficulty in sample preconcentration, and the high ionic strength of the matrix. Therefore, the discussion will focus on the particular developments needed to most appropriately apply various analytical methods to microdialysis sampling.

J. C. Cooley · A. R. Regel · C. E. Lunte (✉)
Ralph N. Adams Institute for Bioanalytical Chemistry and Department of Chemistry, University of Kansas, Lawrence, KS, USA
e-mail: clunte@ku.edu

M. W. Ducey
Ralph N. Adams Institute for Bioanalytical Chemistry and Department of Chemistry, Missouri Western State University, St. Joseph, MO, USA

S. M. Lunte
Department of Pharmaceutical Chemistry and Ralph N. Adams Institute for Bioanalytical Chemistry and Department of Chemistry, University of Kansas, Lawrence, KS, USA

P. Nandi
Department of Pharmaceutical Chemistry and Ralph N. Adams Institute for Bioanalytical Chemistry, University of Kansas, Lawrence, KS, USA

M. Müller (ed.), *Microdialysis in Drug Development*, AAPS Advances in the Pharmaceutical Sciences Series, DOI: 10.1007/978-1-4614-4815-0_3,

3.1 Analytical Challenges of Microdialysis Sampling

Microdialysis sampling poses several analytical challenges. The general issue is that the design of the microdialysis experiment dictates the requirements of the analytical method employed or conversely that the limits/requirements of the analytical method dictate the parameters of the microdialysis experiment (Fig. 3.1). By its nature, microdialysis is used to monitor dynamic processes and therefore the temporal resolution of the sampling experiment is a key parameter. However, the volume of the microdialysis sample is directly proportional to the sampling interval; therefore, the sampling interval must be sufficient to provide the amount of sample volume required by the analytical method. The microdialysis sample volume is also directly proportional to the perfusion flow rate, higher flow rates providing more sample volume over the same sampling interval. However, the relative recovery decreases with increasing perfusion rate, so that although larger sample volumes are collected they are at significantly lower concentrations. Higher perfusion rates require low concentration limits of detection. Finally, the goal of matching the perfusion solution to the extracellular fluid of the sampled tissue results in microdialysis samples being of high ionic strength. These interdependencies of the microdialysis experiment and the associated analytical method require that analytical methods to be used with microdialysis sampling must have low concentration limits of detection, small sample volume requirements, and be able to tolerate high ionic strength sample matrices. A final consideration that is often overlooked is that the typical microdialysis sampling experiment results in a significantly greater number of samples than other sampling approaches, requiring that analytical methods for microdialysis studies should have relatively high sample throughputs.

3.2 Analytical Techniques for Off-Line Analysis

The most common design of a microdialysis experiment is to collect the microdialysis samples and store them for later analysis. The lack of protein in the samples allows for long-term stability of analytes against enzymatic degradation, allowing samples to be stored at −80 °C for extended periods of time. Off-line analysis is typically accomplished using high performance liquid chromatography (HPLC) or capillary electrophoresis (CE), as these techniques lend themselves to small sample volumes of complex mixtures. Using this approach, the dialysate is collected for a specific amount of time, so that the flow rate (typically 1 μL/min) will determine the volume collected (typically 5–10 μL). Sample collection can be manual or automated. While use of an autosampler requires significantly less operator attention and is often more precise than manual collection, many analytes require immediate preservation that requires manual manipulation. While simple and robust, automated sample collection with off-line analysis can be more challenging when high temporal

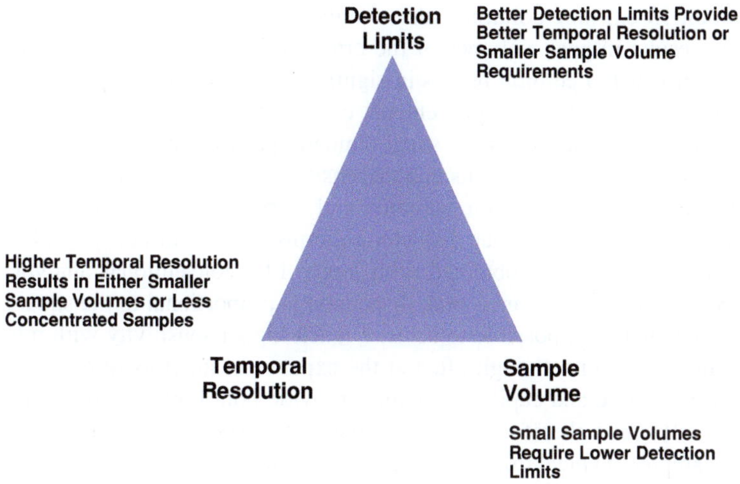

Fig. 3.1 Parameter interactions of temporal resolution, sample volume, and detection limits associated with microdialysis

resolution, very low flow rates, or low volumes are required. Challenges include evaporation for small volumes and loss of temporal resolution for very low flow rates. These issues are further magnified when working with awake, freely moving animals due to the larger dead volumes associated with connecting tubing. In response to these issues, other sample collection methods and online analysis have been developed to improve automation and better handle small volumes. These approaches will be discussed later in this chapter.

3.2.1 Liquid Chromatography

Liquid chromatography (LC) is the most commonly used analytical method for the separation of analytes present in microdialysis samples. The high ionic strength aqueous nature of microdialysis samples and the wide range of analytes of interests make LC the most robust separation technique for microdialysis samples. While the choice of stationary phase to use is dictated by the physiochemical properties of the analytes of interest, reversed phase systems are by far the most widely used for microdialysis samples. The primary reason being that microdialysis sampling selects for water soluble compounds. Other types of stationary phases, such as ion exchange, mixed mode, phenyl, amino, and carbohydrate columns have also been used for unique applications.

The other characteristics of the column (length, particle size, and internal diameter) are determined by the sampling interval desired and the required sensitivity and detection limits. Chromatographic methods are by their nature diluting, the eluting sample zones are in significantly greater volumes than when injected.

A common 15-cm long column with a 4-mm i.d. and packed with 5 μm particles typically results in peak volumes of the order of 500 μL. Injecting small sample volumes into such a column results in significant dilution and requires the use of high sensitivity detectors. Improvements can be made by use of microbore and capillary columns, and the use of small diameter packing materials. Both of these approaches lead to improvements in both sensitivity and temporal resolution due to the need for smaller injection volumes and improved resolution (Schultz and Kennedy 2008). These smaller diameter columns (<2.1 mm i.d.) provide separations equivalent to those obtained with larger I.D. conventional analytical columns. More importantly, since peak dispersion is proportional to the square of the column diameter, it is possible to achieve much higher sensitivity with microbore and capillary columns, though often at the expense of analysis time.

Short microbore and capillary columns provide both high sensitivity and rapid analysis (<10 min); therefore, they are a popular choice for the analysis of microdialysis samples. Typical improvements range from a 20-fold improvement relative to standard 4.6 mm analytical columns to 2,000-fold improvement when capillary columns are employed (Zang 2001). These methods do present several challenges including the need for low flow rate, pulse-free pumps, and minimized dead volumes. Dead volumes can be reduced by minimizing the length and the internal diameter of all tubing. Dead volume can also be reduced by employing low volume flow cells.

Capillary columns have very small internal diameters (150–300 μm) and offer the advantage of high mass sensitivity, low flow rate requirements, and low consumption of sample and reagents. Typical injection volumes are 50 nL and flow rates are of the order of 1–100 μL/min. These aspects make these columns particularly well suited for online microdialysate analysis, (Baseski et al. 2005; Shackman et al. 2006); however, separation times can be relatively long, making use of these columns for online analysis more challenging. In addition, capillary LC columns are not as robust as conventional LC columns (prone to clogging) and require specialized equipment, such as internal loop nanoliter volume injectors, to load samples. It is important to keep in mind that detection methods and separation efficiency are also key parameters of sensitivity, which will be addressed in later sections.

Ultra-high Performance Liquid Chromatography (UPLC) refers to modern LC systems using very small particles (>2 μm), short columns (>5 cm), and high flow rates (<0.5 mL/min) to achieve fast, high efficiency separations using small sample volumes. UPLC requires specialized LC equipment due to the extremely high pressures generated. UPLC should be well suited to the analysis of MD samples. It can separate the same samples as HPLC using identical separation fundamentals, but require smaller volumes (1 μL or less is typical), with greater speed, and improved resolution. To date, however, very few reports have been published describing the use of UPLC with MD sampling.

In one of the earliest reports Huang and coworkers (Huang et al. 2008) described the quantitation of baicilin (a flavone commonly used in Chinese medicine) in rat CSF by MD sampling coupled with UPLC-MS. Baicilin could be separated in less than 1.6 min with limits of detection in CSF of 1.185 ng/mL. In spite of the rapid separation sampling was only performed at 20 min intervals.

Table 3.1 Common detection methods employed with microdialysis coupled with liquid chromatographic analysis

Method	Representative analytes and detection limits
Chemiluminescence	Nitric oxide (nM) ,Yao et al. (2002a), Wang et a. (2001), Raimondi et al. (2007)
	Antioxidant capacity Yao et al. (2002b)
Electrochemical	Dopamine, aromatic amines, and endogenous quinones (nM–pM) Church and Justice 1987, Zhong et al. (1996), Pravda et al. (1997)
	8-oxoG and 8-OHdG (nM) Arnett et al. (2005)
	Glycerol and Mannitol (nM) Tuma et al. (2010)
	3-mercaptopropionic acid (μM) Crick et al. (2007)
Enzyme assay	Glutamate (μM) Obrenovitch and Zilkha (2001), Obrenovitch et al. 1990
	Glucose and lactate (μM) Hopwood et al. 2005, Deeba et al. (2008)
Fluorescence	Hydroxytyrosol (nM to μM) Wu et al. (2009)
	Levofloxacin (nM) Chen et al. (2002)
	Amino acid neutrotransmitters (nM) Sun et al. (2011), Klinker and Bowser (2007)
Mass spectrometry	Amino acid neurotransmitters and aromatic amines (fM to pM) Uutela et al. (2009), Sun et al. (2011, Buck et al. 2009)
	GSH and GSSG (nM) Robin et al. (2011)
	Oxcarbazepine and metabolites (nM) Paglia et al. (2007)
Ultraviolet	Acetaminophen and metabolites (nM-μM) Steele and Lunte (1995)
	Cephalexin (μM) Tsai et al. (2000)
	Nicotine and cotinine (nM) Chang et al. (2005)

Additionally, 5 μL samples were injected into the analytical system. The advantages of UPLC were further demonstrated by Liu et al. (2010) for the analysis of serotonin in mouse brain microdialysate samples. By injecting 500 nL dialysate samples and carrying out the separation at high temperatures and pressures, they were able to separate and quantify serotonin in less than 1 min.

The detection method employed for LC analysis is dependent on the analyte of interest and its relative concentration in the dialysate (Nandi and Lunte 2009; Perry et al. 2009). In general, UV detection has been a popular detection method for online monitoring of drugs for pharmacokinetic studies. Electrochemical detection has been employed for the detection of catecholamines and other redox active compounds of biological interest such as thiols and aromatic amines. Fluorescence detection, including laser-induced fluorescence (LIF) has been widely applied to the analysis of amino acids. Finally, mass spectrometry is becoming increasingly popular for analysis of microdialysis samples following liquid chromatographic separations. A summary of detection methods is identified in Table 3.1 and is discussed in detail in the section on detectors.

3.2.2 Capillary Electrophoresis

CE is a particularly attractive method for the analysis of microdialysis samples as a result of its very low sample volume requirements (1–10 nL injection volumes) and ability to perform extremely fast separations (<1 min). Analysis speed and separation efficiency may be improved in CE with increases in field strength (in the absence of Joule heating). The use of very short capillaries and high field strengths allows for very fast, highly efficient separations, where plate counts in the millions have been achieved.

Analysis of microdialysis samples by CE is not without challenge. One of the primary sample challenges is the high ionic strength associated with MD samples. CE is generally carried out under low ionic strength buffer conditions (10–50 mM is common) where microdialysis samples are, by necessity, in high salt concentrations (approximately 150 mM). The consequence of the high ionic strength is sample destacking (observed as band broadening) under separation conditions. Sample destacking occurs as a result of differences in field strength within the sample in contrast to the background electrolyte (BGE). To counter this band broadening, a variety of sample stacking approaches have been developed. Sample stacking for CE has been reviewed (Shihabi 2000; Mala et al. 2007).

The most general approach to sample stacking in CE is field amplified stacking (Chien and Burgi 1991, 1992) where the sample matrix is of significantly lower ionic strength than the BGE. For microdialysis samples, this involves a significant dilution of the sample with water. The high ionic strength nature of microdialysis samples requires such a significant dilution of the sample so as to make it practically useless for microdialysis samples. A modification of field amplification stacking, termed pH-mediated stacking, has been developed specifically for analyte preconcentration of microdialysis samples (Fig. 3.2) (Gillogly and Lunte 2005;

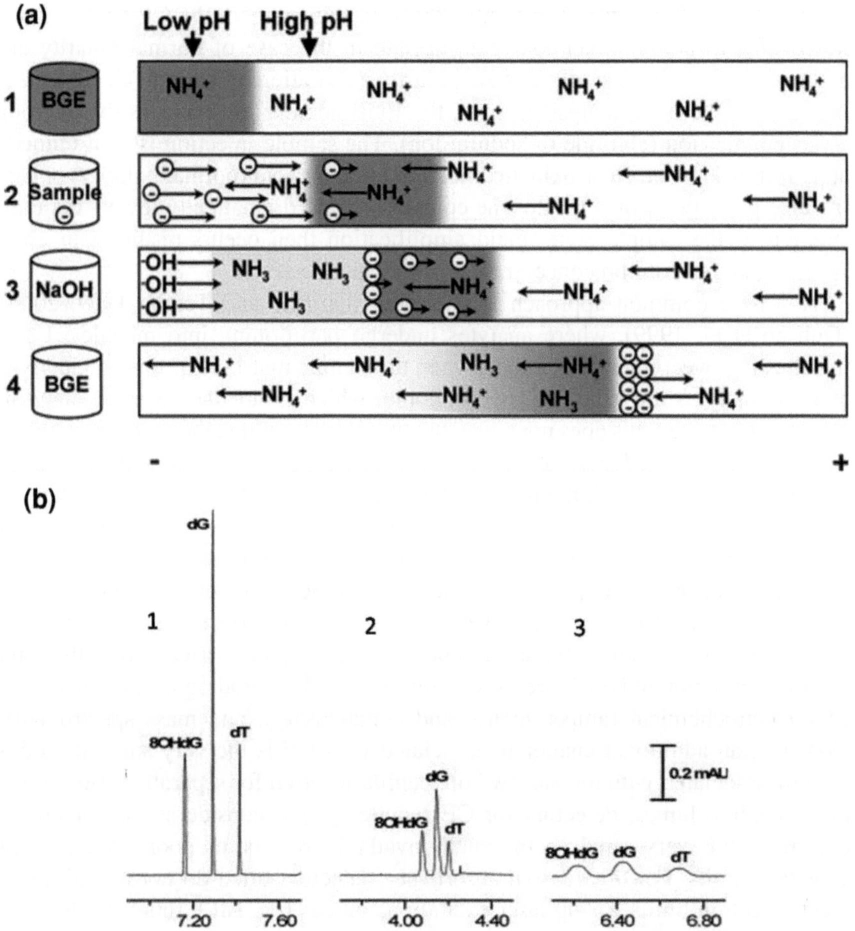

Fig. 3.2 Enhanced pH-mediated stacking of anions (base stacking) with a dynamic pH junction. **a** (*1*) a small plug (*shaded gray*) of lower pH BGE consisting of a weak base is injected hydrodynamically into the capillary filled with higher pH BGE also consisting of a weak base, (*2*) high ionic strength sample is injected electrokinetically and titrable BGE cations migrating toward the cathode replace untitrable sample cations (such as Na^+) in the sample zone, (*3*) basic solution is electrokinetically injected and hydroxide ions titrate BGE cations to form a low conductivity zone in which analyte velocity increases under the higher field strength, when analytes encounter both the high ionic strength BGE zone and the lower pH zone, they are concentrated at the interface due to both the lower field strength and decreased charge state, (*4*) the higher-pH BGE vial is replaced, and the titrated zone ceases to grow, as analytes exit the low pH, low conductivity zone, they regain charge state and can be separated in the BGE based on their electrophoretic mobilities before reaching the detection window. **b** CE-UV electropherograms of nucleoside standards 8-hydroxy-2-deoxyguanosine (*8OHdG*), 2-deoxyguanosine (*dG*), and thymidine (*dT*) in Ringer's solution with (*1*) enhanced base stacking, (*2*) normal base stacking, and (*3*) no base stacking. Adapted with permission from (Arnett and Lunte 2007) (Figs. 3.1 and 3.7)

Hoque et al. 2005; Arnett and Lunte 2007). In this method, the CE BGE is composed of a titrable counter ion (e.g., acetate in the case of normal polarity and ammonium in the case of reverse polarity separations). During electrokinetic injection of the microdialysis sample, the BGE counter ion replaces the microdialysis counter ion (chloride or sodium ion). The sample injection is then immediately followed by an electrokinetic injection of either acid (normal polarity) or base (reverse polarity), which titrates the counter ion and dramatically lowers the ionic strength of the sample zone. Field amplification then occurs resulting in up to several hundred fold preconcentration of the analyte zones.

A second common approach is sweeping (Quirino and Terabe 1999, 2000); (Quirino et al. 1999), where analytes undergo partitioning into micelles. Using micelles for sweeping in CE is ideal when the sample matrix is of low conductivity and the analytes are relatively hydrophobic, which is not the case for dialysate samples. However, this does not mean that the addition of micelles is only useful for separation purposes. Palmer et al. (1999) described how micelles will stack at the detector side of an injection plug when the sample is of higher conductivity than the BEG. In this instance, the micelles stack because of the decrease in electric field at the sample plug/BGE boundary. Essentially, this is field amplification that occurs in the buffer and not in the sample zone. Analytes will then stack at the sample/micelle boundary as a result of the reduced velocity from the high concentration of micelles.

Because CE is a liquid, typically aqueous, phase separation technique, the same detectors common in HPLC are most common for CE, including UV, fluorescence/LIF, electrochemical (amperometric and conductivity), and mass spectrometry. However, an additional challenge associated with CE is the very small detection volume associated with the narrow bore capillaries used for separation. Because of these small volumes, detectors for CE require special consideration. For optical detection, the very small path length available results in poor concentration detection limits. For this reason, absorbance detectors often do not provide sufficient detection limits. Using laser excitation sources (*i.e.* LIF), this limitation can be overcome for fluorescence detection. Electrochemical detection is also well suited for the small detection volumes of CE. For these reasons, LIF and EC detection are typically the preferred approaches for microdialysis experiments.

Finally, the high field strengths employed in CE require careful isolation of the separation circuit from that of the detector. This is especially true in cases where field strengths can impact detector response and noise (electrochemical and mass spectrometry). Specific details associated with isolation of the capillary from the detector are described in more detail below.

3.3 Detection Methods

A wide variety of detectors are available for use with LC and CE systems. Current trends in detection methods employed with microdialysis have been recently reviewed by Guihen and O'Connor (2009). In choosing a detector, the properties of the analyte or its derivative, concentration, and need for selectivity must be considered. Generally, monoamine neurotransmitters such as dopamine are amenable to electrochemical detection, amino acids are readily derivatized for fluorescence detection, and a majority of pharmacological agents are detectable by absorbance, fluorescence, and/or electrochemistry. Finally, nearly all chemically interesting species may be detected by mass spectrometry including metal ions.

3.3.1 UV–VIS Absorbance

Many exogenous species, including many pharmaceutical agents, are quantifiable using ultraviolet absorbance detection. UV–visible absorbance detectors are inexpensive and simple to use, though they suffer from relatively poor selectivity and high detection limits. This is a result of two factors including the low molar absorptivities of typical analytes and the short path length of the detection cell (especially in the case of CE). In spite of these challenges, absorbance detection remains a common detection method, especially when coupled with HPLC separations.

In a typical use of absorbance detection in MD sampling, Lada and Kennedy (1995) described the detection of ascorbate and lactate at 265 nm following dosing of amphetamine, anesthetic cocktail, and elevated potassium infusion. Using this approach they determined that basal level concentrations of ascorbate in the caudate nucleus were 333 μM. Injections of amphetamine were found to increase the ascorbate concentrations, while injections of anesthetic decreased the concentrations.

In a pharmacokinetics study, doses of amoxicillin or ketoprofen were correlated between microdialysis sampling in the skin and straight plasma sampling (Juluru et al. 2012). Both drugs were dosed either *i.v.* or orally, and detected using UV absorbance detection after HPLC separation. By correlating the area under the curve for both sampling methods, it was found that the systemic bioavailability fraction determined by microdialysis sampling was not statistically different from that of plasma (Juluru et al. 2012).

3.3.2 Fluorescence Detection

Fluorescence detection allows for the quantitation of very low concentrations of analyte (nM–pM) assuming an appropriate fluorophore is available. The vast majority of biologically and pharmaceutically interesting compounds do not possess innate fluorescence and must therefore be derivatized prior to detection. Several fluorescent derivatization reagents are commercially available (Table 3.2) for a wide variety of functional groups including amines, thiols, carbohydrates, carboxylic acids, aldehydes, and other functional groups. Fluorescence intensity is proportional to the quantum yield of the fluorophore, analyte concentration, and the intensity of the excitation source. It is through this last variable that very low

Table 3.2 Common fluorescent derivatizing agents

Functional group	Derivatizing agent
Amine/amino acid	3-(4-Carboxybenzoyl)-2-quinolinecarboxaldehyde (CBQCA) Liu et al. 1991, You et al. 1997; Napthalene dicarboxaldehyde/cyanide (NDA) Hernandez et al. 1993, Robert et al. 1995, Bert et al. 1996; Fluorescein isothiocyanate (FITC) Li et al. 2001, Li et al. 2003; 4-fluoro-7-nitro-2,1,3-benzoxadiazole (DBDF) Klinker and Bowser 2007, Zhao et al. 2005
Carbohydrate	dansylhydrazine Perez and Colon 1996; 2-Aminobenzamide (2-AB) Bigge et al. 1995; 2-aminopyridine (2-AP) Hase et al. 1978
Carboxylic acid	4-Bromoethyl-7-methoxycoumarin (Br-MMC) Wolf and Korf 1992; 9-anthryldiazomethane (ADAM) Wolf and Korf 1992, Katsumi et al. 1984, Norio et al. 1984; 3-Bromoethyl-6,7-dimethyl-2(1H)-quinoxalinone (Br-DMEQ) Yamaguchi et al. 1985; 2-(2,3-napthalimino)ethyl trifluoromethanesulfonate (NE-Otf) Yasaka et al. 1993; 4-(N,N-dimethylminosulfonyl)-7-(5-aminopentylamino)-2,1,3-benzoxadiazole (DBDCD) Toyo'oka et al. 1991; and 5-Bromoethylfluorescein (5-BMF) Mukherjee and Karnes 1996
Thiol	Maleimides Toyo'oka 2009 including N-(9-acridinyl)maleimide (NAM), N-(p-(2-(6-dimethylamino)benzofuranylphenyl)maleimide (DBPM), N-[p-(2-benzimidazolyl)phenyl)maleimide (BIPM), and fluorescein-5-maleimide (FM); 5,5′-dithiobis(2-nitrobenzoic acid) DTNB Guan et al. 2003
Nitrite	2,4-dinitrophenylhydrazine (2,4-DNPH) Sun and Stenken 2003; Griess reaction Tsikas 2007; 2,3-diaminonaphthalene (DAN) Misko et al. 1993
Aldehydes	fluorescein 5-thiosemicarbazide Banos and Silva 2011; dansylhydrazine (DNSH) Claeson et al. 2001; BODIPY-aminozide Xiong et al. 2010; FMOC-hydrazine Shangguan et al. 2001

detection limits may be achieved using LIF. A drawback to LIF is that the excitation source is not tunable (as is the case for grating systems using a xenon arc lamp) thus limiting the detector to examination of a few specific fluorophores.

NDA (naphthalene-2,3-dicarboxaldehyde) and OPA (*o*-phthalaldehyde) are the most common reagents employed for the detection of amino acid neurotransmitters and other primary amines in vivo. Both are fluorogenic and exhibit rapid reaction kinetics, making them well suited for precolumn derivatization prior to analysis. Online reaction times for OPA/β-ME and NDA/CN with primary amines have been reported to be 10–30 s and 120–240 s, respectively (Shou et al. 2004).

3.3.3 Chemiluminescence

Applications of chemiluminescence detection in microdialysis are not nearly as common as absorbance and fluorescence methods. This is primarily due to the fact that most analytes do not possess intrinsic chemiluminescence, and many chemiluminescent derivatives are large which can influence chromatographic and electrophoresis properties of the analyte. Chemiluminescence detection does, however, lend itself well to the detection of short life radical species including nitric oxide (NO). With the many important physiological roles played by NO, its quantitation is of interest based on its chemical messenger properties as well as its role in response to tissue damage. NO can be quite challenging to quantify as it has a short solution half-life (445 s) (Hakim et al. 1996) and low physiological concentrations in the brain (1–5 nM) (Yao et al. 2002).

The use of 3-aminopthalhydrazide (luminol) has been described for the detection of NO in rat brain MD samples (Yao et al. 2002; Wang et al. 2001; Yao et al. 2002; Raimondi et al. 2007). In these experiments, luminol is mixed with the MD sample stream (dialysate) and passed immediately into a chemiluminescence detector. Wang et al. (2001) demonstrated that NO concentrations briefly increased to over 400 % of basal levels following traumatic brain injury. NO detection limits as low as 1 nM were reported by Yao and coworkers using a similar approach (Yao et al. 2002).

An indirect method for NO was developed in which the metabolites nitrite and nitrate (NO_2^- and NO_3^-) were determined in microdialysis samples using an ozone-phase chemiluminescence method. Samples are reacted with acidic vanadium to reduce NO_2^- and $NO_3^-$to NO which reacts with ozone to produce chemiluminescence (Jou and Ma 2009; Reynolds et al. 1998).

Luminol may also be employed in precapillary derivatization approaches as well. Wang and coworkers described the use of luminol to derivatize seven monosaccharides including glucose (Wang et al. 2003). The authors were able to separate and quantify glucose in rate brain dialysate using CE separation with chemiluminescence detection in an off-line format. The technique was quite sensitive with detection limits of the order of 2 pmol of glucose.

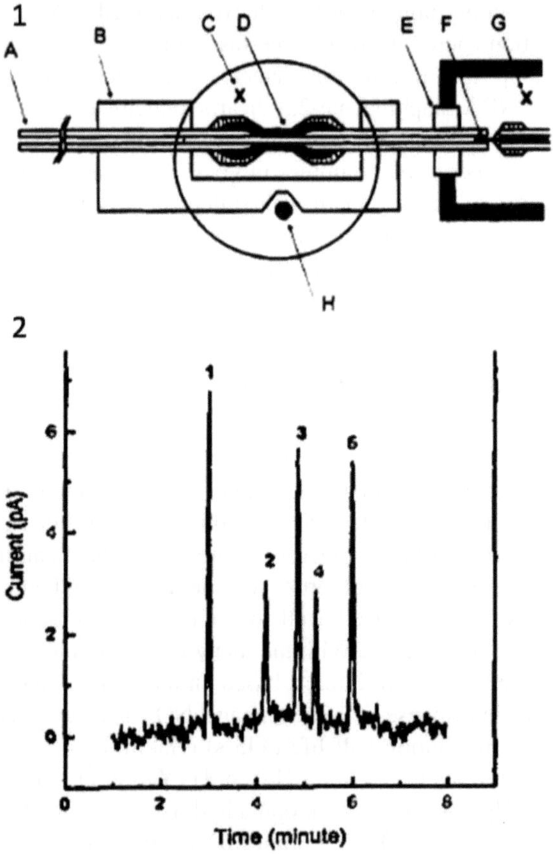

Fig. 3.3 **a** *Top view* of decoupled CE-EC system using Nafion joint, (*A*) Anodic CE capillary end, (*B*) epoxy board, (*C*) cathodic CE buffer reservoir, (*D*) cast Nafion membrane, (*E*) septum, (*F*) carbon fiber working electrode, (*G*) electrochemical cell reservoir, and (*H*) cathode for CE. **b** Electropherogram of 6 nM (*1*) hydroquinone and 12 nM (*2*) cholorogenic acid, (*3*) caffeic acid, (*4*) protocatechuic acid, and (*5*) gentistic acid. Adapted with permission from (Park et al. 1995) (Figs. 3.1 and 3.11a)

3.3.4 Electrochemical

Electrochemical (EC) detection provides several advantages including low detection limits (nM), selectivity, and no need for derivatization of the analyte prior to analysis. In addition, the detector is easily miniaturized for use in capillary LC and CE methods. Typically, amperometric detection is employed, where a working electrode is held at a redox potential specific to the analyte(s) of interest. Neurotransmitters (dopamine, norepinephrine, etc.) are particularly amenable to electrochemical detection. Church and Justice describe the use of electrochemical detection of dopamine in rat brain microdialysis samples separated by small bore LC (Church and Justice 1987). Many pharmaceutical compounds are also natively electroactive at analytically useful potentials. For example, isoproterenol has been determined in microdialysis samples using CE with electrochemical detection (Hadwiger et al. 1997). Similarly, the oxidized nucleotide 8-hydroxy deoxyguanosine has been detected in microdialysis samples by CE with electrochemical detection as a biomarker of DNA oxidation (Arnett et al. 2005).

Additional selectivity may be introduced using dual electrode detection where a second working electrode is placed immediately downstream of the first electrode for redox coupling. For example, 3-nitrotyrosine (3-NT) (a decomposition product of peroxynitrite and biomarker for RNS) oxidation occurs at high potentials (>800 mV vs Pd/H_2) which can lead to interferences from other electroactive compounds, at greater concentrations, recovered from microdialysate. To increase the selectivity of the method, 3-NT was first reduced to 3-aminotyrosine at the upstream electrode (−1,200 mV), and the subsequent oxidation of 3-aminotyrosine was then monitored at the downstream electrode at −100 mV, a much more selective potential (Richards et al. 2006). This method was used to detect 3-NT in liver dialysates at a concentration of 2 nM.

The coupling of electrochemical detection to CE requires that the detection electrode be isolated from the separation circuit by using a decoupler or placing the electrode outside of the capillary. A variety of approaches have been described to carry out this isolation including the introduction of a fracture in the capillary at the ground (Wallingford and Ewing 1987; Huang and Zare 1990; Linhares and Kissinger 1991; Yik et al. 1991; O'Shea et al. 1992; Lu and Cassidy 1994) and joining together two pieces of capillary via a conductive polymer joint (Park et al. 1995). Figure 3.3 shows the experimental set up of a system employed by Park et.al and an electropherogram for a 10 nM injection of dopamine, norepinephrine, and isoprotenol. An additional challenge in the application to CE is the size of the separation capillary. Carbon fiber microelectrodes are generally employed by placing them slightly within the separation capillary and locating the counter and reference immediately downstream (Park et al. 1997).

The use of capacitively coupled contactless conductivity detection (C4D) does not require the decoupling of the separation current and has been used with microdialysis samples (Petsch et al. 2004; Tuma et al. 2010). This type of detection, however, is not as selective as other electrochemical techniques and suffers from poor detection limits.

3.3.5 Mass Spectrometry

Mass spectrometric detection addresses a variety of the analytical challenges associated with MD sampling. Especially, using MS^n techniques, conclusive analyte identification can be accomplished with very low detection limits (low pM). While derivatization is not required for MS detection, a variety of reports describe analyte derivatization to enhance resolution in the separation step (particularly for GC), improve certainty in analyte identification, and to change the analyte properties to make them more amenable to ionization (Miller et al. 2011; Che et al. 1999; Uutela et al. 2009; Harvey 2011; Sun et al. 2011). A limitation of MS analysis for MD samples is ionization suppression due to the high ionic strength sample matrix of the dialysate. The application of mass spectrometry to the analysis of MD samples was recently reviewed by Lanckmans et al. (2008).

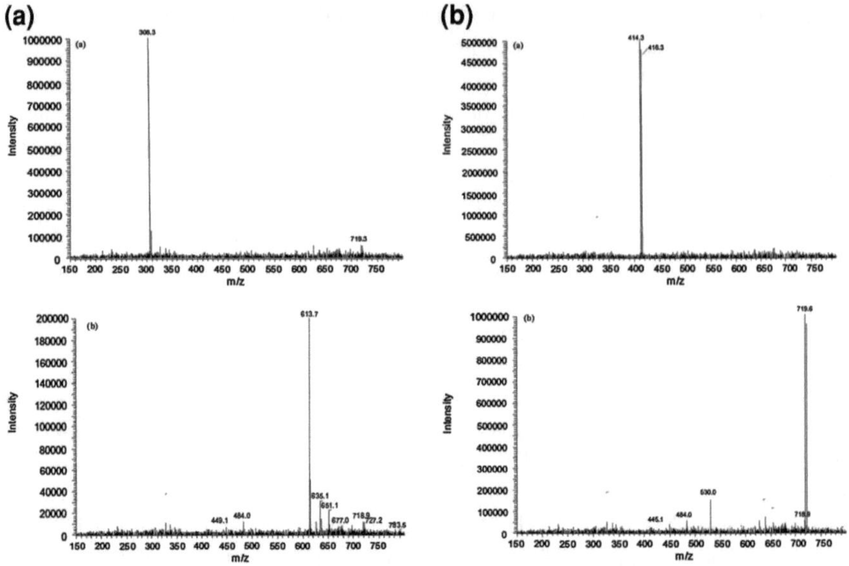

Fig. 3.4 Increased sensitivity for the detection of GSH and GSSG from adding Ag + post column separation. **a** ESI mass spectra of (*a*) GSH and (*b*) GSSG in positive ion mode without $AgNO_3$ post column reagent. **b** ESI mass spectra of (*a*) GSH and (*b*) GSSG in positive ion mode with $AgNO_3$ as post column reagent. Reprinted with permission from (Robin et al. 2011) (Fig. 3.1)

MS has been demonstrated to be particularly useful for the identification and quantitation of peptides due to their low abundance endogenously. Traditional means for quantifying peptides involve radioimmunoassay where significant cross-reactivity (especially between peptide, precursor and metabolites) can be a significant problem (Maidment et al. 1989; Consolo et al. 1994; Blakeman et al. 2001). To improve sensitivity for glutathione (GSH) and glutathione disulfide (GSSG), silver nitrate (100 μM at 0.05 ml/min) was infused as a post column additive to create a $[M + Ag]^+$ ion (Robin et al. 2011). The addition of Ag^+ improved peak intensity by fivefold for both GSH and GSSG decreasing limits of detection to 1.3 and 0.8 nM, respectively (Fig. 3.4). The concentrations of GSH and GSSG in dermal dialysate samples were 50 and 3 ng/ml, respectively.

As a detection method, MS may be coupled with HPLC (especially microbore) and CE. The high salt concentrations typical of MD samples provide an additional challenge requiring a desalting step prior to introduction into the MS. This may be accomplished using an ion exchange column prior to switching the sample into the separation column. Additionally, the separation conditions must be carried out in low salt buffer and high volatility solvents. While MS is useful as a detection method when coupled to HPLC or CE, the separation step is not always necessary. Matrix-assisted laser desorption ionization (MALDI) was used to investigate the metabolism of several opioid peptides including β-endorphin (Reed et al. 2008) and dynorphin A (Reed et al. 2003). Matrices such as N-(1-naphthyl) ethylenediamine dihydrochloride (NEDC) are ideal for the ionization of microdialysis samples because of its high salt tolerance and little interference at low molecular weight region. This matrix was sufficient at producing $[M - H]^-$ and $[M + Cl]^-$ ions in negative ion mode for the detection of submicromole of glucose, sucrose, aspartic acid, and glutamic acid in high ionic strength solutions with no pretreatment or desalting (Chen et al. 2012).

Pickl and Sinner developed an SPE-MS/MS method to study the pharmacokinetics of topically administered ketoprofen (a non-steroidal anti-inflammatory drug) (Pickl et al. 2007). In this method, 10 μl of sample was injected onto the SPE column followed by 50 μl (0.5 min) of water to flush the salt. The ketoprofen was then eluted and detected with a total time from injection to injection of 1.5 min.

3.3.6 Enzyme Assay

The use of enzyme assays (especially couple assays) allow for the rapid, and in some cases continuous, analysis of MD dialysate with very low detection limits, and without the need for a separation step (imparted by the high specificity of the enzyme). To date, most MD-coupled enzyme assays employ nicotinamide adenine dinucleotide (NAD)-dependent enzymes and produce fluorescent or electrochemical active products. These assays have been reviewed by Obrenovitch and Zilkha (2001). In general the use of MD-coupled enzyme assay allows for temporal resolution of the order of 2 min and is amenable to analytes that can serve as substrates for an NAD- or NADP-dependent enzyme.

Obrenovitch and coworkers examined changes in L-glutamate in ischemic rat brains (Obrenovitch et al. 1990). Using glutamate dehydrogenase and NAD^+, NADH was produced and quantified by fluorescence detection. Glutamate concentrations were monitored continuously over a 30-min perfusion of 100 mM potassium ion. When compared to an off-line HPLC method, the authors were able to demonstrate the relationship between glutamate concentration and spreading depression waves with much greater temporal resolution.

Glucose and lactate were detected in dialysate using a dual online enzymatic system (Fig. 3.5) (Hopwood et al. 2005). Outlet tubing was connected to a multiport injector programmed to switch between two reaction cells splitting the dialysate. Each cell consisted of either glucose or lactate oxidase with horseradish peroxidase and an amperometric detector set at −100 mV (vs Ag/AgCl) to reduce the ferricinium ion produced. This method was able to detect glucose and lactate on 15–30 s intervals. This detection method was applied to clinical bowel ischemia (Deeba et al. 2008).

The CMA 600 commercial analyzer uses enzymatic reagents and colorimetric detection for glucose, lactate, pyruvate, glycerol, glutamate, and urea. Up to four analytes can be measured per sample with volumes required between 0.2 and 1 μl/analyte and analysis time of about 2 min per analyte.

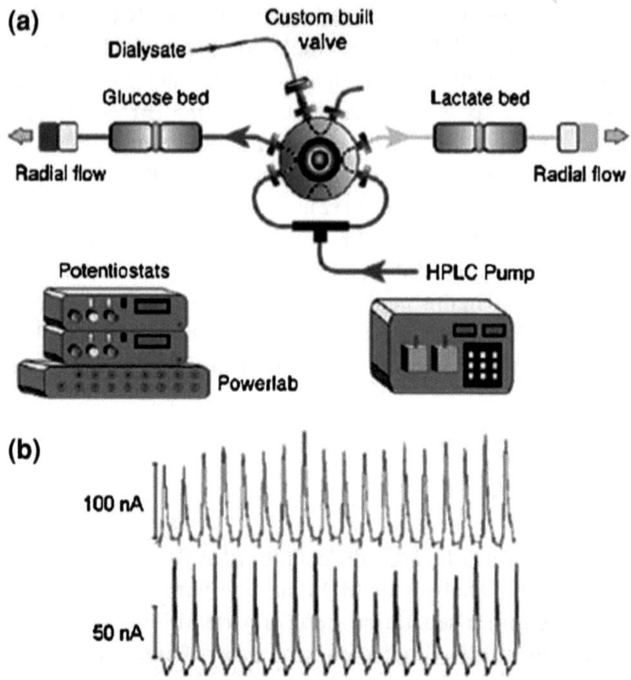

Fig. 3.5 On-line enzymatic assay for glucose and lactate. **a** Schematic of microdialysis assay showing sequential detection of glucose and lactate. Internal loops in the 2-position valve ensure delivery of fresh sample to the appropriate enzyme bed when the valve changes. **b** Example of the glucose and lactate peaks recorded from assay system. Reprinted with permission from Hopwood et al. (2005) (Fig. 3.1) SE, Parkin, MC, Bezzina, EL, Boutelle, MG, and Strong, AJ. J of Cerebral Blood Flow and Metabolism. 25 (2005), 391-401. (Fig. 3.1)

3.3.7 Derivatization Methods

Analyte derivatization is carried out to impart characteristics that improve the detection sensitivity and selectivity (*i.e.*, introduction of a fluorophore or chromophore). Derivatization may occur at a variety of stages in the analysis including prior to introduction into the separation system (CE or HPLC) in either an off- or online format, or carried out following the separation. A number of reviews and guides are available describing all manner of derivatization schemes and available reagents (Harvey 2011; Imai 1987; Bardelmeijer et al. 1997; Lunn and louise 1998; Waterval et al. 2000; Toyo'oka 2009; Szoko and Tabi 2010).

The choice of derivatization mode is dictated by several factors including the stability of both the analyte and the derivative, whether the MD sampling is off-line or online, the required reaction conditions, and the kinetics of the derivatization reaction. Additional factors that impact the choice of mode include the sample size and the presence of any interfering species in the sample matrix.

3.3.7.1 Pre-Capillary/Column Derivatization

Pre-separation derivatization is amenable to derivatizations requiring long reaction times, heating steps, or reaction conditions (pH, solvent conditions, etc.) widely different from the separation conditions. Pre-separation derivatization can be used in conjunction with both off- and online MD sampling. In general, the sample is combined with the derivatizing reagent(s), allowed to react, and injected for separation. In some cases, the resulting derivative may be quite stable and can be stored for days to months prior to analysis.

3.3.7.2 Online Derivatization

While the specific details of online MD sampling will be described below, the term online derivatization here applies to coupling the derivatization process with the separation and detection steps. Online derivatization involves combining and mixing the sample with the derivatizing agent(s) using a mixing T or similar junction immediately prior to the analysis system. The derivatized sample then passes out of the mixing T directly into the separation system. Online derivatization is especially useful for submicroliter samples and cases where high temporal resolution is required.

Lada, Vickory, and Kennedy developed an online MD-CE method for the analysis of glutamate and aspartate in anesthetized rat striatum (Lada et al. 1997). While the analytical method provided a sampling time of 5 s (resolving and quantifying the analytes), the temporal resolution (time to see a response), however, was limited to 12 s. This was attributed to significant extracapillary band broadening occurring with the MD probe and transfer tubing.

3.4 Online Versus Off-Line Analysis of Microdialysis Samples

Direct, "on-line" coupling of the microdialysis sample collection to the analysis method offers a number of advantages over off-line analysis (Westerink and Cremers 2007). Issues related to handling submicroliter sample volumes (sample loss, mislabeling, evaporation, and surface tension) are minimized as the sample is moved directly from collection to introduction into the analysis method. In addition to the minimizing sample handling issues, sample degradation by exposure to air (e.g., ascorbic acid and catecholamines) can be avoided (Tsai et al. 1996; Jin et al. 2008). Direct coupling between the collection and analysis allows for automation and submicroliter samples to be collected and analyzed. This small sample volume can improve temporal resolution, yielding near real-time data if the analysis method is fast.

In the choice of an analytical method, the first parameter to be addressed is whether the method possesses sufficient detection limits for the analyte of interest in the volume of sample generated at the desired sampling rate. For example, if a probe is perfused at a typical rate of 1 μL min^{-1} and 1 min temporal resolutions is desired, then the analytical system must have sufficient mass/concentration sensitivity to detect 1 pmol of analyte if it is at a concentration of 1 μM in the dialysate. If the analytical method is capable of detecting only 10 pmol or higher, then a much larger sample (10 μL) must be collected in order to quantify the analyte of interest, limiting the temporal resolution to 10 min at this flow rate.

In the case of online systems, temporal resolution is generally defined not by the collection rate, but rather by the analysis time. If the analysis time is longer than the duration of the event being measured, then real-time analyte monitoring cannot be done, as the change will appear digital and show up in the next injection. In contrast, when the analysis step is much faster than the event being measured, changes in analyte concentration may be measured and reported as a function of time. For rapid and transient analyte changes, the system rise time must also be considered, where the rise time is defined as the time required for the signal (analyte concentration) to increase from 10 to 90 % of maximum value (Mecker and Martin 2008). For very fast analyses, the rise time becomes limited by the rate of mass transport of the analyte across the probe membrane. In most cases, this is the limiting condition where the ultimate online system sampling rate limit is a function of the system dead volume (tubing, connections, etc.), injection method, and the flow rate of the dialysate.

As noted earlier, analyte recovery is a direct function of flow rate, where recoveries of 100 % can be approached with very low flow rates (*i.e.* 100 nL/min) (Menacherry et al. 1992). There is a direct tradeoff, however, in sample volume for the increase in analyte concentration in the dialysate. In this case, an analytical method that is both sensitive and capable of analyzing submicroliter sample volumes is necessary if 1 min or less temporal resolution is to be obtained. This is one of the driving forces for the use of capillary and microchip electrophoresis (ME) for the analysis of microdialysis samples.

Improved temporal resolution can be extremely useful in the case of neurochemical experiments as many neurochemical events (*i.e.* neurotransmitter release) occur on the second to subsecond time scale. Direct coupling of microdialysis sampling to the analytical method allows sample collection to approach these time scales, and therefore, observe these rapidly changing species. In contrast, with most (though not all) off-line methods these concentration changes would not be resolvable due to dilution and averaging of the signal with longer sampling times.

In an example of high temporal resolution sampling with off-line analysis, Rossell and coworkers described one-second resolution nanoliter microdialysis sampling from the brains of awake rats (Rossell et al. 2003). Using a unique sampling design, the dialysate was immediately combined with fluorescein isothocyante in a 26 nL reactor then collected in a 99 μm id capillary tube, where every 4 mm (30 nL) corresponded to 1 s. Following the sample collection the tube was separated into 4 mm segments with the contents allowed to derivatize for 18 h, followed by CE analysis.

The consequence of the temporal, volume, and concentration limitations generally results in the application of microdialysis systems with online, separation-based analysis for analytes with high in vivo concentrations (*i.e.* amino acid neurotransmitters) and others that may be derivatized with high quantum yield fluorescent tags or may be detectable by other sensitive techniques such as electrochemical methods. In such cases, temporal resolution is dictated by the time required to separate the compounds so that serial analyses can be performed without overlapping the analysis peaks from two different runs.

Online microdialysis systems can also be useful for studies where high temporal resolution is not essential. Pharmacokinetic experiments in which analysis is required every few minutes over a period of hours are a typical example. Because most drugs exhibit pharmacokinetic profiles (absorption-distribution-metabolism-excretion) lasting a few hours to several days, measuring the average concentration over a period of time (10–20 min) is sufficient for pharmacokinetic modeling studies. Tsai and coworkers described the simultaneous monitoring of unbound cephalexin in both blood and brain using MD sampling coupled to microbore-HPLC and UV detection (Tsai et al. 2000). They were able to demonstrate that cephalexin has good blood–brain barrier penetration. Pharmacokinetics was determined by collecting samples at 10 min intervals (1 μL/min perfusion rate) for a 2 h period. Cephalexin was found to exhibit mean residence times of 117 and 84 min in blood and brain, respectively.

Perhaps, the greatest challenge in online microdialysis is developing a robust, low dead volume interface between the microdialysis sampling and analytical systems. Tubing and connectors associated with such an interface can produce an increase in dead volume in the system that leads to a delayed response in online monitoring experiments as well as Taylor dispersion resulting in band broadening. This is especially challenging in experiments involving awake, freely moving animals, as there is significant additional dead volume introduced with the swivel and tubing connecting the probe to the analytical system. In some cases, where the analysis is very rapid, the dialysis membrane and the tubing connecting to the analytical system can be the factor limiting temporal resolution.

3.4.1 Online Coupling of Microdialysis to Liquid Chromatographic Analysis

LC is a well-established analytical method that is used extensively for the analysis of biological samples. The strengths and weaknesses of LC for off-line analysis of microdialysis samples, as well as the mechanism of separation and detection methods were discussed above.

The general instrumentation required for online analysis of microdialysis samples consists of a microdialysis probe connected through the outlet to the injector of an LC system (Westerink and Cremers 2007). Automated sample injectors are commonly used (Tsai et al. 1996; Yang et al. 1999; Sheu et al. 2003). Directly coupling microdialysis sampling to LC systems can simplify sample handling and protect samples from degradation. For example, to establish the pharmacokinetics of omeprazole (an unstable acid pump inhibitor) in the blood, brain, and bile after *i.v.* dosing an online LC-UV method was developed (Cheng et al. 2002).

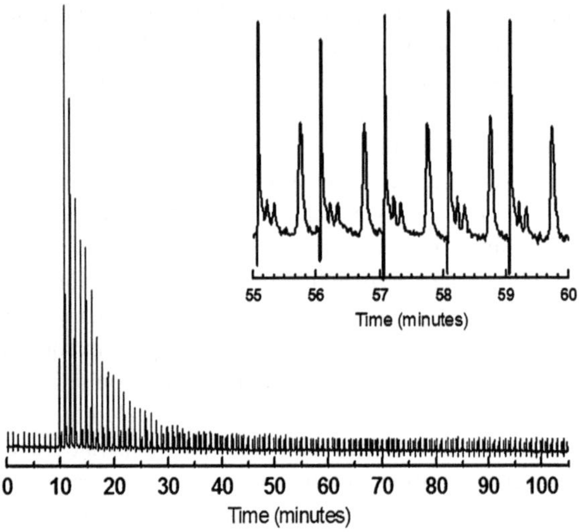

Fig. 3.6 Detector output from online microdialysis sampling with fast microbore-HPLC following plasma concentration of caffeine and its metabolites. Caffeine was administered i.v. (30 mg/kg) at time 10 min. Insert is an expanded scale from 55 to 60 min of individual 60 s HPLC sample runs. Reprinted with permission from (Chen and Lunte 1995)

As discussed in Sect. 3.1 temporal resolution in microdialysis sampling has many determining factors, such as flow rate and the injection volume of sample required. Typical microdialysis volumes needed for LC analysis are 5–10 μL. However, directly interfacing microdialysis sampling with the analysis makes it possible to accurately inject sub-microliters of sample volume (Steele and Lunte 1995). This would be difficult, if not impossible to do using off-line sample collection due to evaporation, surface tension, and other factors. However, as stated in Sect. 3.2 temporal resolution for online analysis differs from off-line analysis in that analysis time must also be considered. This is because a new sample cannot be injected into the system until the previous sample has been analyzed. For example, if 5 μL of sample is required at a 1 μL min^{-1} flow rate but it takes 10 min to analysis a sample then 10 min is the fastest sampling rate possible. In this case, 5 μL of sample will need to be diverted to waste while the analysis takes place.

Steele and Lunte (1995) were able to overcome this limitation by using multiple sample loops. The injection from one sample loop onto the column simultaneously occurs with the filling of another sample loop. In this case, no sample is lost and you obtain integrated concentration data for the time of sampling. Using this approach, the pharmacokinetics of acetaminophen was determined in freely moving rats. This method only works if the time required to fill the injection loop is longer than the analysis time.

Near real-time resolution can be achieved in online microdialysis sampling coupled to LC if small sample volumes (≤1 μL) can be analyzed in a short time period (minutes). Chen and Lunte were able to determine the concentration of caffeine, theobromine, paraxanthine, acetaminophen, and acetaminophen metabolites in vivo in near real time with an awake freely moving animal (Fig. 3.6) (Chen and Lunte 1995). They used a microbore column with 1.5 μL min^{-1} microdialysis flow rate and a 0.5 μL injection volume and an in-house made concentric style probe placed in the jugular of a rat. Under these conditions, the analysis was accomplished within one minute.

The direct coupling of microdialysis to LC can also protect highly reactive analytes from degradation due to exposure to air and light (Yang et al. 2001). Malondialdehye (MDA) is a biomarker for lipid peroxidation, which is indicative of several disease states. Sheu et al. (2003) were able to continuously monitor MDA in the brain, using a concentric style probe in vitro, using online microdialysis coupled to LC. Online microdialysis sampling has also been used to protect ascorbic acid and hydroxyl radicals from degradation prior to analysis (Tsai et al. 1999, 2005).

Baseski et al. (2005) were able couple microdialysis of both anesthetized and freely moving rats to capillary LC using mass spectrometry detection. The probes were implanted in the striatum and the neuropeptides (met)enkephalin and (leu)enkephalin were monitored with detection limits of standards at 2 pM.

3.4.2 Online Coupling of Microdialysis to Capillary Electrophoresis

Analogous to the case of coupling microdialysis online with LC, temporal resolution for online CE is additionally dependent on analysis time and the sensitivity of the analytical method. CE requires only nanoliters of sample and most analyses can be accomplished on the order of minutes.

The first report of online microdialysis sampling with CE was developed by Hogan et al. (1994). The instrument consisted of a rotary microinjection valve and an in-house built CE interface. The rotary microinjection valve injected microdialysis sample into a capillary in 60 nL plugs. The CE interface was a junction that placed the sample plug capillary approximately 50 μm away from the separation capillary. The junction was filled with BGE and a platinum electrode was placed in the solution so that injection and separation voltages could be applied to the separation capillary (Fig. 3.7). The system was used to evaluate the pharmacokinetics of the anticancer drug SR 4233 and yielded a temporal resolution of 90 s (Hogan et al. 1994). Using the same interface design, Zhou et al. (1995) were able to monitor the release of aspartate and glutamate in the striatum of live rats, with a temporal resolution of less than 2 min. Online derivatization has also been accomplished using a modified version of this system (Robert et al. 1998).

An alternative interface developed by Lada et al. which was based on Lemma and Jorgenson's flow gated injection was designed for coupling LC to CZE (Lemmo and Jorgenson 1993). This instrumentation was used to derivatize microdialysis samples from the caudate nucleus in rat and separate primary amines using CE and LIF detection. Lada employed metal plates with an approximate 75 μm space between them to create a BGE flow channel (Fig. 3.8) (Lada et al. 1995). This flow channel is perpendicular to the reaction capillary and the separation capillary, which are each interfaced through the metal plates on opposite sides of the flow channel. The BGE is hydrodynamically pumped through the flow channel and is actuated using a pneumatically controlled valve. Microdialysis sample is pumped through a reaction capillary where derivatizing agents are added

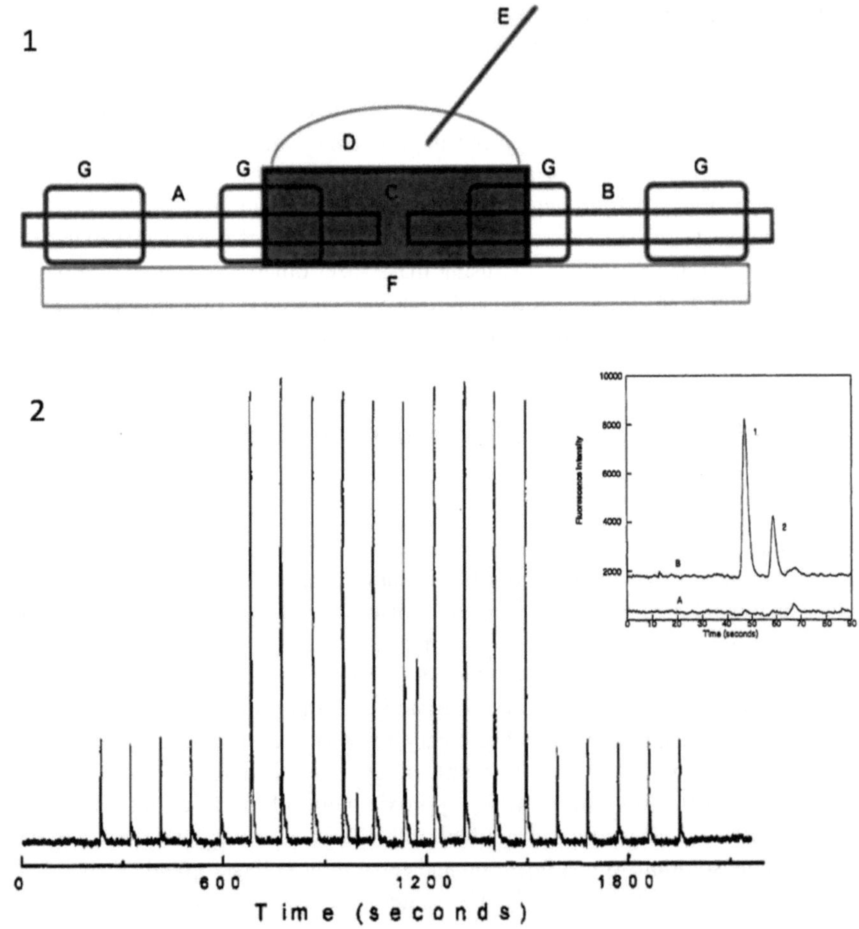

Fig. 3.7 Online microdialysis coupling to CE-flourescence for antineoplastic SR 4233 and metabolite SR 4317. **a** Interface between microdialysis sampling and CE separation, (*A*) Transfer capillary from rotary microinjector, (*B*) separation capillary, (*C*) BGE run buffer reservoir (~50 μm gap), (*D*) BGE, (*E*) platinum CE ground electrode, (*F*) microscope slide, and (*G*) guide tubing. **b** Response of interface with step change concentration of SR 4233 from 0, 2, 10, 2, and 0 ug/mL: insert typical electopherogram from blood dialysate (*A*) blank and (*B*) 10 min after 4 mg/kg SR 4233 dose. Adapted with permission from Hogan BL, and Lunte, SM. Anal Chem 66(5) (Hogan et al. 1994) 596–602

using a T connector. In order for the derivatized sample to be injected into the separation capillary the BGE flow at the interface is stopped. This allows the sample to fill the space between the reaction capillary and the separation capillary. A voltage is then applied to inject sample into the separation capillary. Once sample has been injected, the BGE flow is resumed to sweep away any sample that was not injected and to prevent sample from leaking into the separation capillary (Lada et al. 1995; Bowser and Kennedy 2001). This scheme is compatible with

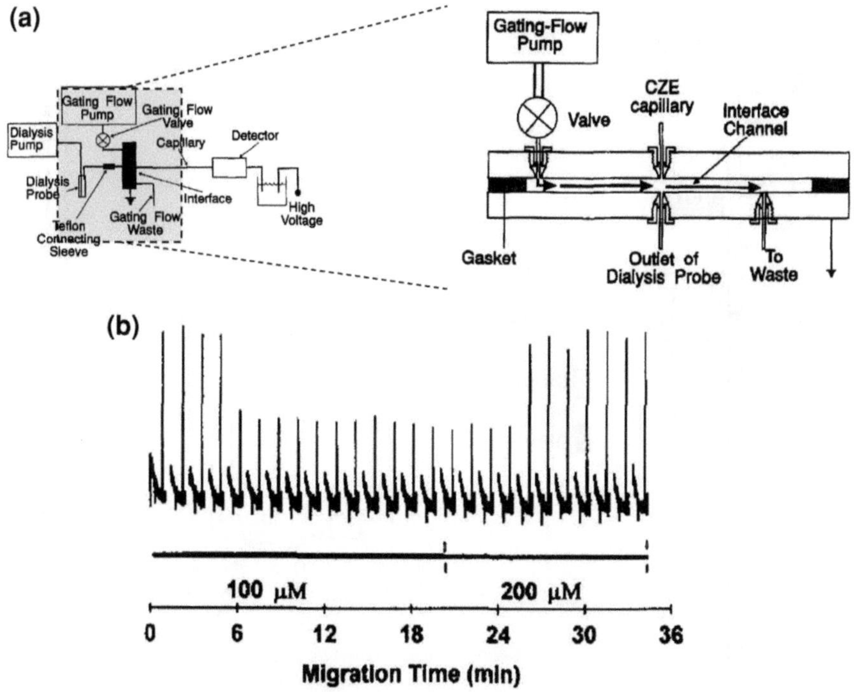

Fig. 3.8 **a** Cross-section of interface between online microdialysis and CZE with flow gated injection. **b** Series of electropherograms illustrating the response of the system with step changes in ascorbic acid with initial concentration of 200 uM with switches indicated by bars. Adapted with permission from Lada et al. (1995)

very low microdialysis sampling flow rates, 40–155 nL min^{-1} (Lada and Kennedy 1995, 1996), which is advantageous for analyte recovery.

This low sample volume requirement for CE means that it is the analysis time that most often determines temporal resolution. As discussed in Sect. 3.2.2, sample destacking due to the high ionic strength of microdialysis samples can cause longer analysis times and band broadening. The off-line solutions to destacking, such as field-amplified stacking and pH-mediated stacking, can be difficult to perform when using online systems. This is because these techniques for sample stacking often utilize different conductivity of leading BGE, following BGE and/or the sample itself. The introduction of different compositions of BGE is more difficult to perform online because the separation capillary cannot merely be moved to and from different vials for fluid loading.

Zhao et al. (1998) were able to perform field-amplified stacking in an online system by switching the highly conductive sample matrix to a low conductivity buffer using a LC guard column (Fig. 3.9). The analytes from the microdialysis sample were retained on the guard column momentarily as the high conductivity matrix was washed away (Fig. 3.9a, b). The analytes were then eluted in a low conductivity buffer and electrokinetically injected into the separation capillary by applying a high

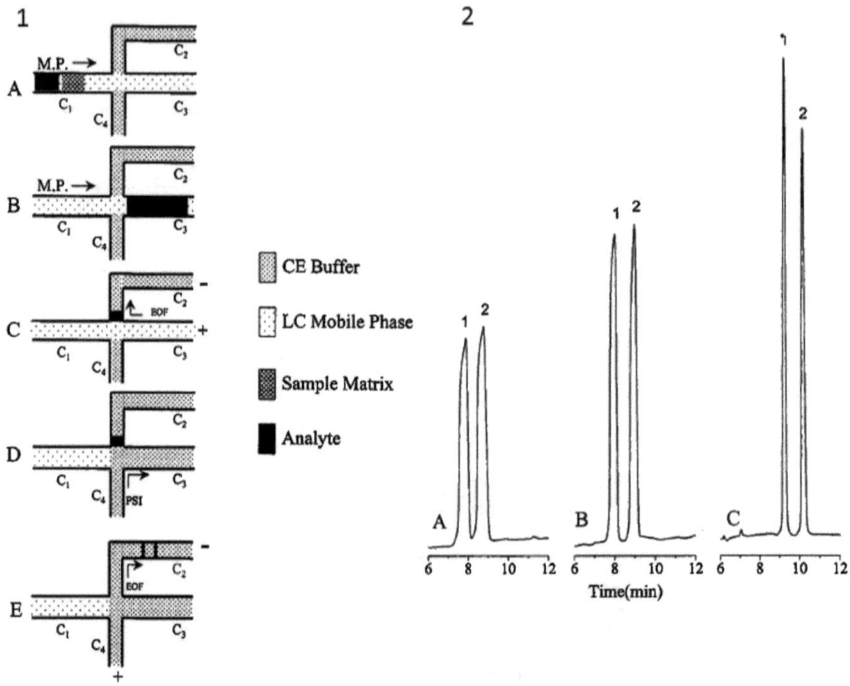

Fig. 3.9 **a** Diagram of matrix switching using LC-CE for microdialysis sampling, (*A*) Injection of sample on to LC column for matrix switching, (*B*) trapping analyte peak in stacking capillary, (*C*) field amplification stacking of analyte, (*D*) flushing LC mobile phase from CE system, and (*E*) CE separation. **b** Effect of ionic strength of LC (lithium citrate pH 2.5/ACN (70:30 v/v)) and CE (lithium acetate pH 4.75) buffers, (*A*) 10 mM LC buffer and 100 mM CE buffer, (*B*) 10 mM LC and 150 mM CE buffers, (*C*) 7.4 mM LC and 150 mM buffers. Reprinted with permission from Zhao et al. (Zhao et al. 1998)

voltage (Fig. 3.9c). The sample, at that time, is sandwiched between the high conductivity BGE in the separation capillary and the low conductivity LC buffer; this causes the sample plug to stack at the beginning of the separation capillary. The LC buffer is removed and replaced with the high conductivity CE BGE using pressure (Fig. 3.9d). Finally, the separation voltage is applied and separation occurs (Fig. 3.9e).

3.4.3 Mass Spectrometry

The online coupling of MS (no prior separation step) with MD sampling is challenging as a result of the high salt content of the perfusate and the complexity of the sample. Despite these challenges, microdialysis sampling with online MS detection has been reported for the detection of pharmaceutical compounds and metal ions. A novel central analgesic R-84760 was sampled via MD from rat blood after *i.v.* administration and analyzed by MS/MS in a quantitative fashion (Kobayashi et al. 2000). Direct coupling of the microdialysis flow to the ionization source is reported and compared to a previously optimized LC–MS/MS method. To eliminate high salt problems and ensure compatibility with the electrospray ionization method, a perfusate of 50:50 ethanol:H_2O was used. The extent of protein-binding of this drug was also determined by LC–MS and compared to results obtained by ultracentrifugation. A temporal resolution of 2 s was achieved with the method.

Chung and coworkers reported on MD sampling coupled to an inductively coupled plasma-mass spectrometer (ICP-MS) (Chung et al. 2007). In this work, the concentration of Cu, Zn, Mn, Co, Ni, Ca, and Mg ions were monitored in rat brain. To allow the perfusate to more closely match the ECF, a desalting membrane was placed between the MD probe and ICP-MS. Limits of detection were in the sub ng/mL range with a temporal resolution of 10 min. The authors also demonstrated the transfer of platinum into the brain ECF following *i.p.* administration of cisplatin.

Coupling MD to MS via a solid phase extraction (SPE) step has also been employed to help desalt the samples. To monitor trace metals in microdialysate samples, an online step using polytetrafluoroethylene (PTFE) as a SPE loop was connected to ICP-MS. The PTFE was used to remove the salt matrix and preconcentrate heavy metals copper, zinc, manganese, nickel, cobalt, cadmium, and lead with detection limits of sub ng/L and temporal resolution of 14 min (Su et al. 2012).

3.5 Microchip Electrophoresis

The analysis of microdialysis samples by ME has some advantages over LC and conventional capillary analysis in regard to temporal resolution. Electrophoretic separations in the microchip format are generally accomplished in seconds as opposed to minutes. Like CE, ME has very low sample volume requirements. When microdialysis is coupled to ME, it is possible to monitor fast changes in analyte concentration. The first coupling of microdialysis to microchip CE was reported in 2004 for monitoring the activity of beta galactosidase in vitro (Huynh et al. 2004).

As with the other separation methods discussed in this chapter, many different detection methods can be employed for ME including mass spectrometry (Wen et al. 2000), electrochemical (Mecker and Martin 2008), and LIF (Cellar and Kennedy 2006).

Several different approaches have been employed for coupling microdialysis to ME. These include the use of gated injection, pneumatic valves, and droplet microfluidics. Gated injection has been employed for the online analysis of enzyme reactions and the in vivo analysis of neurotransmitters (Huynh et al. 2004; Nandi et al. 2010). In this case, the chip has a separate channel for the microdialysis flow. Sample is injected from the flowing microdialysis stream by floating the voltage and allowing sample to leak into the separation channel (Fig. 3.10).

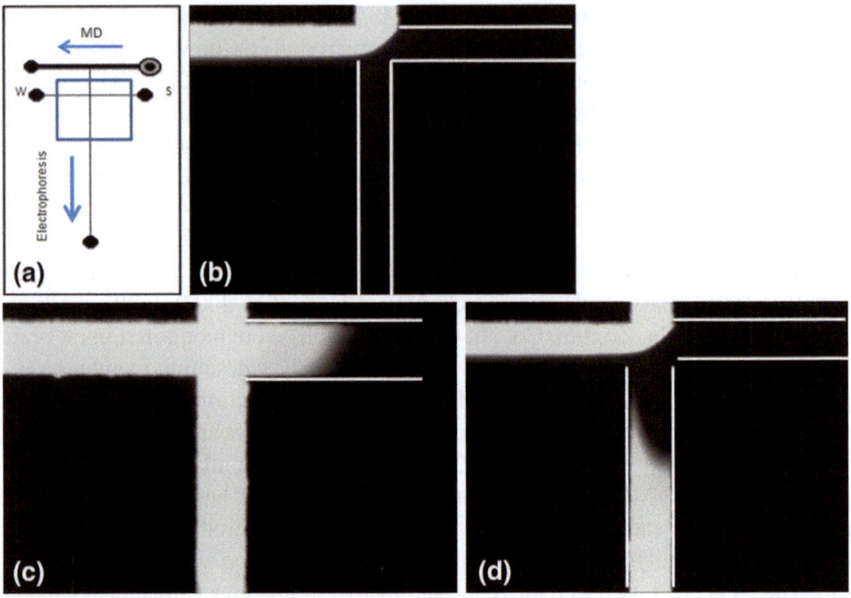

Fig. 3.10 Gated injection approach for introducing microdialysis samples intothe separation channel . (**a**) Schematic of chip used for microdialysis sampling. The microdialysis sample contains fluorescein for visualization. (**b**) microdiaysis sample flows to waste reservoir by a combination of hydrodynamic and electroosmotic flow (**c**) voltage is floated at the sample reservoir causing sample to leak into the separation channel (**d**) voltage is reapplied to the sample reservoir and injected sample plug is separated by electrophoresis

Gated injection can be used to make very fast injections from a flowing microdialysis stream but it can exhibit an electrokinetic bias of the analytes. An alternative method of injection was described by Martin's group using a PDMS pneumatic valve (Fig. 3.11) (Li and Martin 2007; Johnson et al. 2011). This method works extremely well but the pneumatically controlled valves can be difficult to fabricate and require a gas source for activation. Also, ridged substrates cannot be used, such as glass or hard plastics.

The most recent approach to coupling microdialysis online with ME is the use of droplet-based microfluidic systems. In this case, a segmented flow consists of a continuous phase and disperse phase differentiated because the two liquids are immiscible. In the case of microdialysis sampling, a hydrophobic liquid is the continuous phase and the aqueous microdialysis sample is the disperse phase (Fig. 3.12) (Wang et al. 2008, 2010). These droplets can be made to have a very controlled and small volume, which means each droplet is in essence a single precise time point. Droplets employed in the microfluidic systems commonly exhibit volumes in the fL–nL range. If a 1 μL min^{-1} flow rate is used, then each droplet translates into time points in the 60 ns–60 ms range. At this time scale, temporal resolution is actually limited by diffusion of analytes across the dialysis membrane, which can take a few seconds. The droplets can also be generated very

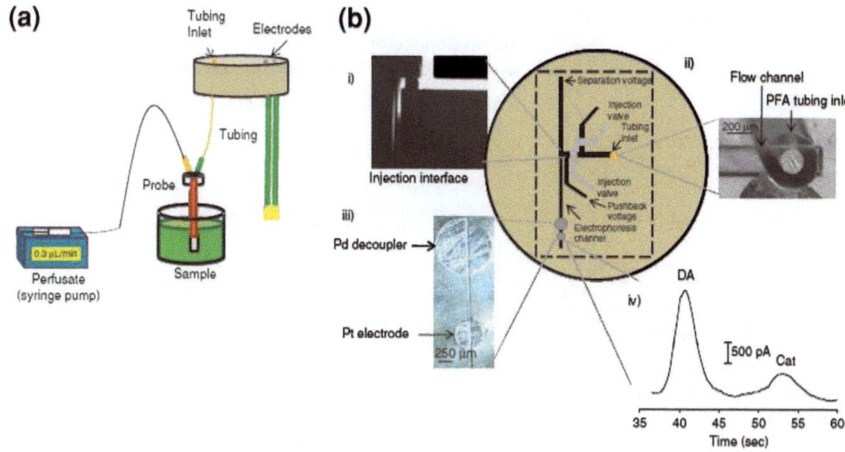

Fig. 3.11 Use of epoxy-embedded electrodes and tubing to enable the injection, electrophoresis, and amperometric detection of dopamine and catechol sampled through microdialysis. **a** Schematic depicting microdialysis set up and solution transfer through epoxy base. **b** *Top-down* view of PDMS microchip and pneumatic valves sealed on epoxy base. (*i*) Micrograph of a pressure injection using on chip valves into the separation channel, (*ii*) micrograph of microfluidic flow channel reversibly sealed over epoxy-encapsulated PFA tubing, (*iii*) micrograph of 1 mm decoupler and platinum working electrode, (*vi*) electropherogram of dopamine (*DA*) and catechol (*Cat*). Reprinted with permission from Johnson et al. (Johnson et al. 2011)

close to the point of sampling, which minimizes sample dispersion and dilution due to diffusion. Droplets can be collected during the experiment and analyzed at a later time or at another location without sacrificing temporal resolution. Wang and Kennedy used segmented flow to collect and derivatize 2 nL samples off-line (Wang et al. 2010). The samples were then analyzed using microchip with fluorescence detection with a temporal resolution of 2 s (limited by mass transport across the membrane).

3.6 Summary

As shown in this chapter, a wide variety of analytical techniques can be used in combination with microdialysis sampling. Most of these methods are also commonly used for other biological samples, such as blood, urine, and tissues. However, when using microdialysis, the analysis and sampling must be viewed as a unified experiment where a parameter directly affecting one aspect will indirectly affect all the other components of the experiment. This is shown most clearly in Fig. 3.1 which shows the interaction between the analytical parameters and the microdialysis sampling parameters. It is therefore almost never appropriate to simply apply an analytical method developed for a different sampling system to be

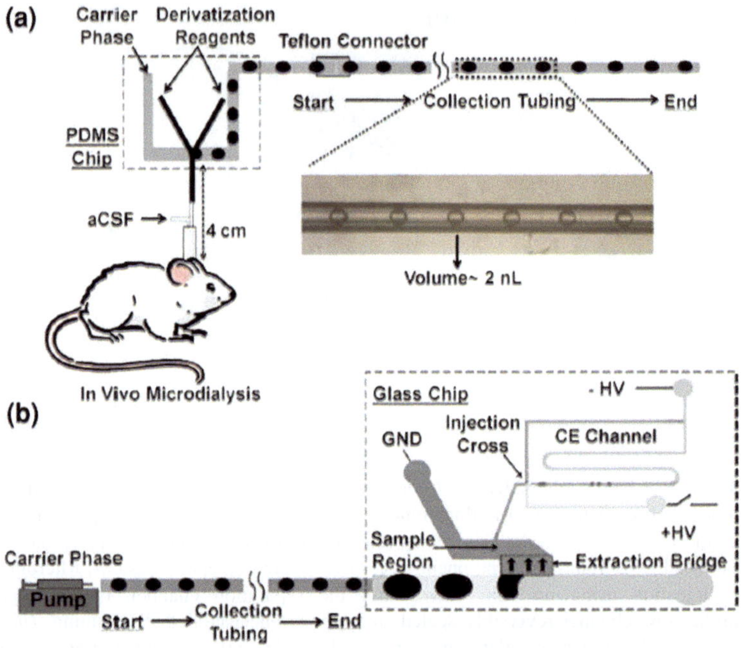

Fig. 3.12 Collection of samples using segmented flow with off-line CE separation. **a** In vivo dialysate derivatized online and stored in collection tubing, each plug is about 2 nL. **b** Plugs in the tubing were pumped to microfluidic chip for electrophoretic analysis. Reprinted with permission from Wang et al. (2010)

directly applied to microdialysis sampling. In the authors' opinion, trying to apply an analytical method developed for urine or plasma samples is the most common reason for microdialysis sampling to "not work". The analyst must develop a method with the recognition of the small sample volumes available, the difficulty in sample preconcentration, and the high ionic strength of the matrix. As new analytical techniques emerge with decreased sample volume requirements and improved concentration detection limits, microdialysis sampling will find utility in even more cases than at present.

References

Anna SL, Bontoux N, Stone HA (2003) Appl Phys Lett 82(3):364–366

Arnett SD, Lunte CE (2007) Electrophoresis 28(20):3786–3793

Arnett SD, Osbourn DM, Moore KD, Vandaveer SS, Lunte CE (2005) J Chromatogr B 827:16–25

Banos CE, Silva M (2011) J Chromatogr B 879(17–18):1412–1418

Bardelmeijer HA, Waterval JCM, Lingeman H, Hof Rvt, Bult A, Underberg WJM (1997) Electrophoresis 18:2214–2227

Baseski HM, Watson CJ, Cellar NA, Shackman JG, Kennedy RT (2005) J Mass Spectrom 40(2):146–153
Bert L, Robert F, Denoroy L, Stoppini L, Renaud B (1996) J Chromatogr A 755:99–111
Bigge JC, Patel TP, Bruce JA, Goulding PN, Charles SM, Parekh RB (1995) Anal Biochem 230:229–238
Blakeman KH, Wiesenfeld-Hallin Z, Alster P (2001) Exp Brain Res 139:354–358
Bowser MT, Kennedy RT (2001) Electrophoresis 22(17):3668–3676
Buck K, Voehringer P, Ferger B (2009) J Neurosci Methods 182(1):78–84
Cellar NA, Kennedy RT (2006) Lab Chip 6(9):1205–1212
Chang YL, Tsai PL, Chou YC, Tien JH, Tsai TH (2005) J Chromatogr A 1088(1–2):152–157
Che F-Y, Song J-F, Zeng R, Wang K-Y, Xia Q-C (1999) J Chromatogr A 858(2):229–238
Chen AQ, Lunte CE (1995) J Chromatogr A 691(1–2):29–35
Chen FC, Tsai TR, Chen YF, Hung LC, Tsai TH (2002) J Chromatogr A 961(1):131–136
Chen R, Xu W, Xiong C, Zhou X, Xiong S, Nie Z, Mao L, Chen Y, Chang HC (2012) Anal Chem 84(1):465–469
Cheng FC, Ho YF, Hung LC, Chen CF, Tsai TH (2002) J Chromatogr A 949(1–2):35–42
Chien R-L, Burgi DS (1991) J Chromatogr 559:141–152
Chien R-L, Burgi DS (1992) Anal Chem 64:489A–496A
Chung YT, Ling YC, Yang CS, Sun YC, Lee PL, Lin CY, Hong CC, Yang MH (2007) Anal Chem 79:8900–8910
Church WH, Justice JB (1987) Anal Chem 59:712–716
Claeson K, Thorsen G, Karlberg B (2001) J Chromatogr B 763(1–2):133–138
Consolo S, Baldi G, Russi G, Civenni G, Bartfai T, Vezzani A (1994) Proc Natl Acad Sci U S A 91:8047–8051
Crick EW, Osorio I, Bhavaraju NC, Linz TH, Lunte CE (2007) Epilepsy Res 74(2–3):116–125
Deeba S, Corcoles EP, Hanna GB, Pareskevas P, Aziz O, Boutelle MG, Darzi A (2008) Dis Colon Rectum 51(9):1408–1413
Gillogly JA, Lunte CE (2005) Electrophoresis 26(3):633–639
Guan X, Hoffman B, Dwivedi C, Matthees DP (2003) J Pharm Biomed Anal 31:251–261
Guihen E, O'Connor WT (2009) Electrophoresis 30(12):2062–2075
Hadwiger ME, Park S, Torchia SR, Lunte CE (1997) J Pharm Biomed Anal 15(5):621–629
Hakim T, Sugimori K, Camporesi E, Anderson G (1996) Physiol Meas 17(4):267–277
Harvey DJ (2011) J Chromatogr B 879:1196–1225
Hase S, Ikenaka T, Matsushima Y (1978) Biochem Biophys Res Commun 85:257–263
Hernandez L, Joshi N, Murzi E, Verdeguer P, Mifsu JC, Guzman N (1993) J Chromatogr A 652(2):399–405
Hogan BL, Lunte SM, Stobaugh JF, Lunte CE (1994) Anal Chem 66(5):596–602
Hopwood SE, Parkin MC, Bezzina EL, Boutelle MG, Strong AJ (2005) J Cereb Blood Flow Metab 25(3):391–401
Hoque ME, Arnett SD, Lunte CE (2005) J Chromatogr B 827(1):51–57
Huang X, Zare RN (1990) Anal Chem 62(5):443–446
Huang H, Zhang Y, Yang R, Tang X (2008) J Chromatogr B 874:77–83
Huynh BH, Fogarty BA, Martin RS, Lunte SM (2004) Anal Chem 76(21):6440–6447
Imai K (1987) Adv Chromatogr 27:215–245
Jin G, Cheng Q, Feng J, Li F (2008) J. Chrom Sci 46:276–287
Johnson AS, Selimovic A, Martin RS (2011) Electrophoresis 32(22):3121–3128
Jou NT, Ma SX (2009) Microcirculation 16(5):434–443
Juluru R, Shukla C, Yin H, Stagni G (2012) J Pharm Sci 101(1):405–413
Katsumi H, Jiro K, Koichi Y, Katsutoshi A, Kazuro T (1984) Anal Biochem 136(2):314–320
Klinker CC, Bowser MT (2007) Anal Chem 79:8747–8754
Kobayashi N, Kazui M, Ikeda T (2000) J Pharm Biomed Anal 21:1233–1242
Lada MW, Kennedy RT (1995) J Neurosci Methods 63:147–152
Lada MW, Kennedy RT (1996) Anal Chem 68(17):2790–2797

Lada MW, Schaller G, Carriger MH, Vickroy TW, Kennedy RT (1995) Anal Chim Acta 307(2–3):217–225
Lada MW, Vickroy TW, Kennedy RT (1997) Anal Chem 69(22):4560–4565
Lanckmans K, Sarrea S, Smoldersa I, Michotte Y (2008) Talanta 74(4):458–469
Lemmo AV, Jorgenson JW (1993) Anal Chem 65(11):1576–1581
Li MW, Martin RS (2007) Electrophoresis 28(14):2478–2488
Li Y-M, Qu Y, Vandenbussche E, Arckens L, Vandesande F (2001) J Neurosci Methods 105:211–215
Li H, Wang H, Chen JH, Wang LH, Zhang HS, Fan Y (2003) J Chromatogr B 788:93–101
Linhares MC, Kissinger PT (1991) Anal Chem 63:2076–2078
Liu J, Hsieh Y-Z, Wiesler D, Novotny M (1991) Anal Chem 63:408–412
Liu U, Zhang J, Xu X, Zhao MK, Andrews AM, Weber SG (2010) Anal Chem 82:9611–9616
Lu W, Cassidy RM (1994) Anal Chem 66:200–204
Lunn GH, Louise C (1998) Albert Wiley, New York, p 1632
Maidment NT, Brumbaugh DR, Rudolph VD, Erdelyi E, Evans CJ (1989) Neuroscience 33:549–557
Mala Z, Krivankova L, Gebauer P, Bocek P (2007) Electrophoresis 28:243–253
Mecker LC, Martin RS (2008) Anal Chem 80(23):9257–9264
Menacherry S, Hubert W, Justice JB Jr (1992) Anal Chem 64(6):577–583
Miller BF, Ellis D, Robinson MM, Rivera JD, Kjaer M, Langberg H (2011) Scand J Med Sci Sports 21(6):e1–e8
Misko TP, Schilling RJ, Salvemini D, Moore WM, Currie MG (1993) Anal Biochem 214(1):11–16
Mukherjee PS, Karnes HT (1996) Analyst 121:1573
Nandi P, Lunte SM (2009) Anal Chim Acta 651(1):1–14
Nandi P, Desai DP, Lunte SM (2010) Electrophoresis 31(8):1414–1422
Norio I, Kyoko N, Chiaki S, Hisashi K, Ken O (1984) J Chromatogr 295:463–469
Obrenovitch TP, Zilkha E (2001) Methods 23:63–71
Obrenovitch TP, Sarna GS, Millan MH, Lok S-Y, Kamauchi M, Ueda Y, Symon L, Krieglstein J, Oberpichler J (1990) (eds) Wissenschaftliche verlagsgesellschaft: stuttgart, pp 23–31
O'Shea TJ, Greenhagen RD, Lunte SM, Lunte CE (1992) J Chromatogr 593:305–312
Paglia G, D'Apolito O, Garofalo D, Scarano C, Corso G (2007) J Chromatogr B 860(2):153–159
Palmer J, Munro NJ, Landers JP (1999) Anal Chem 71(9):1679–1687
Park S, Lunte SM, Lunte CE (1995) Anal Chem 67:911–918
Park S, McGrath MJ, Smyth MR, Diamond D, Lunte CE (1997) Anal Chem 69:2994–3001
Perez SA, Colon LA (1996) Electrophoresis 17(2):352–358
Perry M, Li Q, Kennedy RT (2009) Anal Chim Acta 653(1):1–22
Petsch M, Mayer-Helm BX, Sauermann R, Joukhadar C, Kenndler E (2004) Electrophoresis 25(14):2292–2298
Pickl KE, Magnes C, Bodenlenz M, Pieber TR, Sinner FM (2007) J Chromatogr B Analyt Technol Biomed Life Sci 850(1–2):432–439
Pravda M, Bogaert L, Sarre S, Ebinger G, Kauffmann J-M, Michotte Y (1997) Anal Chem 69:2354–2361
Quirino JP, Terabe S (1999) Anal Chem 71:1638–1644
Quirino JP, Terabe S (2000) Anal Chem 72:1023–1030
Quirino JP, Terabe S, Otsuka K, Vincent JB, Vigh G (1999) J Chromatogr A 838:3–10
Raimondi L, Alfarano C, Pacini A, Livi S, Ghelardini C, DeSiena G, Pirisino R (2007) Br J Pharmacol 150:1003–1010
Reed B, Zhang Y, Chait BT, Kreek MJ (2003) J Neurochem 86:815–823
Reed B, Bidlack JM, Chait BT, Kreek MJ (2008) J Neuroendocrinol 20:606–616
Reynolds JD, Zeballos GA, Penning DH, Kimura KA, Atkins B, Brien JF (1998) J Pharmacol Toxicol Methods 39(3):125–128
Richards DA, Silva MA, Devall AJ (2006) Anal Biochem 351(1):77–83
Robert F, Bert L, Denoroy L, Renaud B (1995) Anal Chem 67:1838–1844

Robert F, Bert L, Parrot S, Denoroy L, Stoppini L, Renaud B (1998) J Chromatogr A 817(1–2):195–203
Robin S, Leveque N, Courderot-Masuyer C, Humbert P (2011) J Chromatogr B Analyt Technol Biomed Life Sci 879(30):3599–3606
Rossell S, Gonzalez LE, Hernandez L (2003) J Chromatogr B 784:385–393
Schultz KN, Kennedy RT (2008) Annu Rev Anal Chem 1(1):627–661
Shackman HM, Shou M, Cellar NA, Watson CJ, Kennedy RT (2006) J Neurosci Methods 159(1):86–92
Shangguan D, Han HW, Zhao R, Zhao YX, Xiong SX, Liu GQ (2001) J Chromatogr A 910(2):367–372
Sheu J-Y, Ku H-P, Tseng W-C, Chen M-T, Tsai L-Y, Huang Y-L (2003) Anal Sci 19(4):621–624
Shihabi ZK (2000) J Chromatogr A 902:107–117
Shou M, Smith AD, Shackman JG, Peris J, Kennedy RT (2004) J Neurosci Methods 138(1–2):189–197
Steele KM, Lunte CE (1995) J Pharm Biomed Anal 13(2):149–154
Su C-K, Lin Y-L, Sun Y-C (2012) J Anal At Spectrom 27(1):56
Sun L, Stenken JA (2003) J Chromatogr B 796(2):327–338
Sun Z, You J, Li G, Zhao XE, Suo Y, Wang X (2011) J Chromatogr B 879:1367–1374
Szoko E, Tabi T (2010) J Pharm Biomed Anal 53:1180–1192
Toyo'oka T (2009) J Chromatogr B 877:3318–3330
Toyo'oka T, Ishibashi M, Takeda Y, Imai K (1991) Analyst 116:609–613
Tsai P-J, Wu J-P, Lin N-N, Kuo J-S, Yang C-S (1996) J Chromatogr B Biomed Sci Appl 686(2):151–156
Tsai TH, Cheng FC, Hung LC, Chen CF (1999) J Chromatogr B: Biomed Sci Appl 734(2):277–283
Tsai TH, Hung LC, Chang YL, Shum AYC, Chen CF (2000) J Chromatogr B 740:203–209
Tsai PJ, Chen WY, Tzeng SF, Liang WM, Yang CS (2005) Clin Chim Acta 362(1–2):94–100
Tsikas D (2007) J Chromatogr B 851(1–2):51–70
Tuma P, Malkova K, Wedellova Z, Samcova E, Stulik K (2010) Electrophoresis 31(12):2037–2043
Uutela P, Ketola RA, Piepponen P, Kostiainen R (2009) Anal Chim Acta 633:223–231
Wallingford RA, Ewing AG (1987) Anal Chem 59:1762–1766
Wang J, Lu M, Yang F, Zhang X, Bayens WRG, Campana AMG (2001) Anal Chim Acta 428:173–181
Wang X, Wang Q, Chen Y, Han H (2003) J Chromatogr A 992:181–191
Wang M, Roman GT, Schultz K, Jennings C, Kennedy RT (2008) Anal Chem 80(14):5607–5615
Wang M, Slaney T, Mabrouk O, Kennedy RT (2010) J Neurosci Methods 190(1):39–48
Waterval JCM, Lingerman H, Bult A, Underberg WJM (2000) Electrophoresis 21:4029–4045
Wen J, Lin YH, Xiang F, Matson DW, Udseth HR, Smith RD (2000) Electrophoresis 21(1):191–197
Westerink BHC, Cremers TIFH (2007) Elsevier, Dusseldorf, Vol. 16, p 697
Wolf JH, Korf J (1992) J Pharm Biomed Anal 10:99–107
Wu Y-T, Lin L-C, Tsai T-H (2009) J Chromatogr A 1216:3501–3507
Xiong XJ, Wang H, Rao WB, Guo XF, Zhang HS (2010) J Chromatogr A 1217(1):49–56
Yamaguchi M, Hara S, Matsunaga R, Nakamura M, Ohkura Y (1985) J Chromatogr 346:227–236
Yang C-S, Tsai P-J, Chen W-Y, Tsai W-J, Kuo J-S (1999) J Chromatogr B Biomed Sci Appl 734(1):1–6
Yang CS, Tsai PJ, Chen WY, Kuo JS (2001) J Chromatogr B 752(1):33–38
Yao D, Vlessidis AG, Evmiridis NP, Evangelou A, Karkabounas S, Tsampalas S (2002a) Anal Chim Acta 458:281–289
Yao D, Vlessidis AG, Evmiridis NP (2002b) Analytica Chimica Acta 467:133–144
Yasaka Y, Tanaka M, Shono T, Tetsumi T, Ji Katakawa (1993) J Chromatogr 508:133–140
Yik YF, Lee HK, Li SFY, Khoo SB (1991) J Chromatogr 585(1):139–144
You WW, Haugland RP, Ryan DK, Haugland RP (1997) Anal Biochem 244(2):277–282

Zang W (2001) San Jose State University, San Jose
Zhao YP, McLaughlin K, Lunte CE (1998) Anal Chem 70(21):4578–4585
Zhao S, Youan H, Xiao D (2005) J Chromatogr B 822:334–338
Zhong M, Zhou J, Lunte SM, Zhao G, Giolando DM, Kirchhoff JR (1996) Anal Chem 68:203–207
Zhou SY, Zuo H, Stobaugh JF, Lunte CE, Lunte SM (1995) Anal Chem 67(3):594–599

Chapter 4
Regulatory Aspects of Microdialysis: A United States Food and Drug Administration Perspective

Chinmay Shukla, Jing Li, Robert Lionberger and Edward D. Bashaw

Abstract The purpose of this chapter is to illustrate the potential applications of microdialysis (MD) in drug discovery and development with the aim of meeting Food and Drug Administration (FDA) requirements for evaluating drug bioavailability (BA) and/or establishing bioequivalence (BE). This chapter also focuses on the importance of optimizing the MD technique in order to achieve reproducible results. The effects of physicochemical properties of the drug, technical aspects of MD sampling, and physiological factors of the target tissue that affect MD recovery are illustrated with regard to optimizing drug recovery. Ultimately, this chapter provides a concise overview of the applications of MD in preclinical and clinical setting in line with FDA's critical path initiative and regulatory requirements for BA and BE assessment.

The views expressed in this chapter are those of the authors and do not necessarily reflect the opinion of the FDA. No official support or endorsement by FDA is intended or should be inferred.

C. Shukla (✉) · E. D. Bashaw
Office of Clinical Pharmacology, Center for Drug Evaluations and Research, United States Food and Drug Administration, 10903 New Hampshire Avenue, Silver Spring, MD, USA
e-mail: Chinmay.Shukla@fda.hhs.gov

J. Li · R. Lionberger
Office of Generic Drugs, Center for Drug Evaluations and Research, United States Food and Drug Administration, Rockville, MD, USA

M. Müller (ed.), *Microdialysis in Drug Development*, AAPS Advances in the Pharmaceutical Sciences Series, DOI: 10.1007/978-1-4614-4815-0_4,

4.1 Background

Introduction of new technologies into the drug development process is challenging because they must not only provide valuable scientific information but also must interact with the existing regulatory framework. The purpose of this chapter is to illustrate the different areas where microdialysis (MD) sampling could potentially be applied during drug discovery and development and be useful in meeting the requirements of Food and Drug Administration (FDA) for evaluating bioavailability (BA) and/or establishing bioequivalence (BE).

FDA recognized that the technological advances in basic science were not being adequately used in drug development science and hence launched its Critical Path Initiative in 2004 (FDA 2004). The purpose of the Critical Path Initiative was to encourage drug developers to implement new technologies early on during development at the preclinical and early phase clinical stages and use the data generated to improve the predictability and efficiency of the clinical trials. This represented a recognition by the FDA that completes characterization of the pharmacokinetic (PK), pharmacodynamic (PD), and toxicological properties of a new molecular entity (NME) early on, could yield better developmental decisions during the Phase II and Phase III studies (Hocht 2011; Hughes et al. 2011). An example of one such technique that can be utilized early on to determine some of the delivery characteristics of a compound to or near a physical target is MD.

Following the launch of the critical path initiative focusing on implementing new technologies in the development of new drugs in 2004 (FDA 2004), FDA released critical path for generic drugs in May 2007 and identified challenges in the development of generic drugs (FDA 2007). Under the critical path for generic drug development, several areas were identified where scientific progress could accelerate the development and approval of generic drugs. One such area is the development of methods for the assessment of BE of locally acting drugs under which MD is identified as a direct sampling method of target tissue for establishing BE when blood concentrations are not appropriate (FDA 2007; Lionberger 2008; Chaurasia et al. 2007).

Basically, MD is a sampling technique which involves insertion of a MD probe in the tissue with the goal of measuring concentrations of the drug(s), metabolite(s), and biomarker(s) in the tissue of interest. A MD probe is a hollow tube incorporating a semipermeable membrane. The concept of MD is based on the passive diffusion of the analyte(s) across the semipermeable membrane driven by a concentration gradient. The probe is usually perfused with a physiological solution called the "perfusate", and the resulting "dialysate" that exits the probe contains the analyte(s) that has diffused into the probe through the semipermeable membrane. This dialysate is analyzed to determine the tissue concentrations of the analyte(s) using a suitable validated bioanalytical method.

Along with illustrating different areas where MD can be applied, this chapter also delineates various experimental factors that would need to be optimized when this technique is applied. For additional information on the feasibility of MD sampling for the analyte of interest, the authors of this chapter encourage readers to consult the original literature.

4.2 Role of Microdialysis in Assessing Bioavailability

According to Code of Federal Regulations (CFR) Title 21 Part 320 (CFR Title 21 Part 320 2011), BA is defined as the rate and extent to which the active ingredient or active moiety is absorbed from a drug product and becomes available at the site of action (Guidance for industry 2003). The CFR also states that with the submission of a new drug application (NDA) or a supplemental new drug application (sNDA) to the FDA, evidence measuring the in vivo BA of the drug product or information to permit FDA to waive the submission of evidence measuring in vivo BA is required to be submitted (CFR Title 21 Part 320 2011).

For drugs acting systemically, BA is currently assessed by developing a systemic exposure profile, which can be obtained by measuring the concentrations of active moieties and, when appropriate, the active metabolite(s) over time in samples collected from the systemic circulation (CFR Title 21 Part 314 2011). In this case, blood concentrations are usually linked to both assessments of safety and efficacy, with the underlying assumption that the drug concentrations at the target site would correspond to those measured in the blood. Recently, Stagni et al. have demonstrated estimation of systemic BA fraction using skin MD concentrations of two model drugs, amoxicillin and ketoprofen, by using continuous and intermittent MD sampling in the dermis in a preclinical setting (Juluru et al. 2012).

This classic concept of estimating BA by using systemic concentrations would become a complicated approach to be applied for drugs administered topically because drug products are delivered directly to or near the target site in high concentrations. Topical BA is a result of interplay between the drug substance, formulation, and the effect of the disease on the barrier function of the skin. Systemic drug absorption is not normally desired and systemic BA testing in this case becomes a safety assessment, while the actual BA at the target site is unknown (Cho and Bashaw 2011). As a result of this, PK-PD relationships for most topical products are poorly defined. MD sampling in the diseased skin could prove as a valuable tool in evaluating topical BA. Linking this concentration to drug effect might aid in improving our current understanding of PK-PD relationship for therapies to treat dermatological diseases (Cho and Bashaw 2011). Garcia Ortiz et al. have demonstrated the application of cutaneous microdialysis to determine the penetration of metronidazole from a cream formulation in adult subjects with atopic dermatitis. This study demonstrated that metronidazole concentrations were higher in skin with atopic dermatitis compared to uninvolved skin (Garcia Ortiz et al. 2009). Benfeldt et al. demonstrated an increase in cutaneous penetration of salicylic acid with perturbed skin barrier compared to normal skin (1999) . MD has also been applied to quantify biomarkers in the skin like histamine quantification in human psoriatic plaques (Guihen et al. 2012), nerve growth factor in atopic dermatitis (Papoiu et al. 2011), with the aim of understanding the disease mechanics.

Also, a few other examples where target site BA estimation will prove to be of value for evaluating PK/PD relationship include anti-infective drugs, neuroactive agents, intraocular agents, intra-tumoral agents, etc. In cases where the site of drug action is extravascular, unbound tissue concentrations would provide a more meaningful approach to the evaluation of the relationship between drug concentration and efficacy and/or toxicity and their time course (Schmidt et al. 2008). Assuming drug concentrations at the target site to be proportional to those measured in the blood might increase the risk of underestimating or overestimating drug efficacy and/or toxicity by solely evaluating systemic BA.

In case of anti-infective drugs, most of their target site of action are in the tissue (Shukla et al. 2009; Brunner et al. 2005; Liu et al. 2002). Hence, linking unbound tissue concentrations obtained using MD to drug effect seems a more scientific approach. The efficacy of neuroactive compounds on the other hand is based on their ability to cross the blood–brain barrier (BBB) and remain in the brain interstitial tissues in sufficient concentrations for adequate duration of time. Hence, in this case blood concentrations of the analyte might poorly predict efficacy (Alavijeh and Palmer 2010; McAdoo and Wu 2008). Along with the tight junctions at the blood–brain barrier, drug delivery to the brain is further complicated by the presence of numerous efflux and influx transporters and the non-specific affinity to the brain tissue making exposure to the brain challenging to predict from blood concentrations alone (Hammarlund-Udenaes 2010). This makes MD in the brain an attractive tool. Intraocular drug disposition has always been challenging to evaluate and ocular MD has recently gained popularity in a preclinical setting to evaluate drug concentrations in relevant intraocular regions to elucidate drug PK/PD effect (Duvvuri et al. 2005; Boddu et al. 2010). MD has also been applied to assess intratumoral drug disposition to evaluate the unbound pharmacologically active fraction in the tumor tissue due to poor correlation of plasma drug and/or metabolite(s) concentrations with efficacy and toxicity (Kitzen et al. 2006; Wei et al. 2009; Blakeley and Portnow 2010).

Hence, MD is a unique technique that permits monitoring free extracellular drug(s), metabolite(s), and/or biomarker(s) concentration that are available for pharmacological activity at or near the site of action. The few examples considered so far are in line with the definition of BA as described in 21 CFR 320. In such cases, systemic BA estimation should not be overruled as it may be required for the assessment of systemic safety. Further, multiple MD probes can be placed in different tissues as this would allow monitoring drug concentrations in the same animal supplying valuable drug distribution information for physiologically based PK models and PK-PD models which could be further used during drug development to aid in the prediction of clinical outcome (Hocht 2011; Hocht et al. 2004). This would justify the basis of implementing MD during drug development for studies such as proof of concept (Chaurasia et al. 2007).

4.3 Microdialysis in Bioequivalence Assessment

BE is defined in 21 CFR 320 as the absence of a significant difference in the rate and the extent to which the active ingredient or active moiety in pharmaceutical equivalents or pharmaceutical alternatives becomes available at the site of drug action when administered at the same molar dose under similar conditions in an appropriately designed study (CFR Title 21 Part 320 2011; Guidance for industry 2003). For abbreviated new drug application (ANDA) to the FDA, evidence demonstrating that the drug product is bioequivalent to the reference listed drug is required to be submitted. A reference listed drug is defined in 21 CFR 314 as the listed drug identified by FDA as the drug product upon which an applicant relies in seeking approval of its ANDA (CFR Title 21 Part 314 2011).

Currently, plasma drug concentrations are used to establish bioequivalence of systemically administered drugs (Guidance for industry 2003). This approach becomes challenging for topically administered drugs because only a fraction of drug reaches the systemic blood circulation which might be difficult to quantify and also might not reflect the concentration at the target site, the skin, which from a regulatory standpoint is the criteria of interest (Chaurasia et al. 2007; Holmgaard et al. 2010). With the exception of topical corticosteroids where multipoint skin blanching assay is currently used to establish BE (Guidance for industry 1997), and very few individual topical products (e.g., lidocaine topical patch) where systemic PK is used for BE assessment (Draft guidance on lidocaine 2007), most of the topical formulations require the establishment of BE with efficacy endpoints (i.e., clinical trials). These clinical trials often require large number of subjects and are demanding in terms of time and cost. In addition, due to the route of administration (topical) the trial must be done in subjects with diseased skin and as a parallel design, as even a placebo emollient treatment can affect the skin. Recently there has been a quest towards exploring alternative approaches to establishing BE following topical administration. These approaches include in vitro drug penetration experiments, dermatopharmacokinetic approach using tape stripping technique (Shah 2005), skin biopsies and suction blister technique (Benfeldt et al. 1999), dermal microdialysis, etc. These approaches have been discussed at multiple FDA Advisory Committee meetings (Anti-infective drugs advisory committee meeting 1998), and national professional society meetings such as American Association of Pharmaceutical Scientists (AAPS) (Chaurasia et al. 2007; Bioequivalence of topical drugs 2008), American College of Clinical Pharmacology (ACCP) (ACCP frontiers symposia 2010), International Pharmaceutical Federation (FIP) (Bioequivalence of topical drugs 2008), etc.

Among the aforementioned techniques, MD is the only technique currently available that would provide a means to evaluate topical drug penetration of different formulations by continuous sampling and thus providing real-time PK. Most other techniques produce limited information in the form of a single data point from healthy subjects (2007).

MD drug penetration data with the test and reference could be obtained in the same subject thus reducing the intrasubject variability and such an approach would likely reduce the number of subjects that would be required to establish topical BE compared to the number of subjects that would be required using the current approach of establishing BE using comparative clinical endpoints between the generic drug and the listed drug (Narkar 2010, 2007).

The application of MD for BE estimation or formulation comparison following topical administration is fairly recent. The earliest mention of this technique in the literature was in 2001 by Kreilgaard et al. 2001 who investigated the BE of lidocaine applied in a topical novel microemulsion formulation compared to oil-in-water emulsion formulation in eight healthy male volunteers. In this study, the authors demonstrated improved dermal delivery of lidocaine using MD and with the aid of an appropriate PK model, would provide increased sensitivity to BE estimation. Tegeder et al. used microdialysis to compare skin concentrations after topical bath and cream formulations of 8-methoxypsoralen (Tegeder et al. 2002). Benfeldt et al. evaluated the relationship between dermal MD and dermatopharmacokinetic methods when employed simultaneously for BE investigations of topical lidocaine cream and ointment in eight healthy human volunteers. With this study the authors demonstrated that with MD the intrasubject variability was 19 % between the probes and 20 % between the two penetration areas. The intersubject variability was high and was reported as 61 %. The authors further showed that with MD, BE estimation could be feasible with approximately 18 subjects (Benfeldt et al. 2007). Further, McCleverty et al. demonstrated that in order to achieve BE (within 80–125 % confidence limits) using MD sampling following topical administration, each subject should act as their own control thus removing the element of subject-to-subject variability. In addition, the authors also recommended that measurements using MD be made in duplicates to further reduce variability. From calculations of sample size, the authors recommended that a subject population of approximately 20 would be adequate to demonstrate BE (McCleverty et al. 2006). On the other hand, Agabeyoglu et al. demonstrated a failed approach to achieve BE with dermal MD for two commercial topical ointment formulations of oxytetracycline HCl in 12 healthy volunteers with the 90 % confidence interval for AUC and C_{max} reported as 89–135 % and 82–161 %, respectively (Incecayir et al. 2011). Kanfer et al. demonstrated bioequivalence of ketoprofen gel in 18 subjects using only the AUC estimation from concentration time profiles from pairs of application sites in each subject. The intrasubject variability was 10 %, while the intersubject variability was 68 % (Tettey-Amlalo et al. 2009).

4.4 Experimental Factors Affecting the Outcome of Microdialysis

The applicability of microdialysis data depends on the ability to produce reproducible analyte recovery and reliable estimation of tissue concentrations. Recovery rate is defined as the ratio of the concentration of the analyte in the dialysate and the concentration of the analyte in the "periprobe" fluid. Periprobe fluid is the tissue fluid surrounding the probe. Many experimental factors that influence the recovery rate of an analyte and the feasibility of applying MD sampling are discussed in the following sections.

Adequate optimization and validation of these factors will be required before the technique is applied in the clinic and MD sampling may or may not be the best approach for your analyte of interest. Several literature references are available on the technical aspects of MD and the authors of this chapter encourage the readers to refer to this wealth of information to aid in their experimental design. The experimental factors can be broadly classified into physicochemical properties of the analyte, factors related to the MD technique and physiological factors. Several of these are briefly discussed below.

4.4.1 Physicochemical Properties of the Analyte

Molecular weight: One of the biggest advantages of MD sampling is the ability to produce protein-free samples which can be directly analyzed using a well-validated analytical method without requiring any further sample treatment. Hence sampling of large molecular weight compounds such as biologic proteins and peptides was traditionally not possible. With the availability of high molecular weight cutoff (MWCO) membranes (e.g., polycarbonate membranes with MWCO 300 kDa and maximum pore size of 0.3 μm), microdialysis sampling of biologic proteins and peptides, especially cytokines, is now possible (Clough 2005; Ao and Stenken 2006). Also, recently the literature describes the application of a new "vent" probe with push–pull perfusion technology for sampling neuropeptides and other macromolecules in the mouse brain (Takeda et al. 2011). Additional information could be found in the chapter on microdialysis and high molecular weight compounds in this book as well as the other literature references.

Lipophilicity: Traditionally, MD technique was mostly applied to sample hydrophilic moieties of low molecular weight and lipophilic molecules were challenging to sample and led to poor recovery due to physiological perfusate and/or non-specific binding to the probe materials (Duo et al. 2006). Improvement in probe design and modification of perfusate has made possible recovery of lipophilic compounds. Substances such as albumin (Carneheim and Stahle 1991; Klede et al. 2005), lipid emulsions such as Intralipid® and Encapsin® (Ward et al. 2003), chemical binding affinity agents such as cyclodextrins (Duo et al. 2006; Sun and Stenken 2003) etc., have been used in or as perfusate to improve recovery of lipophilic substances.

4.4.2 Factors Related to the Microdialysis Technique

Microdialysis probe membrane materials and membrane window: Ideally, probe material should be inert and should not bind or react with the analyte of interest as this can have direct effect on the recovery. Also, the thickness of the membrane will decrease recovery (de Lange et al. 2000). MD recovery is directly proportional to the membrane length. A near 100 % recovery is possible with longer membrane windows. However, the tissue where the MD sampling is desired will be crucial when deciding on the length of the membrane window (e.g., when sampling in specific areas in the brain, shorter membrane window is preferred) (Chaurasia 1999). In vitro estimation of any loss of the analyte on the probe materials (including membrane, probe arms and tubing) should be evaluated and optimized prior to in vivo MD sampling.

Perfusate composition: Ideally, the composition, ionic strength, osmotic value, and pH of the perfusate should be as close as possible to the extracellular fluid of the dialyzed tissue. Most often isotonic saline or lactated ringer solution is used. However, with the aim of improving recovery of hydrophilic analytes, organic substances such as ethanol, propylene glycol, dimethylsulfoxide, and arachidonic acid have been added to the perfusate (Holmgaard et al. 2010), and for lipophilic analytes substances such as albumin and cyclodextrins have been added to the perfusate (Carneheim and Stahle 1991; Klede et al. 2005; Duo et al. 2006; Sun and Stenken 2003). Lipid emulsions such as Intralipid® and Encapsin® (Ward et al. 2003; Au et al. 2012) have also been used to improve recovery of lipophilic analytes. The use of antibodies and antibody-immobilized beads have been reported to improve the recovery of cytokines by MD (Duo et al. 2006). The use of modified perfusate might impose additional analytical and calibration challenges.

Perfusate flow rate and sampling interval: The flow rate of a perfusate through a microdialysis probe will affect recovery. Slower flow rates produce higher relative recovery of the analyte with longer sampling interval, but this will result in the loss of temporal resolution (de Lange et al. 2000; Chaurasia 1999; de Lange et al. 1997). These variables should be standardized during development.

Recovery and calibration techniques: The applicability of MD for PK/PD depends on the ability to achieve reliable and reproducible analyte recovery from the tissue of interest. Basically, MD is used to measure the extracellular tissue concentration and MD sampling takes place under non-equilibrium conditions with perfusate continuously flowing through the probe at a predetermined rate. Hence, the concentrations in the dialysate will be proportional to the periprobe concentrations, but will not be exactly equal to the extracellular tissue concentrations of the analyte. When quantitative information on the extracellular tissue concentrations is desired, calibration of the probe is required. Probes with longer membrane window and slow perfusate flow rates might approach recovery close to 100 %. However, clinically with the use of shorter membrane windows make probe calibration an integral part of the MD experiment. Most frequently used calibration methods in the literature are retrodialysis by drug or by calibrator methods (de Lange et al. 2000; Wang et al. 1993; Bouw and Hammarlund-Udenaes 1998), no-net-flux method (Lonnroth et al. 1987), dynamic-no-net-flux method (Olson and Justice 1993) and flow rate method (Jacobson et al. 1985; Cano-Cebrian et al. 2005). Adequate choice and standardization of calibration method will be required during drug development to produce reliable results.

For additional information refer to the chapter on recovery and calibration techniques in this book and also to the literature references.

Bioanalytical method consideration: According to Chaurasia et al., usually MD sampling is "volume neutral", because sampling is carried out using very low perfusate flow rates to minimize the removal of fluid volume from the extracellular space of the tissue of interest (Chaurasia et al. 2007). The low flow rates are also intended to improve relative recovery. This results in very low volume of the dialysate collected per MD sample. Higher volume collection will require longer sampling duration, but this will result in loss of temporal resolution. Higher perfusate flow rates result in lower relative recovery. Hence, highly sensitive and well-validated analytical methods are required to analyze MD samples in order to produce reproducible results (Guihen and O'Connor 2009). With improved probe design and experimental conditions permitting monitoring of bioactive proteins and peptides as well as lipophilic molecules, further development and validation of analytical methodology will be required. For further information, refer to the chapter on analytical aspects in this book and also refer to the several literature references available. For more information on bioanalytical method validation, refer to Guidance for Industry—Bioanalytical Method Validation (Guidance for industry 2001).

Stability of the microdialysis samples: Stability of the analyte in the dialysate should be evaluated to support the stability of the MD samples for the entire period of storage and analysis.

Temperature: There are reports of in vivo recovery being affected by temperature. Hence, it is recommended that in vitro probe recovery be conducted at the temperature of the tissue of interest (Chaurasia et al. 2007).

Sterilization of the probe: MD is an invasive technique. Hence, it would be necessary to maintain aseptic conditions to prevent secondary infection when implanting MD probes into a tissue. Sterilization of probes would be required before implantation and a suitable sterilization method should be evaluated (Hocht et al. 2006).

Training of personnel: MD is a specialized technique and adequate training of personnel is required to achieve low variability in experimental factors such as probe manufacture, probe insertion, probe calibration, etc.

4.4.3 Physiological Factors

Probe insertion tissue trauma and equilibration: MD probe implantation is shown to cause tissue trauma in tissues such as dermis, muscle, tumor, and liver and this could influence recovery (Hocht et al. 2004; Bungay et al. 2003; Chen 2006; de Lange et al. 1997). Adequate time should be allowed for tissue equilibration for the trauma to subside before initiating MD sampling (Chaurasia et al. 2007). The maximum tolerable insertion time for the MD probe should also be considered as long-term implantation might cause inflammatory responses, cell changes, fibrosis, etc. (Herkenne et al. 2008).

Matrix tortuosity: In order for an analyte to be recovered by a MD probe, the nature of the sample matrix will have a direct effect on the recovery as analytes must first diffuse through the tissue surrounding the probe. Tissues with high tortuosity coupled with non-hydrodynamic nature will make in vitro probe calibration challenging to infer in vivo tissue concentrations. In order to minimize the effect of matrix tortuosity, probe calibration should be performed in a matrix that closely represents the tissue (Chaurasia 1999).

Effect of anesthesia: Recovery might be affected by anesthetics since they interfere with numerous physiological processes (de Lange et al. 1997). Hence the effect of an anesthetic agent on recovery should be adequately evaluated.

4.5 Conclusions

This chapter provides a concise overview of the applications of MD in preclinical and clinical setting in lines with FDA critical path initiative and regulatory requirements for BA and BE assessment. Optimization of the MD technique will be critical in order to obtain reproducible and reliable results.

FDA's Center for Devices and Radiological Health has approved two devices utilizing MD principle. These include a Cerebral Tissue Monitoring System to monitor biochemical markers of ischemia in the brain (CMA Cerebral tissue monitoring system 2002) and the GlucoDay diabetes monitoring device (Klonoff 2005). Investigational use of other unapproved microdialysis devices in humans should be conducted under the supervision of an Institutional Review Board (IRB) consistent with the Investigational Device Exemption (IDE) regulations (Medical devices 2012).

Preclinical microdialysis data have been accepted to support the mechanism of action of the anti-seizure drug zonisamide and for providing preclinical in vivo evidence to evaluate the drug CPP-109 in the treatment of cocaine addiction (Chaurasia et al. 2007). Also, AAPS and FDA held a joint workshop titled "Microdialysis Principles, Application and Regulatory Perspectives" in Nashville, TN in 2005 and published a white paper in 2007 (Chaurasia et al. 2007).

In conclusion, from a regulatory point of view, FDA does not require MD studies at this point of time, but the Agency is receptive of including MD during drug development and including MD study results in product labeling. One of the questions often asked to the FDA is "When will MD data be included in a package insert?"; when the question should be, "When will Sponsors pursue the development of MD as a tool for drug development", as the FDA cannot include data it does not have in package inserts. Sponsors interested in applying this technique should engage in an open dialog with the Agency during the process of development. Formal meetings can be requested between the FDA and the Sponsor and you are referred to Guidance for Industry—Formal Meetings between the FDA and the Sponsors or Applicants (Guidance for industry 2009). Currently, there are no individual product

BE Guidances that recommend MD studies for BE. However, protocols for the use of MD to demonstrate BE can be submitted to FDA's Office of Generic Drugs for evaluation of their suitability for submission in an ANDA.

References

ACCP frontiers symposia: 6th international symposium on microdialysis in drug research and development, New Orleans (2010) Last accessed on 01/18/2012. Available from http://accp1.org/college_activities.shtml

Alavijeh MS, Palmer AM (2010) Measurement of the pharmacokinetics and pharmacodynamics of neuroactive compounds. Neurobiol Dis 37(1):38–47 Epub 2009/10/13

Anti-infective drugs advisory committee meeting (64th Meeting) (1998) (Last accessed on 01/18/2010) Available from http://www.fda.gov/ohrms/dockets/ac/98/transcpt/3435t3.pdf

Ao X, Stenken JA (2006) Microdialysis sampling of cytokines. Methods 38(4):331–341 Epub 2006/02/21

Au WL, Skinner MF, Benfeldt E, Verbeeck RK, Kanfer I (2012) Application of dermal microdialysis for the determination of bioavailability of clobetasol propionate applied to the skin of human subjects. Skin Pharmacol Physiol 25(1):17–24 Epub 2011/08/24

Benfeldt E, Serup J, Menne T (1999a) Effect of barrier perturbation on cutaneous salicylic acid penetration in human skin: in vivo pharmacokinetics using microdialysis and non-invasive quantification of barrier function. Br J Dermatol 140(4):739–748 Epub 1999/05/08 http://www.ncbi.nlm.nih.gov/pubmed/10233334

Benfeldt E, Serup J, Menne T (1999b) Microdialysis vs. suction blister technique for in vivo sampling of pharmacokinetics in the human dermis. Acta Derm-Venereol 79(5):338–342 Epub 1999/09/24

Benfeldt E, Bashaw ED, Shah VP (2007) Assessment of topical bioequivalence using microdialysis and other techniques. In: Roberts M, Walters K (eds) Dermal absorption and toxicity assessment. 2nd edn, New York, Taylor and Francis, pp 537–548

Benfeldt E, Hansen SH, Volund A, Menne T, Shah VP (2007a) Bioequivalence of topical formulations in humans: evaluation by dermal microdialysis sampling and the dermatopharmacokinetic method. J Invest Dermatol 127(1):170–178 Epub 2006/07/29

Bioequivalence of topical drugs: alternative approaches (Bio-international conference, 2008). Last accessed on 01/18/2012 Available from www.apgi.org/documents/Bio-International 2008FINALWEB.pdf

Blakeley J, Portnow J (2010) Microdialysis for assessing intratumoral drug disposition in brain cancers: a tool for rational drug development. Expert Opinion Drug Metabol Toxicol 6(12):1477–1491 Epub 2010/10/26

Boddu SH, Gunda S, Earla R, Mitra AK (2010) Ocular microdialysis: a continuous sampling technique to study pharmacokinetics and pharmacodynamics in the eye. Bioanalysis 2(3):487–507 Epub 2010/11/19

Bouw MR, Hammarlund-Udenaes M (1998) Methodological aspects of the use of a calibrator in in vivo microdialysis-further development of the retrodialysis method. Pharm Res 15(11):1673–1679 Epub 1998/12/02

Brunner M, Derendorf H, Muller M (2005) Microdialysis for in vivo pharmacokinetic/pharmacodynamic characterization of anti-infective drugs. Curr Opin Pharmacol 5(5):495–499 Epub 2005/08/09

Bungay PM, Newton-Vinson P, Isele W, Garris PA, Justice JB (2003) Microdialysis of dopamine interpreted with quantitative model incorporating probe implantation trauma. J Neurochem 86(4):932–946 Epub 2003/07/31

Cano-Cebrian MJ, Zornoza T, Polache A, Granero L (2005) Quantitative in vivo microdialysis in pharmacokinetic studies: some reminders. Curr Drug Metab 6(2):83–90 Epub 2005/04/28
Carneheim C, Stahle L (1991) Microdialysis of lipophilic compounds: a methodological study. Pharmacol Toxicol 69(5):378–380 Epub 1991/11/01
Chaurasia CS (1999) In vivo microdialysis sampling: theory and applications. Biomed chromatogr : BMC 13(5):317–332 Epub 1999/07/29
Chaurasia CS, Muller M, Bashaw ED, Benfeldt E, Bolinder J, Bullock R et al (2007) AAPS-FDA workshop white paper: microdialysis principles, application and regulatory perspectives. Pharm Res 24(5):1014–1025 Epub 2007/04/27
Chen KC (2006) Effects of tissue trauma on the characteristics of microdialysis zero-net-flux method sampling neurotransmitters. J Theor Biol 238(4):863–881 Epub 2005/09/01
Cho S, Bashaw ED (2011) Clinical pharmacology for development of topical dermatological products: present and future opportunities for safety and efficacy. Clin Pharmacol Ther 89(2):167–169 Epub 2011/01/22
Clough GF (2005) Microdialysis of large molecules. AAPS J 7(3):E686–E692 Epub 2005/12/16
CMA Cerebral tissue monitoring system (2002) Available from http://www.accessdata.fda.gov/cdrh_docs/pdf2/k020285.pdf
Code of federal regulations Title 21 Part 314 (2011) Available from http://www.accessdata.fda.gov/scripts/cdrh/cfdocs/cfcfr/CFRsearch.cfm?CFRPart=314
Code of federal regulations Title 21 Part 320 (2011) Available from http://www.accessdata.fda.gov/scripts/cdrh/cfdocs/cfcfr/cfrsearch.cfm?cfrpart=320
de Lange EC, Danhof M, de Boer AG, Breimer DD (1997) Methodological considerations of intracerebral microdialysis in pharmacokinetic studies on drug transport across the blood-brain barrier. Brain Res Brain Res Rev 25(1):27–49 Epub 1997/11/25
de Lange EC, de Boer AG, Breimer DD (2000) Methodological issues in microdialysis sampling for pharmacokinetic studies. Adv Drug Deliv Rev 45(2–3):125–148 Epub 2000/12/08
Draft Guidance on Lidocaine (2007). Available from http://www.fda.gov/downloads/Drugs/GuidanceComplianceRegulatoryInformation/Guidances/ucm086293.pdf
Duo J, Fletcher H, Stenken JA (2006) Natural and synthetic affinity agents as microdialysis sampling mass transport enhancers: current progress and future perspectives. Biosens Bioelectron 22(3):449–457 Epub 2006/06/27
Duvvuri S, Rittenhouse KD, Mitra AK (2005) Microdialysis assessment of drug delivery systems for vitreoretinal targets. Adv Drug Deliv Rev 57(14):2080–2091 Epub 2005/11/08
FDA (2004) Challenge and opportunity on the critical path to new medical products. Available from http://www.fda.gov/ScienceResearch/SpecialTopics/CriticalPathInitiative/CriticalPathOpportunitiesReports/ucm077262.htm
FDA (2007) Critical path opportunities for generic drugs. Available from http://www.fda.gov/ScienceResearch/SpecialTopics/CriticalPathInitiative/CriticalPathOpportunitiesReports/ucm077250.htm
Garcia Ortiz P, Hansen SH, Shah VP, Menne T, Benfeldt E (2009) Impact of adult atopic dermatitis on topical drug penetration: assessment by cutaneous microdialysis and tape stripping. Acta Derm-Venereol 89(1):33–8. Epub 2009/02/07
Guidance for Industry—Bioanalytical Method Validation (2001). Available from: http://www.fda.gov/downloads/Drugs/GuidanceComplianceRegulatoryInformation/Guidances/ucm070107.pdf
Guidance for Industry—Bioavailability and Bioequivalence Studies for Orally Administered Drug Products—General considerations (2003) Available from: http://www.fda.gov/downloads/Drugs/GuidanceComplianceRegulatoryInformation/Guidances/ucm070124.pdf
Guidance for Industry—Formal Meetings Between the FDA and Sponsors or Applicants (2009) Available from http://www.fda.gov/downloads/drugs/GuidanceComplianceRegulatoryInformation/guidances/ucm153222.pdf
Guidance for industry—Topical Dermatologic Corticosteroids in vivo Bioequivalence (1997) Available from http://www.fda.gov/downloads/Drugs/GuidanceComplianceRegulatoryInformation/Guidances/ucm070234.pdf

Guihen E, O'Connor WT (2009) Current separation and detection methods in microdialysis the drive towards sensitivity and speed. Electrophoresis 30(12):2062–2075 Epub 2009/07/08

Guihen E, Ho WL, Hogan AM, O'Connell ML, Leahy MJ, Ramsay B et al (2012) Rapid quantification of histamine in human psoriatic plaques using microdialysis and ultra high performance liquid chromatography with fluorescence detection. J Chromatogr, B: Anal Technol Biomed Life Sci 880(1):119–124 Epub 2011/12/20

Hammarlund-Udenaes M (2010) Active-site concentrations of chemicals—are they a better predictor of effect than plasma/organ/tissue concentrations? Basic Clin Pharmacol Toxicol 106(3):215–220 Epub 2010/01/07

Herkenne C, Alberti I, Naik A, Kalia YN, Mathy FX, Preat V et al (2008) In vivo methods for the assessment of topical drug bioavailability. Pharm Res 25(1):87–103 Epub 2007/11/07

Hocht C (2011) Microdialysis in drug discovery. In: Tsai T-H (ed) Applications of microdialysis in pharmaceutical science, 1st edn. John Wiley and Sons, pp 7–30

Hocht C, Opezzo JA, Taira CA (2004) Microdialysis in drug discovery. Curr Drug Discov Technol 1(4):269–285 Epub 2006/02/14

Hocht C, Opezzo JA, Bramuglia GF, Taira CA (2006) Application of microdialysis in clinical pharmacology. Curr Clin Pharmacol 1(2):163–183 Epub 2008/08/01

Holmgaard R, Nielsen JB, Benfeldt E (2010) Microdialysis sampling for investigations of bioavailability and bioequivalence of topically administered drugs: current state and future perspectives. Skin Pharmacol Physiol 23(5):225–243 Epub 2010/05/21

Hughes JP, Rees S, Kalindjian SB, Philpott KL (2011) Principles of early drug discovery. Br J Pharmacol 162(6):1239–1249 Epub 2011/11/26

Incecayir T, Agabeyoglu I, Derici U, Sindel S (2011) Assessment of topical bioequivalence using dermal microdialysis and tape stripping methods. Pharm Res 28(9):2165–2175 Epub 2011/04/19

Jacobson I, Sandberg M, Hamberger A (1985) Mass transfer in brain dialysis devices—a new method for the estimation of extracellular amino acids concentration. J Neurosci Methods 15(3):263–268 Epub 1985/11/01

Juluru R, Shukla C, Yin H, Stagni G (2012) Skin microdialysis-based estimation of systemic bioavailability fraction. J Pharm Sci 101(1):405–413 Epub 2011/09/22

Kitzen JJ, Verweij J, Wiemer EA, Loos WJ (2006) The relevance of microdialysis for clinical on:cology. Curr Clin Pharmacol 1(3):255–263 Epub 2008/08/01

Klede M, Schmitz H, Goen T, Fartasch M, Drexler H, Schmelz M (2005) Transcutaneous penetration of toluene in rat skin a microdialysis study. Exp Dermatol 14(2):103–108 Epub 2005/02/01

Klonoff DC (2005) Continuous glucose monitoring: roadmap for 21st century diabetes therapy. Diabetes Care 28(5):1231–1239 Epub 2005/04/28

Kreilgaard M, Kemme MJ, Burggraaf J, Schoemaker RC, Cohen AF (2001) Influence of a microemulsion vehicle on cutaneous bioequivalence of a lipophilic model drug assessed by microdialysis and pharmacodynamics. Pharm Res 18(5):593–599 Epub 2001/07/24

Medical devices: how to market your device. Last accessed on 01/18/2012 Available from http://www.fda.gov/MedicalDevices/DeviceRegulationandGuidance/HowtoMarketYourDevice/InvestigationalDeviceExemptionIDE/default.htm

Lionberger RA (2008) FDA critical path initiatives: opportunities for generic drug development. AAPS J 10(1):103–109 Epub 2008/05/01

Liu P, Muller M, Derendorf H (2002) Rational dosing of antibiotics: the use of plasma concentrations versus tissue concentrations. Int J Antimicrob Agents 19(4):285–290 Epub 2002/04/30

Lonnroth P, Jansson PA, Smith U (1987) A microdialysis method allowing characterization of intercellular water space in humans. Am J Physiol 253(2 Pt 1):E228–E231 Epub 1987/08/01

McAdoo DJ, Wu P (2008) Microdialysis in central nervous system disorders and their treatment. Pharmacol Biochem Behav 90(2):282–296 Epub 2008/04/26

McCleverty D, Lyons R, Henry B (2006) Microdialysis sampling and the clinical determination of topical dermal bioequivalence. Int J Pharm 308(1–2):1–7 Epub 2005/12/21

Narkar Y (2010) Bioequivalence for topical products–an update. Pharm Res 27(12):2590–2601 Epub 2010/09/23

Olson RJ, Justice JB Jr (1993) Quantitative microdialysis under transient conditions. Anal Chem 65(8):1017–1022 Epub 1993/04/15

Papoiu AD, Wang H, Nattkemper L, Tey HL, Ishiuji Y, Chan YH et al (2011) A study of serum concentrations and dermal levels of NGF in atopic dermatitis and healthy subjects. Neuropeptides 45(6):417–422 Epub 2011/09/07

Schmidt S, Banks R, Kumar V, Rand KH, Derendorf H (2008) Clinical microdialysis in skin and soft tissues: an update. J Clin Pharmacol 48(3):351–364 Epub 2008/02/21

Shah VP (2005) IV-IVC for topically applied preparations–a critical evaluation. Eur J Pharm Biopharm: official J Arbeitsgemeinschaft Fur Pharmazeutische Verfahrenstechnik eV. 60(2):309–314 Epub 2005/06/09

Shukla C, Patel V, Juluru R, Stagni G (2009) Quantification and prediction of skin pharmacokinetics of amoxicillin and cefuroxime. Biopharm Drug Dispos 30(6):281–293 Epub 2009/07/11

Sun L, Stenken JA (2003) Improving microdialysis extraction efficiency of lipophilic eicosanoids. J Pharm Biomed Anal 33(5):1059–1071 Epub 2003/12/06

Takeda S, Sato N, Ikimura K, Nishino H, Rakugi H, Morishita R (2011) Novel microdialysis method to assess neuropeptides and large molecules in free-moving mouse. Neuroscience 186:110–119 Epub 2011/05/03

Tegeder I, Brautigam L, Podda M, Meier S, Kaufmann R, Geisslinger G et al (2002) Time course of 8-methoxypsoralen concentrations in skin and plasma after topical (bath and cream) and oral administration of 8-methoxypsoralen. Clin Pharmacol Ther 71(3):153–161 Epub 2002/03/22

Tettey-Amlalo RN, Kanfer I, Skinner MF, Benfeldt E, Verbeeck RK (2009) Application of dermal microdialysis for the evaluation of bioequivalence of a ketoprofen topical gel. Eur J Pharm Sci : official J Eur Fed Pharm Sci 36(2–3):219–225 Epub 2008/10/14

Wang Y, Wong SL, Sawchuk RJ (1993) Microdialysis calibration using retrodialysis and zero-net flux: application to a study of the distribution of zidovudine to rabbit cerebrospinal fluid and thalamus. Pharm Res 10(10):1411–1419 Epub 1993/10/01

Ward KW, Medina SJ, Portelli ST, Mahar Doan KM, Spengler MD, Ben MM (2003) Enhancement of in vitro and in vivo microdialysis recovery of SB-265123 using intralipid and encapsin as perfusates. Biopharm Drug Dispos 24(1):17–25 Epub 2003/01/08

Wei YH, Xu LZ, Shen Q, Li FZ (2009) Microdialysis: a technique for pharmacokinetic-pharmacodynamic studies of oncological drugs. Curr Pharm Biotechnol 10(6):631–640 Epub 2009/07/22

Part II
Specific Topics

Chapter 5
Microdialysis in CNS PKPD Research: Unraveling Unbound Concentrations

Margareta Hammarlund-Udenaes

Abstract The pharmacokinetic (PK) processes of drug movements between blood and tissues and the drug concentration profiles within a tissue over time can be described with the help of microdialysis sampling. In addition, microdialysis uniquely allows measurement of the concentrations of unbound, pharmacologically active drug; it is not possible to obtain these directly with any other method. These are the most valuable outcomes of the microdialysis technique for pharmacokinetic/pharmacodynamic (PKPD) studies, and are the basis of the principles of drug distribution and PKPD in the central nervous system (CNS) presented in this chapter. This chapter will cover microdialysis studies investigating the PK and PKPD relationships of drugs in the CNS, and will also summarize alternative methods of achieving these results when microdialysis does not work or is too slow, for example in drug discovery processes.

Abbreviations

5-OH-DPAT	5-hydroxy-2-(N,N,-di-n-propylamino)tetralin
$A_{tot,brain}$	Total amount of drug in brain tissue per g tissue
AUC	Area under the concentration–time curve in blood, plasma, or brain
$AUC_{u,blood}$	Unbound AUC in blood
$AUC_{u,brain}$	Unbound AUC in brain ISF
BBB	Blood–brain barrier
BCRP	Breast cancer resistance protein
C_{blood}	Concentration of drug in blood
CL_{in}	Influx clearance from blood to brain across the BBB
CL_{out}	Efflux clearance from brain to blood across the BBB

M. Hammarlund-Udenaes (✉)
Department of Pharmaceutical Biosciences, Uppsala University, Uppsala, Sweden
e-mail: mhu@farmbio.uu.se

M. Müller (ed.), *Microdialysis in Drug Development*, AAPS Advances in the Pharmaceutical Sciences Series, DOI: 10.1007/978-1-4614-4815-0_5,

CNS	Central nervous system
$C_{u,\text{blood}}$	Unbound concentration of a drug in blood
$C_{u,\text{brain}}$	Unbound concentration of a drug in brain ISF
EC50	Concentration resulting in half-maximal effect
ECF	Extracellular fluid (same use as ISF here)
f_u	Fraction of unbound to total drug concentration in plasma
$f_{u,\text{brain}}$	Fraction of unbound to total drug concentration in brain tissue
GHB	Gamma-hydroxybutyric acid
hCMEC	Human brain endothelial capillary cell line
ISF	Interstitial fluid (same use as ECF here)
Km	Michaelis Menten constant (substrate concentration at half-maximal rate of metabolism)
Ki	Equilibrium dissociation constant
K_p	Ratio of total brain to total plasma concentrations
$K_{p,\text{uu}}$	Unbound concentration ratio between brain ISF and blood
$K_{p,\text{uu,cell}}$	Ratio of unbound intracellular concentration (including lysosomal and other organelle distribution) to unbound ISF concentration
LC–MS/MS	Liquid chromatography mass spectrometry
MPTP	N-methyl-4-phenyl-1,2,3,6-tetrahydropyridine
MPP+	1-methyl-4-phenylpyridinium
NMDA	N-methyl-D-aspartate
PET	Positron emission tomography
P-gp	P-glycoprotein
PD	Pharmacodynamic
PK	Pharmacokinetic
PKPD	Pharmacokinetic/pharmacodynamic
pKa	−10logKa, the acid dissociation constant
PS	Permeability surface area product
V_{blood}	Physiological volume of blood in brain per g tissue
$V_{u,\text{brain}}$	Unbound volume of distribution in the brain, in ml × g_brain^{-1} tissue

5.1 Introduction

The microdialysis technique has allowed the measurement of unbound drug concentrations at relevant receptor sites within the central nervous system (CNS). As such, microdialysis results have created the basis for understanding the quantitative role of blood–brain barrier (BBB) transport for drug action in the CNS. In addition, microdialysis has provided an ethical advantage over the sampling of whole tissue in that it results in rich data from fewer animals.

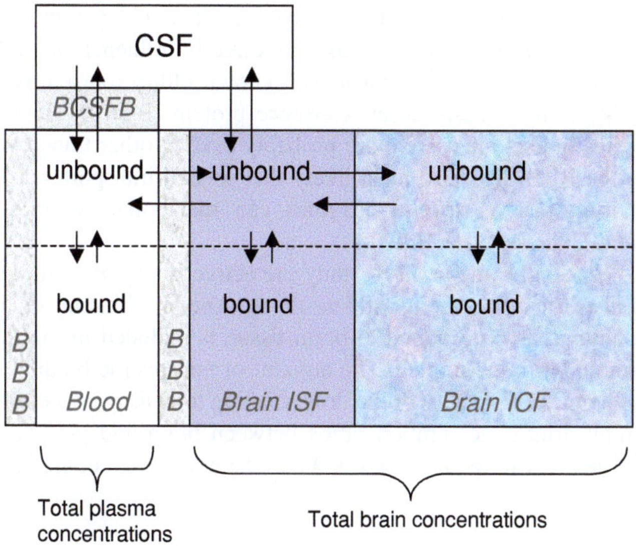

Fig. 5.1 A schematic drawing of the equilibration of drug concentrations among blood, brain, and CSF. BBB is the blood–brain barrier, BCSFB is the blood CSF barrier, brain ISF is the brain interstitial fluid and brain ICF is the brain intracellular fluid. Plasma concentrations include both bound and unbound drug, and total brain concentrations include bound and unbound drug both in brain ISF and intracellularly. From Hammarlund-Udenaes et al. (2008) with permission from the publisher

The free drug hypothesis states that it is the concentration of free (unbound) drug at the site of action that is related to the drug's action. Pharmacokinetic (PK) principles also tell us that unbound drug is the only form of the drug that can traverse membranes and that the concentration of unbound drug reaches equilibrium via random movements across all the membranes in the body, as the drug moves between blood and tissue interstitial fluid (ISF) across the capillary walls and between ISF and intracellular fluid across the cellular membranes (Fig. 5.1). The distribution of the drug is also influenced by differences between the pH of these compartments and the pKa of the drug. The equilibration of drug bound to plasma proteins or tissue components are separate equilibria, both from each other and from the equilibration across the membranes of the body.

As knowledge of the role of efflux and influx transporters at some of the membrane surfaces of the body has increased, it has become apparent that the principle of similar unbound drug concentrations for all tissues will not hold true, as the transporters actively influence transport of the drug in one or the other direction across the membrane. Active transport can cause differences in the concentrations of unbound drug between tissue ISF and blood. This is especially true in the CNS, where it has been clearly shown that the concentrations of unbound drug in the brain are better determinants of the pharmacodynamics (PD) of the drug in the brain than the total drug concentrations in the brain or the total or unbound concentrations in the plasma (Watson et al. 2009).

The function and efficiency of the transporters have a very important effect on the transport of drugs into the CNS, as they directly influence the unbound drug concentrations in the CNS. The most well-known efflux transporters are P-glycoprotein (P-gp) and Breast cancer resistance protein (BCRP). However, microdialysis studies indicate that there are probably several other transporters that act on drugs at the BBB. It is also most likely that several transporters can act on an individual drug, just as several enzymes can metabolize an individual drug (Hammarlund-Udenaes et al. 2008).

It is thus especially important to study the active transport of drugs across the BBB. It is also important to investigate the nonspecific binding of drugs to the brain parenchyma, as drug bound to brain tissue is included in the measurement when the whole brain is sampled. The amount of nonspecific binding differs with the physicochemical properties of the drug, making total drug concentrations in the brain and total drug concentration ratios between brain and plasma less able to predict drug delivery to the brain and PD parameters (Friden et al. 2009b; Watson et al. 2009).

5.2 Experimental Prerequisites for Microdialysis PKPD Studies

There are several prerequisites for microdialysis to function as a method of studying drug distribution in general and specifically to the CNS. Some of these issues are covered in more depth in other chapters, but will be briefly mentioned here.

5.2.1 In Vivo Recovery

The importance of measuring the recovery of the drug in vivo cannot be stressed enough. Microdialysis measures a fraction of the extracellular drug concentration in the tissue in which the microdialysis probe is placed; the higher the flow rate through the microdialysis probe, the lower the fraction recovered. Recovery of the drug can vary according to the tissue properties and active transport mechanisms present, thus giving false concentration values if in vivo recovery is not assessed at the site of the measurement. This is a prerequisite for comparing drug concentrations measured by microdialysis in different tissues or when comparing concentrations in a specific tissue before and after an intervention. The recovery of a drug will probably be changed if an efflux transporter is blocked, as it is mainly influenced by elimination from the studied tissue (Olson and Justice 1993; Olson Cosford et al. 1996; de Lange et al. 1998; Xie et al. 1999). Performing a drug distribution study without performing an in vivo recovery will therefore decrease the value of the study significantly, giving results that cannot be interpreted in a scientifically sound way.

5.2.2 In Vitro Recovery and Sticking to Tubings

The main disadvantage of microdialysis is that lipophilic substances often stick to the tubing and/or microdialysis probe. In vitro drug recovery and sticking to the tubings should be checked before starting an in vivo PK or PKPD microdialysis study. Problems with drug attachment to tubing can be handled by using tubing that will not adsorb the drug, or by adding low concentrations (approximately 0.5 %) of bovine serum albumin to the perfusate (Chaurasia et al. 2007). If these interventions do not solve the problem, it is better not to perform an in vivo microdialysis study, as the data cannot be interpreted correctly.

Several things need to be checked:

- *Tubing properties:* the tubing can be checked for adsorption by perfusing it with low concentrations of the drug, sampling from the end of the tubing every 10 min at a flow rate of 1 μl/min, changing the perfusion fluid to blank buffer after an hour, and then sampling every 10 min for another hour. A rapid change in the outflow concentrations with the changed perfusion fluid ("square wave") means that the drug does not stick and the tubing can be used. Conversely, if the outflow concentration decreases during the drug perfusion and then slowly returns to the inflow concentration with the buffer (a "mirror" profile), the tubing is not suitable. If the drug does adsorb to the tubing, the test can be repeated with the addition of albumin. If the tubing can be used with this addition, the concentrations should change more rapidly.
- *Probe properties:* any loss or gain across the microdialysis probe in vitro can be checked after connecting the probe to the tubing, switching between perfusates with and without drug, and switching between surrounding solutions with and without drug (4 periods). Again, a "square wave" should be observed when switching between perfusates with and without drug (Chaurasia et al. 2007).
- *High albumin binding:* if the drug is highly protein bound and albumin is required in the perfusate, the obtained concentrations and the recovery measurements may be wrong. The loss may be different from the gain, and recovery calculations using loss of drug will not allow the true extracellular fluid (ECF) concentrations from gain of drug from the surrounding fluid to be assessed (unpublished experience from our laboratory).

Thus, the in vitro drug recovery test is not important in itself, as it cannot be used to calculate drug recovery from the in vivo data. The value of performing these pretests is to find out whether the drug is dialyzable and whether it will be possible to interpret the data from the in vivo studies.

5.2.3 *Study Design*

The mode of drug administration should of course be related to the question under investigation. For a PK study with simultaneous measurement of PD, it is an advantage to infuse the drug so that the concentrations increase at first and then decrease, within the time frame of an experiment. It is recommended that at least three samples be collected along with the PD measurements as the concentrations are increasing, and at least three as they decrease, in order to relate the PD properties to the PK properties (Tunblad et al. 2004b; Bostrom et al. 2008). Thus, a medium length infusion is preferred. This will show if there is any delay in the PD response in relation to the PK. Modeling of the information in this way is a very valuable method of refining the analysis of the data (Tunblad et al. 2004b; Bostrom et al. 2006).

If, on the other hand, the concentrations of unbound drug in the brain are to be related to the results of an in vitro receptor binding assay, it may be preferable to use steady-state infusions. The purpose in this context is more to obtain a reliable relationship between the concentrations of unbound drug in the brain and those in the blood ($K_{p,uu}$), so that the unbound drug blood concentrations can later be correlated with PD results and dose requirements (Hammarlund-Udenaes et al. 2008; Hammarlund-Udenaes 2010; Kielbasa and Stratford 2012). Depending on the half-life of the drug, an initial short, more rapid infusion will result in reaching steady state more quickly, within the time frame of a study. An optimal initial infusion should be based on the known PK of the drug in the species studied.

Combining a steady-state infusion with mapping the fall in drug concentrations in the brain and blood after the end of the infusion provides more information for modeling the data (Tunblad et al. 2004b; Bostrom et al. 2006). Changes in the brain concentrations with time, relative to the plasma concentrations, will also show if there are any restrictions on how the drug leaves the brain. If, however, the concentration profiles are parallel, the plasma concentration–time profile is the slowest step.

5.3 Measuring the Distribution of Drugs in the Brain with Microdialysis

Microdialysis has been important in revealing the in vivo role of transporters at the BBB. As it is possible to measure the concentrations of unbound drug on each side of the BBB by applying one probe in the brain parenchyma and the other in a blood vessel, the steady-state equilibrium will directly describe any deviation from equal unbound drug concentrations on each side (i.e. passive transport across the BBB), and subsequently describe the quantitative in vivo influence of the transporters (Hammarlund-Udenaes et al. 2008).

5.3.1 $K_{p,uu}$, the Unbound Drug Concentration Ratio Between Brain Interstitial Fluid and Blood

There are three measurements of brain drug distribution that can be obtained with microdialysis; the most easily obtainable is the ratio of unbound drug concentrations in the brain to those in blood, $K_{p,\mathrm{uu}}$ (Gupta et al. 2006). This ratio can be determined at steady state as

$$K_{p,\mathrm{uu}} = \frac{C_{u,\mathrm{brainISF}}}{C_{u,\mathrm{blood}}} \tag{5.1}$$

$C_{u,\mathrm{brain}}$ is the concentration of unbound drug in brain ISF and $C_{u,\mathrm{blood}}$ the corresponding concentration of unbound drug in blood (plasma). Alternatively, $K_{p,\mathrm{uu}}$ can be determined by calculating the area under the concentration–time curve (AUC) for unbound drug in the brain ISF vs that for unbound drug in blood: $AUC_{u,\mathrm{brain}}/AUC_{u,\mathrm{blood}}$. The unbound drug concentration in blood can be directly measured with microdialysis or can be calculated from regular blood sampling with correction for protein binding.

The value of estimating $K_{p,\mathrm{uu}}$ lies in that it is a direct in vivo measure of the role of transporters for the studied drug. This can help drug discovery laboratories to select the optimal drug for further development, instead of basing the decision on total brain and plasma drug concentrations. It can also explain a lack of PD response if the efflux is too efficient in relation to the doses administered. $K_{p,\mathrm{uu}}$ gives a direct estimation of the net in vivo activity of efflux and/or influx transporters on the drug studied. This parameter can be used to determine whether a drug is suitable for the treatment of CNS disorders or whether the ratio is too low for effective drug dosage. It can also be used to discover whether CNS side effects can be avoided; if this is an issue, a low $K_{p,\mathrm{uu}}$ is preferable. As $K_{p,\mathrm{uu}}$ relates the concentration of unbound drug in plasma to its counterpart in the brain, knowing the value of $K_{p,uu}$ makes it easier to use plasma concentration data for effect correlations. The relationship of unbound drug brain concentrations to in vitro receptor occupancy or related PD measurements can also be estimated without measuring $C_{u,\mathrm{brain}}$ in all experiments.

5.3.2 $V_{u,brain}$, the Unbound Drug Volume of Distribution in Brain Tissue

The second easiest parameter to measure with microdialysis is the intrabrain distribution, $V_{u,\mathrm{brain}}$, i.e., the ratio of the total concentration of drug in the brain to the concentration of unbound drug in the brain ISF,

$$V_{u,\mathrm{brain}} = \frac{A_{\mathrm{tot,brain}} - V_{\mathrm{blood}} \times C_{\mathrm{blood}}}{C_{u,\mathrm{brainISF}}} \tag{5.2}$$

$V_{u,\mathrm{brain}}$ estimation requires the total amount of drug in the brain ($A_{\mathrm{tot,brain}}$) to be corrected for the amount of drug in the blood in the whole-brain sample. To compensate for the amount in the blood, it is necessary to estimate the physiological volume of blood in the brain, V_{blood}, and to obtain the concentration of the drug in the blood, C_{blood}. It has been shown that, for drugs with low nonspecific binding in the brain, the values used for V_{blood} and C_{blood} are crucial; in fact, V_{blood} is generally lower than the 3 % often used to represent the volume of blood as a percentage of the volume of tissue in the brain (Friden et al. 2010). The total amount of drug in the brain is best measured at steady state in parallel with measurement of the unbound drug concentrations. This can be done by sampling at the end of a constant rate infusion.

The value of estimating $V_{u,\mathrm{brain}}$ lies in the resultant ability to estimate how much of the drug is nonspecifically bound or dissolved in brain parenchymal tissue. This value can be used to compare total drug concentrations in the brain to the concentrations of unbound active drug. Studies using microdialysis and the brain slice method have shown that $V_{u,\mathrm{brain}}$ can vary between 0.2 and 3000 ml × g_brain tissue^{-1} (Friden et al. 2009b).

5.3.3 Permeability

The most difficult parameter to obtain from microdialysis studies is the permeability of the BBB to the drug at the BBB, or expressed differently the rate of transport of drug into and out of the brain. From a PK viewpoint, it is the clearance into and clearance out of the brain (CL_{in} and CL_{out}) that are measured. In other methods, the net CL_{in} parameter that is measured is called the permeability surface area product, PS. PS is often measured in μl × min^{-1} × g^{-1} of brain in both microdialysis and in situ brain perfusion studies (Dagenais et al. 2000). It is necessary to model the microdialysis data in order to obtain these parameters (Tunblad et al. 2004b; Bostrom et al. 2006). The reliability of the estimation is dependent on the study design; both the increases and decreases in the profile should be followed for both brain and blood.

The relationship between CL_{in} and CL_{out} is related to $K_{p,\mathrm{uu}}$ as

$$K_{p,\mathrm{uu}} = \frac{C_{u,\mathrm{brainISF}}}{C_{u,\mathrm{blood}}} = \frac{CL_{\mathrm{in}}}{CL_{\mathrm{out}}} \tag{5.3}$$

Thus, if $CL_{\mathrm{in}} = CL_{\mathrm{out}}$, passive transport across the BBB dominates while active efflux and influx cancel each other out, and $K_{p,\mathrm{uu}}$ will be = 1. When $CL_{\mathrm{in}} > CL_{\mathrm{out}}$, $K_{p,uu}$ will be >1 and transport across the BBB is dominated by active uptake. If

$CL_{in} < CL_{out}$, $K_{p,uu}$ will be <1, and active efflux dominates (Hammarlund-Udenaes et al. 2008).

From a PD perspective, the permeability is important if rapid effects are wanted, for example in anesthesia. However, for treatment of chronic diseases, the permeability of the drug at the BBB seems to be of less importance. If the permeability of the drug is high enough for gastrointestinal absorption, the BBB is permeable enough to allow subsequent distribution into the brain and thereby for central effects.

5.4 Alternative Methods to Microdialysis to Determine Brain Distribution of Drugs

Although microdialysis is an excellent method, it does not always work, as discussed above with, for example, compounds that stick to the tubings and the probe. Microdialysis also requires time and is labor intensive. Other methods are therefore required in drug discovery processes, where there is a need to study compounds in a high-throughput manner but not for the exact results obtained with microdialysis.

5.4.1 Estimation of $K_{p,uu}$

One way of obtaining this parameter with techniques other than microdialysis is to combine in vivo estimation of K_p, the ratio of total brain to total plasma drug concentrations, with estimations of plasma protein binding and brain tissue binding (Friden et al. 2009a).

K_p can be estimated from a steady-state experiment or from the AUC after other dosage strategies. In practice, a single sample taken 1 h after drug dosage is often used in drug discovery procedures. However, this might not give the same answer as that obtained from a steady-state relationship and it should be tested more in-depth for drugs with lower permeabilities.

Plasma protein binding can be estimated using equilibrium dialysis to give f_u, the fraction of unbound drug in plasma. Brain tissue binding can be estimated either with the brain slice method, using $V_{u,\text{brain}}$ (Wang and Welty 1996; Kakee et al. 1997; Friden et al. 2007, 2009a, 2010), or with the brain homogenate method, where the fraction of unbound drug in brain tissue, $f_{u,\text{brain}}$, is calculated (Kalvass and Maurer 2002; Mano et al. 2002).

$$K_{p,\text{uu}} = \frac{K_p \times f_{u,\text{brain}}}{f_{u,\text{plasma}}} \tag{5.4}$$

$$K_{p,\mathrm{uu}} = \frac{K_p}{V_{u,\mathrm{brain}} \times f_{u,\mathrm{plasma}}} \tag{5.5}$$

It has been shown that the brain homogenate and brain slice methods do not give the same results for nonspecific binding. The advantage of the slice method is that the cells and cell organelles remain mainly intact, so that intracellular distribution is also captured. Thus, the trapping of basic drugs in lysosomes that is measured with the brain slice method, is not detected with the brain homogenate method. It has been recommended that the pH partitioning equation should be used to adjust brain homogenate results to better fit brain slice to in vivo distribution (Friden et al. 2011).

5.4.2 Estimation of $V_{u,brain}$ or $f_{u,brain}$

$V_{u,\mathrm{brain}}$ can be estimated in vitro by utilizing the brain slice or brain homogenate techniques. The brain slice technique has been developed for high-throughput processes and can be performed in cassette mode, where from five to possibly ten compounds per cassette can be studied simultaneously (Friden et al. 2009a, Friden et al. 2010).

The brain homogenate technique for estimating $f_{u,\mathrm{brain}}$ can be easily performed with frozen tissue (Kalvass and Maurer 2002; Mano et al. 2002). It is, however, important to compensate brain homogenate results for pH-dependent partitioning into lysosomes to more correctly estimate in vivo partitioning (Friden et al. 2011).

5.4.3 One Step Further: Estimation of $K_{p,uu,cell}$

$K_{p,\mathrm{uu,cell}}$ gives the ratio of the concentration of unbound drug in the cell, including that in lysosomes and other organelles, to the unbound ISF concentration. Intracellular drug distribution cannot be estimated by microdialysis, which only measures extracellular concentrations. The intracellular partitioning of unbound drug can be estimated with a combination of brain slice and brain homogenate methods (Friden et al. 2009a, 2011). It assumes that the nonspecific binding is the same for the two methods. Alternatively, $K_{p,\mathrm{uu,cell}}$ can be theoretically determined from homogenate studies alone, using physiological values and pKa (Friden et al. 2011).

5.5 Examples of Brain PK and PKPD Microdialysis Studies from the Literature

This section presents some recent studies that have used microdialysis to study brain PK, BBB transport of drugs, and CNS PKPD. Modeling of experimental microdialysis results has increased over recent years, which may be advantageous for the interpretation of the results but can also blur the overall picture.

5.5.1 PKPD Relationships and Biophase Concentrations

Several studies have shown that unbound brain concentrations are better correlated with effect parameters than total brain concentrations, unbound plasma concentrations, or total drug concentrations. In a study by Kalvass et al. the intrinsic potency, measured as the equilibrium dissociation constant (K_i), of various opioids in the CNS was best related to half the maximum effective concentration (EC50) of the unbound drug in mouse brains (Kalvass et al. 2007). Large et al. found that the best correlation was between the in vitro K_i and the concentrations of unbound antiepileptic drugs in the rat brain (Large et al. 2009). Watson and colleagues compared the unbound plasma, brain, and CSF concentrations with in vitro D_2-receptor occupancy for six antipsychotic drugs and found the best correlation to be between brain unbound drug concentrations and receptor occupancy (Watson et al. 2009).

When oxycodone and morphine PKPD were studied in rats using microdialysis, the potency comparison between the two drugs was very different depending on whether plasma or unbound drug brain concentrations were used (Bostrom et al. 2008). This was due to the very different BBB transport properties of oxycodone and morphine, with active uptake of oxycodone and active efflux of morphine dominating, resulting in a 10-fold difference in unbound brain concentrations at given unbound plasma concentrations (Fig. 5.2).

Microdialysis is suitable for three purposes: (1) evaluation of the biophase pharmacokinetics of drugs, (2) evaluation of the drug response in the form of neurotransmitter release, and (3) evaluation of the BBB transport properties of the drug in specific situations, like disease states. A review article on how microdialysis has been used to monitor the biophase concentrations of antiepileptic drugs was published by Clinckers et al. (2009). They state that *"intracerebral microdialysis is an indispensable tool for the assessment of the PKs of antiepileptic drugs"* and that it is very well suited together with *"mechanism-based PKPD modeling for dose selection and optimization of the therapeutic regimen for novel compounds"*. Clinckers and colleagues have published several important papers on BBB transport, and brain PK and PD, for antiepileptic drugs. They showed that epileptic seizures clearly increased the concentrations of the oxcarbamazepine metabolite 10,11-dihydro-10-hydroxy-carbamazepine in

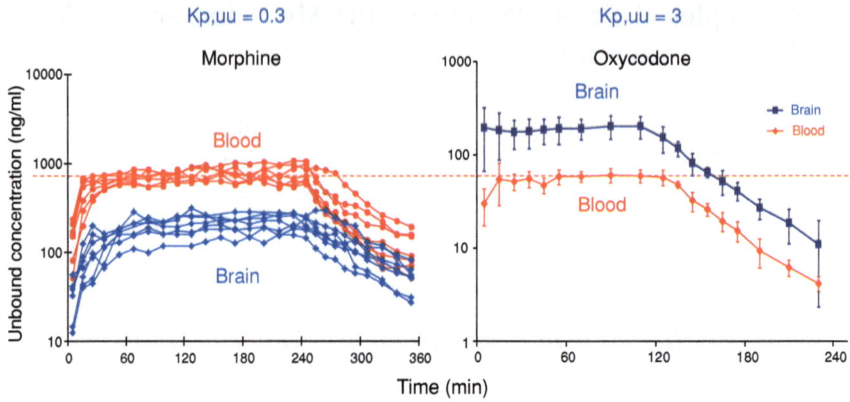

Fig. 5.2 Unbound blood (*red*) and brain (*blue*) concentrations of morphine (*left*) and oxycodone (*right*) measured with microdialysis in rats. The $K_{p,uu}$ of morphine is 0.3 and of oxycodone 3, making the concentration differences in brain be 10-fold for a defined unbound blood concentration. It can also be observed that oxycodone brain concentrations are much more rapidly obtained than those of morphine. The influx clearance, CL_{in}, is 10 μl × min^{-1} × g_brain^{-1} for morphine and 1910 μl × min^{-1} × g_brain^{-1} for oxycodone (Bostrom et al. 2006). The oxycodone concentration–time curve (Bostrom et al. 2006) is published with permission from the publisher

the rat brain (Clinckers et al. 2008). The same paper showed that co-administration of verapamil also increased brain concentrations of this metabolite.

Early studies by Wang and Welty on the PKPD relationships of gabapentin using microdialysis (Wang and Welty 1996) indicated that the PD of gabapentin lags behind the concentration profile in rat brain ECF (Welty et al. 1993).

Fluvoxamine PKPD relationships have been studied in rats using regular blood and brain sampling in combination with microdialysis at several dose levels (Geldof et al. 2008). A relationship between the PKPD and serotonin transporter (SERT) occupancy was found; the EC50 was 0.22 ng/ml in brain ECF and 14.8 ng/ml in total brain tissue, indicating extensive nonspecific binding of fluvoxamine in the brain.

Wegener et al. used microdialysis to study brain concentrations of the noncompetitive N-methyl-D-aspartate (NMDA) receptor antagonist MK-801 in relation to the behavior of rats (Wegener et al. 2011). They found that estimates of total brain concentrations were of little relevance for interpreting the behavioral effects and the extent of MK-801 receptor binding in vitro, and that the doses administered resulted in extracellular concentrations of unbound drug in the brain close to those linked to the known affinity of MK-801 for NMDA receptors.

Joosen et al. used microdialysis to show that pre-administration of tariquidar improved the treatment effect of the oxime HI-6 on nerve agent poisoning in rats by doubling the HI-6 brain ECF concentrations (Joosen et al. 2011). While doubling the dose of HI-6 would have done the same job, peripheral side effects could preclude using higher doses.

Ackaert studied the PKPD of (S)-5-hydroxy-2-(N,N,-di-npropylamino)tetralin (5-OH-DPAT) after transdermal iontophoresis and iv infusion in rats, and modeled the data (Ackaert et al. 2011). Striatal dopamine concentrations, measured by microdialysis, were used as the PD marker, and the 5-OH-DPAT concentrations were measured by microdialysis in the contralateral striatum with a dual probe setup. It was not reported whether drug recovery was measured but brain ECF concentrations were reported. The dopamine response was evaluated as the percentage change in concentration from baseline, which is the usual way of presenting microdialysis transmitter results. The authors found good PKPD correlations between plasma 5-OH-DPAT concentrations and dopamine response after adding an effect compartment to the plasma concentrations. Through modeling, they found indications that the delivery rate of the drug could influence the dopamine response. The authors stated that *"PK–PD modeling suggests that target distribution causes the delay between plasma concentration and drug effect"*. It is interesting that the striatal concentration–time profile for 5-OH-DPAT was not used in the model in this instance, although it could have supported the conclusions.

The importance of connecting measurements of PK and PD was demonstrated in a study of a new triple re-uptake inhibitor for treatment of major depression (Prins et al. 2011). Microdialysis was only used for PD sampling via transmitter analysis. Whole-brain and plasma drug concentrations were determined at three time points in rats. The study found that there was no lasting antidepressant-like behavioral effect in the absence of drug from the brain, and that the 24 h PD sampling regularly used was too late to detect any effect of the drug. In this case, connecting PK and PD measurements provided some very important explanations and offered essential directions for study design; this kind of help can be crucial for the success of drug discovery and development.

Roiko et al. measured the brain concentrations after three doses of gamma-hydroxybutyric acid (GHB) in rats using microdialysis and established the PKPD relationship (Roiko et al. 2012). This study is an elegant example of how microdialysis can be used to study BBB transport and PKPD relationships. It took about half an hour for GHB to distribute into frontal cortex ECF and equilibrate with plasma, to reach a steady-state relationship of 10 % ($K_{p,\mathrm{uu}}$ 0.1). The extent of partitioning into the brain did not differ among the three doses. While in vivo partitioning across the BBB was linear, the uptake of GHB in RBE4 and hCMEC/D3 cells was concentration-dependent. The in vitro K_m values of 18.1 and 23 mM, respectively, for these cells were higher than the peak plasma concentration after the highest dose in vivo, which was 16 mM. GHB is a substrate for monocarboxylate transporter 1 (MCT1) at the BBB. The authors concluded that measurement of the ISF concentrations in the frontal cortex provided a suitable method for estimating the potential sedative/hypnotic effect of GHB. Whole-brain GHB concentrations were about double the ISF concentrations, indicating that GHB is transported into brain parenchymal cells and/or is bound to tissue components.

Effects of pramipexole after administration as a constant infusion using Alset osmotic minipumps were compared with those after intermittent dosing in a

Parkinson's disease model in rats (Ferger et al. 2010). The difference in effect on dopamine levels was obvious as were differences in early morning akinesia, with better results after the constant infusion regimen. This study provides preclinical evidence of the value of using controlled release dosage forms in humans.

Morphine BBB transport has been studied using microdialysis at ultra-low concentrations with accelerator mass spectrometry detection. The BBB transport of the drug did not differ with differing concentrations (Sadiq et al. 2011b).

5.5.2 Drug Interactions at the BBB

Since both oxycodone and diphenhydramine are actively influxed at the BBB, with $K_{p,\mathrm{uu}}$ ratios of 3 and 5, respectively, in the rat (Bostrom et al. 2006; Sadiq et al. 2011a), a potential interaction between the drugs at the transporter level in the BBB after co-administration was studied (Sadiq et al. 2011a). Although in vitro studies showed clear competitive inhibition between the two drugs, in vivo microdialysis studies could not detect significant changes in $K_{p,\mathrm{uu}}$. This was mainly because the clinically relevant and highest tolerated concentrations were much lower than the K_i of the compounds. The maximal in vivo plasma concentrations of unbound diphenhydramine and oxycodone were 22 and 2.3 μM, respectively; the brain ISF concentrations of unbound drug were 5.5- and 3.3-fold higher. The in vitro K_i of oxycodone on diphenhydramine BBB transport was 106 μM, and the in vitro K_i of diphenhydramine on oxycodone transport was 35 μM. This is a caveat for in vitro studies of drug interactions or other transporter-related issues, as the relevant concentrations are often much lower than those used for in vitro studies.

N-methyl-4-phenyl-1,2,3,6-tetrahydropyridine (MPTP) and its interaction with amantadine at the BBB were investigated by Lin et al. in an extensive study utilizing cell cultures, positron emission tomography (PET) and microdialysis in both mice and rats. They demonstrated the presence of the organic anion transporters OAT1 and OAT2 on the luminal side of the BBB (Lin et al. 2010). MPTP is taken up into the brain by the organic cation transporters OCT1 and OCT2 to form MPP+. The co-administration of amantadine extensively decreased the brain concentrations of MPTP, demonstrating a significant influence on uptake. The authors do not mention if in vivo drug recovery was estimated.

O'Brien et al. studied the transport of imipramine and desipramine across the BBB in rats, with and without co-administration of verapamil and cyclosporine A (O'Brien et al. 2012). They found that imipramine is a drug that sticks to the tubings and that it was therefore not possible to estimate in vivo drug recovery. It is highly likely that this also influenced the resulting dialysate concentrations, although brain dialysate profiles showed the expected decline with time. The addition of drugs that influence efflux from the brain, such as verapamil and cyclosporine A, probably also change the recovery (Olson and Justice 1993;

Olson Cosford et al. 1996; de Lange et al. 1998; Xie et al. 2000). However, because inhibition of efflux from the measured site will reduce the recovery, the dialysate values were probably relatively lower than they would have been otherwise, as discussed by the authors. Thus, cyclosporine A might actually increase the brain concentrations more than the 84 % observed. The non-significant increase after verapamil of 44 % could also be higher.

5.5.3 Disease Models in Animals

The effect of rosiglitazone, an agonist at the peroxisome proliferator-activated receptor (PPAR-gamma), on brain trauma and glucose was studied in gerbils by Sheu et al. (2011). Rosiglitazone and glucose were measured in the brain, using microdialysis sampling from the cortex, by splitting the samples to liquid chromatography with tandem mass spectrometry detection and continuous monitoring approach glucose-monitoring equipment. Estimation of rosiglitazone recovery was not mentioned in the paper. Rosiglitazone had a clear effect on both brain glucose and cerebral infarct volume.

In an interesting study, quinidine was administered to rats with kainic acid-induced epilepsy and control rats, seven days after induction of status epilepticus (Syvanen et al. 2012). The K_p for quinidine was low but the brain ISF concentrations increased 7.2-fold after administering tariquidar to block P-gp, as measured with microdialysis, while the total brain concentration increased about 40-fold. Thus, the increase in intrabrain distribution was greater than that in BBB transport after blocking P-gp with tariquidar. The data clearly indicated the presence of the P-gp transporter in the brain parenchyma after induction of the disease. Unfortunately, direct $K_{p,\mathrm{uu}}$ and $V_{u,\mathrm{brain}}$ values could not be extracted from the information in the paper, despite modeling.

In a rat model investigated by the same group, the animals were studied two weeks after unilateral induction of Parkinson's disease by rotenone (Ravenstijn et al. 2012). The authors reported no changes in the BBB transport of L-dopa as a result of the disease. However, lower basal levels of 3.4-dihydroxyphenyl-acetic acid and homovanillic acid were observed in the diseased hemisphere.

5.5.4 Studies of Human Brain Penetration

Microdialysis was used to study the penetration of meropenem into the brains of two patients with acute brain injury (Dahyot-Fizelier et al. 2010). The investigators compared the AUC values in the brain ISF and plasma and $K_{p,\mathrm{uu}}$ values of 0.73 and 0.14 were obtained for the two patients. The data had been modeled to obtain reliable parameters. Any clinical explanation as to why the $K_{p,\mathrm{uu}}$ values were different was not provided in this preliminary study.

Valproate was studied in four patients with epilepsy who were undergoing presurgical evaluation with subdural EEG monitoring (Lindberger et al. 2001). It was found that the CSF, brain ISF, and peripheral subcutaneous concentrations were similar (the CSF value was 0.78 of the plasma value), and that subcutaneous plasma concentrations could be used for monitoring valproate concentrations, although there was a delay in CSF concentrations compared to brain ISF or subcutaneous plasma in the concentration–time profiles.

Poeppl and coworkers measured dialysate doripenem concentrations in five patients with acute brain injury in an observational study (Poeppl et al. 2012). Because calibration of the microdialysis probe was not possible, they obtained a recovery value for the brain probe of 38 % from a soft tissue study. The ratio of brain ISF AUC to total plasma AUC using this recovery value was 0.17 in one patient but only 0.01 in the other four. The authors stated that CSF measurements from the literature are in good agreement with the brain concentrations obtained in their study. They also concluded that the concentrations obtained in the brain were probably too low for an antimicrobial effect against most pathogens.

The distribution of morphine in the brains of patients with brain trauma, measured using microdialysis, showed a $K_{p,\mathrm{uu}}$ of around 0.5, with higher values in the penumbra zone closer to the trauma site (Ederoth et al. 2004). Comparing different species, this value is quite similar to rat and pig data, but is not similar to sheep BBB transport (Bouw et al. 2000; Tunblad et al. 2003, 2004a; Bostrom et al. 2008; Bengtsson et al. 2009).

A modeling approach was used by Kielbasa and Stratford for measuring atomoxetine distribution into the brain and CSF of rats (Kielbasa et al. 2009). They found a $K_{p,\mathrm{uu}}$ of 0.7 across the BBB, and a $K_{p,\mathrm{uu,CSF}}$ of 1.7. Their model did not include different influx and efflux clearances at the BBB, restricting it to drugs with mainly passive transport across the BBB. The model did, however, have separate clearances for the intrabrain distribution into and out of parenchymal cells. The authors used this information to then predict the human brain distribution of atomoxetine and duloxetine through a translational modeling approach (Kielbasa and Stratford 2012). Strong directional transport at the BBB in rats was not seen for duloxetine. The simulation results were compared to results from studies using clinical doses and in vitro receptor occupancy and PET data with good correlations.

5.6 Conclusions

In conclusion, several important papers on the PK and PKPD of drugs with brain effects have been published. Microdialysis has been instrumental in revealing in vivo BBB transport of drugs and the role of transporters for this transport. There is room for many more microdialysis studies investigating PKPD in the CNS, in which both PK and PD is measured simultaneously, where drug recovery is estimated according to established principles, and where the attachment of drugs to tubings and membranes has been taken into consideration to prevent erroneous conclusions. Modeling should be an accepted part of this work.

References

Ackaert OW, De Graan J, Shi S, Vreeken R, Pasqua OE, Dijkstra D, Westerink BH, Danhof M, Bouwstra JA (2011) The pharmacokinetics and pharmacological effect of (S)-5-OH-DPAT following controlled delivery with transdermal iontophoresis. J Pharm Sci 100:2996–3009

Bengtsson J, Ederoth P, Ley D, Hansson S, Amer-Wahlin I, Hellstrom-Westas L, Marsal K, Nordstrom CH, Hammarlund-Udenaes M (2009) The influence of age on the distribution of morphine and morphine-3-glucuronide across the blood-brain barrier in sheep. Br J Pharmacol 157:1085–1096

Bostrom E, Hammarlund-Udenaes M, Simonsson US (2008) Blood-brain barrier transport helps to explain discrepancies in in vivo potency between oxycodone and morphine. Anesthesiology 108:495–505

Bostrom E, Simonsson US, Hammarlund-Udenaes M (2006) In vivo blood-brain barrier transport of oxycodone in the rat: indications for active influx and implications for pharmacokinetics/pharmacodynamics. Drug Metab Dispos: Biol Fate Chem 34:1624–1631

Bouw MR, Gardmark M, Hammarlund-Udenaes M (2000) Pharmacokinetic-pharmacodynamic modelling of morphine transport across the blood-brain barrier as a cause of the antinociceptive effect delay in rats–a microdialysis study. Pharm Res 17:1220–1227

Chaurasia CS, Muller M, Bashaw ED, Benfeldt E, Bolinder J, Bullock R, Bungay PM, DeLange EC, Derendorf H, Elmquist WF, Hammarlund-Udenaes M, Joukhadar C, Kellogg DL Jr, Lunte CE, Nordstrom CH, Rollema H, Sawchuk RJ, Cheung BW, Shah VP, Stahle L, Ungerstedt U, Welty DF, Yeo H (2007) AAPS-FDA workshop white paper: microdialysis principles, application and regulatory perspectives. Pharm Res 24:1014–1025

Clinckers R, Smolders I, Michotte Y, Ebinger G, Danhof M, Voskuyl RA, Della Pasqua O (2008) Impact of efflux transporters and of seizures on the pharmacokinetics of oxcarbazepine metabolite in the rat brain. Br J Pharmacol 155:1127–1138

Clinckers R, Smolders I, Vermoesen K, Michotte Y, Danhof M, Voskuyl R, Della Pasqua O (2009) Prediction of antiepileptic drug efficacy: the use of intracerebral microdialysis to monitor biophase concentrations. Expert Opin Drug Metab Toxicol 5:1267–1277

Dagenais C, Rousselle C, Pollack GM, Scherrmann JM (2000) Development of an in situ mouse brain perfusion model and its application to mdr1a P-glycoprotein-deficient mice. J Cereb Blood Flow Metab 20:381–386

Dahyot-Fizelier C, Timofeev I, Marchand S, Hutchinson P, Debaene B, Menon D, Mimoz O, Gupta A, Couet W (2010) Brain microdialysis study of meropenem in two patients with acute brain injury. Antimicrob Agents Chemother 54:3502–3504

de Lange EC, de Bock G, Schinkel AH, de Boer AG, Breimer DD (1998) BBB transport and P-glycoprotein functionality using MDR1A (−/−) and wild-type mice. Total brain versus microdialysis concentration profiles of rhodamine-123. Pharm Res 15:1657–1665

Ederoth P, Tunblad K, Bouw R, Lundberg CJ, Ungerstedt U, Nordstrom CH, Hammarlund-Udenaes M (2004) Blood-brain barrier transport of morphine in patients with severe brain trauma. Br J Clin Pharmacol 57:427–435

Ferger B, Buck K, Shimasaki M, Koros E, Voehringer P, Buerger E (2010) Continuous dopaminergic stimulation by pramipexole is effective to treat early morning akinesia in animal models of Parkinson's disease: a pharmacokinetic-pharmacodynamic study using in vivo microdialysis in rats. Synapse 64:533–541

Friden M, Bergstrom F, Wan H, Rehngren M, Ahlin G, Hammarlund-Udenaes M, Bredberg U (2011) Measurement of unbound drug exposure in brain: modeling of pH partitioning explains diverging results between the brain slice and brain homogenate methods. Drug Metab Dispos: Biol Fate Chem 39:353–362

Friden M, Ducrozet F, Middleton B, Antonsson M, Bredberg U, Hammarlund-Udenaes M (2009a) Development of a high-throughput brain slice method for studying drug distribution in the central nervous system. Drug Metab Dispos: Biol Fate Chem 37:1226–1233

Friden M, Gupta A, Antonsson M, Bredberg U, Hammarlund-Udenaes M (2007) In vitro methods for estimating unbound drug concentrations in the brain interstitial and intracellular fluids. Drug Metab Dispos: Biol Fate Chem 35:1711–1719

Friden M, Ljungqvist H, Middleton B, Bredberg U, Hammarlund-Udenaes M (2010) Improved measurement of drug exposure in the brain using drug-specific correction for residual blood. J Cereb Blood Flow Metab 30:150–161

Friden M, Winiwarter S, Jerndal G, Bengtsson O, Wan H, Bredberg U, Hammarlund-Udenaes M, Antonsson M (2009b) Structure-brain exposure relationships in rat and human using a novel data set of unbound drug concentrations in brain interstitial and cerebrospinal fluids. J Med Chem 52:6233–6243

Geldof M, Freijer JI, van Beijsterveldt L, Langlois X, Danhof M (2008) Pharmacokinetic-pharmacodynamic modelling of fluvoxamine 5-HT transporter occupancy in rat frontal cortex. Br J Pharmacol 154:1369–1378

Gupta A, Chatelain P, Massingham R, Jonsson EN, Hammarlund-Udenaes M (2006) Brain distribution of cetirizine enantiomers: comparison of three different tissue-to-plasma partition coefficients: K(p), K(p, u), and K(p, uu). Drug Metab Dispos: Biol Fate Chem 34:318–323

Hammarlund-Udenaes M (2010) Active-site concentrations of chemicals—are they a better predictor of effect than plasma/organ/tissue concentrations? Basic Clin Pharmacol Toxicol 106:215–220

Hammarlund-Udenaes M, Friden M, Syvanen S, Gupta A (2008) On the rate and extent of drug delivery to the brain. Pharm Res 25:1737–1750

Joosen MJ, van der Schans MJ, van Dijk CG, Kuijpers WC, Wortelboer HM, van Helden HP (2011) Increasing oxime efficacy by blood-brain barrier modulation. Toxicol Lett 206:67–71

Kakee A, Terasaki T, Sugiyama Y (1997) Selective brain to blood efflux transport of para-aminohippuric acid across the blood-brain barrier: in vivo evidence by use of the brain efflux index method. J Pharmacol Exp Ther 283:1018–1025

Kalvass JC, Maurer TS (2002) Influence of nonspecific brain and plasma binding on CNS exposure: implications for rational drug discovery. Biopharm Drug Dispos 23:327–338

Kalvass JC, Olson ER, Cassidy MP, Selley DE, Pollack GM (2007) Pharmacokinetics and pharmacodynamics of seven opioids in P-glycoprotein-competent mice: assessment of unbound brain EC50, u and correlation of in vitro, preclinical, and clinical data. J pharmacol exp ther 323:346–355

Kielbasa W, Kalvass JC, Stratford R (2009) Microdialysis evaluation of atomoxetine brain penetration and central nervous system pharmacokinetics in rats. Drug Metab Dispo: Biol Fate Chem 37:137–142

Kielbasa W, Stratford RE (2012) Exploratory translational modeling approach in drug development to predict human brain pharmacokinetics and pharmacologically relevant clinical doses. Drug Metab Dispo: Biol Fate Chem

Large CH, Kalinichev M, Lucas A, Carignani C, Bradford A, Garbati N, Sartori I, Austin NE, Ruffo A, Jones DN, Alvaro G, Read KD (2009) The relationship between sodium channel inhibition and anticonvulsant activity in a model of generalised seizure in the rat. Epilepsy Res 85:96–106

Lin CJ, Tai Y, Huang MT, Tsai YF, Hsu HJ, Tzen KY, Liou HH (2010) Cellular localization of the organic cation transporters, OCT1 and OCT2, in brain microvessel endothelial cells and its implication for MPTP transport across the blood-brain barrier and MPTP-induced dopaminergic toxicity in rodents. J Neurochem 114:717–727

Lindberger M, Tomson T, Wallstedt L, Stahle L (2001) Distribution of valproate to subdural cerebrospinal fluid, subcutaneous extracellular fluid, and plasma in humans: a microdialysis study. Epilepsia 42:256–261

Mano Y, Higuchi S, Kamimura H (2002) Investigation of the high partition of YM992, a novel antidepressant, in rat brain—in vitro and in vivo evidence for the high binding in brain and the high permeability at the BBB. Biopharm Drug Dispos 23:351–360

O'Brien FE, Clarke G, Fitzgerald P, Dinan TG, Griffin BT, Cryan JF (2012) Inhibition of P-glycoprotein enhances transport of the antidepressant imipramine across the blood-brain barrier: microdialysis studies in the conscious freely moving rat. Brit J pharmacol

Olson Cosford RJ, Vinson AP, Kukoyi S, Justice JB (1996) Quantitative microdialysis of serotonin and norepinephrine: pharmacological influences on in vivo extraction fraction. J Neurosci Meth 68:39–47

Olson R, Justice J (1993) Quantitative microdialysis under transient conditions. Anal Chem 65:1017–1022

Poeppl W, Zeitlinger M, Donath O, Wurm G, Muller M, Botha F, Illievich UM, Burgmann H (2012) Penetration of doripenem in human brain: an observational microdialysis study in patients with acute brain injury. Int J Antimicrob Agents 39:343–345

Prins J, Westphal KG, Korte-Bouws GA, Quinton MS, Schreiber R, Olivier B, Korte SM (2011) The potential and limitations of DOV 216,303 as a triple reuptake inhibitor for the treatment of major depression: a microdialysis study in olfactory bulbectomized rats. Pharmacol Biochem Behav 97:444–452

Ravenstijn PG, Drenth HJ, O'Neill MJ, Danhof M, De Lange EC (2012) Evaluation of blood-brain barrier transport and CNS drug metabolism in diseased and control brain after intravenous L-DOPA in a unilateral rat model of Parkinson's disease. Fluids Barriers CNS 9:4

Roiko SA, Felmlee MA, Morris ME (2012) Brain uptake of the drug of abuse gamma-hydroxybutyric acid in rats. Drug Metab Dispos: Biol Fate Chem 40:212–218

Sadiq MW, Borgs A, Okura T, Shimomura K, Kato S, Deguchi Y, Jansson B, Bjorkman S, Terasaki T, Hammarlund-Udenaes M (2011a) Diphenhydramine active uptake at the blood-brain barrier and its interaction with oxycodone in vitro and in vivo. J Pharm Sci 100:3912–3923

Sadiq MW, Salehpour M, Forsgard N, Possnert G, Hammarlund-Udenaes M (2011b) Morphine brain pharmacokinetics at very low concentrations studied with accelerator mass spectrometry and liquid chromatography-tandem mass spectrometry. Drug Metab Dispos: Biol Fate Chem 39:174–179

Sheu WH, Chuang HC, Cheng SM, Lee MR, Chou CC, Cheng FC (2011) Microdialysis combined blood sampling technique for the determination of rosiglitazone and glucose in brain and blood of gerbils subjected to cerebral ischemia. J Pharm Biomed Anal 54:759–764

Syvanen S, Schenke M, van den Berg DJ, Voskuyl RA, de Lange EC (2012) Alteration in P-glycoprotein functionality affects intrabrain distribution of quinidine more than brain entry-a study in rats subjected to status epilepticus by kainate. AAPS J 14:87–96

Tunblad K, Ederoth P, Gardenfors A, Hammarlund-Udenaes M, Nordstrom CH (2004a) Altered brain exposure of morphine in experimental meningitis studied with microdialysis. Acta Anaesthesiol Scand 48:294–301

Tunblad K, Hammarlund-Udenaes M, Jonsson EN (2004b) An integrated model for the analysis of pharmacokinetic data from microdialysis experiments. Pharm Res 21:1698–1707

Tunblad K, Jonsson EN, Hammarlund-Udenaes M (2003) Morphine blood-brain barrier transport is influenced by probenecid co-administration. Pharm Res 20:618–623

Wang Y, Welty DF (1996) The simultaneous estimation of the influx and efflux blood-brain barrier permeabilities of gabapentin using a microdialysis-pharmacokinetic approach. Pharm Res 13:398–403

Watson J, Wright S, Lucas A, Clarke KL, Viggers J, Cheetham S, Jeffrey P, Porter R, Read KD (2009) Receptor occupancy and brain free fraction. Drug Metab Dispos: Biol Fate Chem 37:753–760

Wegener N, Nagel J, Gross R, Chambon C, Greco S, Pietraszek M, Gravius A, Danysz W (2011) Evaluation of brain pharmacokinetics of (+)MK-801 in relation to behaviour. Neurosci Lett 503:68–72

Welty DF, Schielke GP, Vartanian MG, Taylor CP (1993) Gabapentin anticonvulsant action in rats: disequilibrium with peak drug concentrations in plasma and brain microdialysate. Epilepsy Res 16:175–181

Xie R, Bouw MR, Hammarlund-Udenaes M (2000) Modelling of the blood-brain barrier transport of morphine-3-glucuronide studied using microdialysis in the rat: involvement of probenecid-sensitive transport. Br J Pharmacol 131:1784–1792

Xie R, Hammarlund-Udenaes M, de Boer AG, de Lange EC (1999) The role of P-glycoprotein in blood-brain barrier transport of morphine: transcortical microdialysis studies in mdr1a (−/−) and mdr1a (+/+) mice. Br J Pharmacol 128:563–568

Chapter 6
Microdialysis in Antibiotic Research

Markus Zeitlinger, Sandrine Marchand, William Couet, Aline Barth and Hartmut Derendorf

Abstract Traditionally, the PK of different anti-infective drugs is evaluated based on the plasma drug concentrations. However, in some cases those concentrations are not identical to those detected at the target site of the infection, i.e., in the affected tissue. In this context, microdialysis is a valuable technique applied for the quantification of free drug in the interstitial space fluid of a great variety of tissues. Considering that generally the infections are located in interstitial space fluid and only the free drug fraction is able to exert the anti-infective effect, it is clear that this approach has the necessary features to quantify the active fraction at the site of action. Using this information for consecutive pharmacokinetic/pharmacodynamic models is increasingly becoming state-of-the-art for optimizing dosing regimens of antibiotics.

6.1 Introduction

The minimal inhibitory concentration (MIC) has been used to define susceptibility breakpoints for anti-infective drugs. A range of pharmacokinetic/pharmacodynamic (PK/PD) parameters are composed by this PD component associated with

M. Zeitlinger
Department of Clinical Pharmacology, Medical University of Vienna, Vienna, Austria

S. Marchand · W. Couet (✉)
School of Medicine and Pharmacy, University of Poitiers, Pharmacologie des Anti-Infectieux, Pôle Biologie Santé, Poitiers cedex, France
e-mail: william.couet@univ-poitiers.fr

A. Barth · H. Derendorf
Department of Pharmaceutics, University of Florida, College of Pharmacy, Gainesville, FL, USA

M. Müller (ed.), *Microdialysis in Drug Development*, AAPS Advances in the Pharmaceutical Sciences Series, DOI: 10.1007/978-1-4614-4815-0_6,

different PK parameters to generate indexes such as free time above the MIC (t > MIC) that relates to the time that the drug remain above the MIC; free maximum concentration above the MIC (C_{max}/MIC), that is the relation of the maximum concentration and the MIC value; as well as free area under the curve above the MIC (AUC/MIC), that is the ratio of the area under the curve over 24 h and the MIC value (Sahre et al. 2011).

The correlation with the antimicrobial effect and these PK/PD parameters varies according to the anti-infective class (Adembri and Novelli 2009). Some examples of anti-infective classes/drugs which effect is related to the t > MIC are beta-lactams, linezolid, and tigecyclin. These drugs should be kept above MIC concentrations for a relatively prolonged time and also the effect is somewhat related to the concentration, as a faster bactericidal activity is detected with a high AUC/MIC. For fluoroquinolones, aminoglycosides, and daptomycin, the effect is best correlated with the concentration (AUC/MIC or C_{max}/MIC) (Adembri and Novelli 2009).

Different PK-PD targets were established for diverse classes of anti-infective drugs. As an example, for beta-lactams the suggested PK-PD target is 40 % of t > MIC, although this threshold might slightly vary between individual drugs (Schmidt et al. 2008a). On the other hand, for aminoglycosides a C_{max}/MIC target of 10 was established as evaluated from clinical patients (Adembri and Novelli 2009). The establishment of these PK-PD parameters can be the funding to the improvement of dosing regimens objectifying an effective dose increasing side effects.

It is important to mention that the MIC is a static concentration that does not consider important factors related to the patient, pathogen or drug that could affect the therapy (Andes et al. 2004). In vivo, the effect is a result of the dynamic exposure of the free drug to the infectious agent at the target site (Brunner et al. 2005). Therefore, even though the MIC is an extensively used PD parameter to predict efficacy for anti-infective drugs it has some disadvantages. No information on the bactericidal activity is generated or if different concentrations can increase this rate. The MIC depends on the growth and death rate and therefore different rates can generate the same MIC. Additionally, there is no information on the persistent anti-infective activity that remains after the exposition to the drug, sometimes referred to as post-antibiotic effect (Muller et al. 2004).

A more appropriate approach is to link the anti-infective effect with the dynamic time-versus-concentration profile. To achieve this goal, in vitro simulations of the pharmacodynamic effect are done based on in vivo PK determination using MD. The generated data is evaluated with a model that links the PK of the free drug to the bacterial kill curves (Muller et al. 2004; Brunner et al. 2005). In this sense, microdialysis (MD) combines the advantage of the dynamic quantification of the free drug in the site of action, demonstrating to be a useful sampling tool to establish the PK/PD parameters (Schmidt et al. 2008a).

Although PK/PD models based on plasma concentrations deliver valuable information during drug development most infections do not occur in blood but in tissues. Thus, the US Food and Drug Administration (FDA) and the European Medicines Agency (EMA) have recommended to include determination of tissue pharmacokinetics into the application dossier of antimicrobial agents, however, the choice of the method for investigating tissue pharmacokinetics remains challenging for individual antibiotics, indications, and localizations (FDA 1998; EMEA 2000). Indeed it is emphasized that no single "optimal" technique exists but the tool for assessment of target site drug levels should be chosen carefully on a case by case base.

In rare cases, the bacteria may be localized in cells or even within specific intracellular compartments such as phago-lysosomes, causing intracellular infections. Examples of obligate or facultative intracellular microorganisms include: *Mycobacterium tuberculosis, Mycoplasma pneumonia, Chlamydia trachomatis, Salmonella spp,* etc. Additionally, there are some bacteria that can survive in the intracellular environment under certain conditions, as is the case for *Bacillus anthracis, Escherichia coli, Helicobacter pylori, and Staphylococcus aureus*, among others; these are the so-called opportunistic bacteria (Van Bambeke et al. 2006). However, for most microorganisms the site of tissue infections, i.e., the location of the bacteria, and thereby the target site of antimicrobial action is represented by the interstitial space fluid (ISF) of tissues and other body fluids such as pleural fluid, bronchial fluid, epithelial lining fluid, middle ear fluid, and cerebrospinal fluid (Ryan 1993).

Not only the site of infection but also the composition of the compartment or organ of interest might become decisive when choosing the appropriate method, since concentrations of antimicrobials may vary within the different compartments of complex organs like lung or brain (Kearney and Aweeka 1999; Zeitlinger et al. 2005). In case of e.g., respiratory tract infections concentrations may be measured by sampling of sputum, respiratory secretions or pleural fluid and surgical collection of whole lung tissue and bronchial mucosa, each yielding concentrations for different compartments of the respiratory system. Yet, none of these techniques necessarily provides adequate information about the site of infection as for pneumonia bacteria may be located in the epithelial lining fluid, bronchial mucosa, the ISF of lung and within cells (Zhanel et al. 2001; Zeitlinger et al. 2005).

As a consequence, a range of different techniques have been used to determine tissue concentrations of antibiotics over several decades including tissue biopsies, saliva sampling or the skin-blister/skin-window fluid technique, PET as well as direct sampling of body fluids like salvia, pleural effusion, ascetic fluid, bile or pus (Thys et al. 1988; Reckziegel et al. 2001; Sauermann et al. 2005b; Ikawa et al. 2008).

6.2 Differences Between Methods to Determine Tissue PK of Antibiotics

Tissue biopsy was one of the very first techniques employed for determination of tissue PK of antimicrobials. Tissue biopsy is relatively easy to perform and there is no special knowledge required to apply it, however, for ethical reasons, only a limited number of biopsies can be taken from a subject, yielding only a limited temporal resolution. In contrast to other methods, measurements from tissue biopsy yield total drug concentrations, an admixture of extracellular and intracellular drug levels. Consequently, total tissue drug concentrations might result in overestimation of antimicrobial activity of agents that accumulate intracellularly or in underestimation of antimicrobials that equilibrate solely with the ISF. Hence, total drug concentrations are increasingly considered of minor importance (Muller et al. 2004).

The skin-blister technique is a method that was used for decades to measure drug concentrations in tissue (Korting et al. 1987; Brunner et al. 1998). Blister formation is induced by applying suction or chemically by administration of cantharidin to the skin. Thereby, a fluid filled cavity between epidermis and dermis is formed. However, blister formation is difficult to standardize and might result in an inflammatory aqueous environment that contains a considerable amount of proteins, possibly enabling protein bound drugs to cumulate in skin-blister fluid thereby distorting results (Philip-Joet et al. 1992).

Most recently, imaging techniques like the positron emission tomography (PET) were established in the field of drug distribution measurement (Fischman et al. 1997; Pien et al. 2005). PET is a nuclear imaging technique that uses compounds labeled with positron-emitting radionuclides. Originally, PET was introduced into clinical medicine for diagnostic purposes, mainly to study tissue metabolism, however, PET suites itself perfectly for PK research. PET offers a non-invasive method to measure drug distribution from plasma into anatomically defined regions and might be used to visualize PK processes simultaneously, virtually in tissues of the whole body (Fischman et al. 1998; Brunner et al. 2004). PET is non-invasive, has a high temporal as well as spatial resolution with 30s and 1–5 mm and can measure drug concentrations at multiple sites at the same time. However, like tissue biopsy, PET reflects the total tissue drug concentration. In addition, the need to radiolabel the drug of interest and the required equipment, i.e., an onsite radiochemistry, a cyclotron, and a PET camera make the technique expensive as restricted to specialized centers (Langer and Muller 2004).

Bronchoalveolar lavage (BAL) is a widely spread technique to measure target site drug concentration in epithelial lining fluid (ELF) of the lung (Capitano et al. 2004; Kikuchi et al. 2007, 2008). A flexible fiber optic bronchoscope is inserted in the lung and small amounts of saline solution are instilled sequentially and immediately aspired each into a separate trap. Before the concentration of the study drug in ELF can be assessed, the concentration has to be corrected for dilution. Usually, this is performed by use of the urea dilution method, a method which has often been criticized for its poor accuracy (Rennard et al. 1986; Gotfried et al. 2001). ELF is a complex mixture of a variety of molecules and cells including alveolar macrophages (AM). Therefore, fast proceeding of BAL fluid is required to avoid lyses of cells which might falsify the free drug concentrations in ELF. Again, this technique yields only limited temporal resolution because of ethical reasons and the fact that concentrations in all but the first ELF sample might be impacted by the previous sampling procedure.

Microdialysis was initially used to evaluate neurotransmission in an animal model in the 1970s (Ungerstedt and Pycock 1974). The recognition as a reliable and advantageous technique to quantify endogenous and/or exogenous compounds in tissues is being reinforced over the time, considering the increased use in diverse areas (Brunner and Derendorf 2006; Li et al. 2006a). Also, regulatory agencies, as the Food and Drug Administration (FDA) and the Committee for Proprietary Medicinal Products (CPMP) recommend the use of the procedure (Schmidt et al. 2008). This minimally invasive catheter based technique is becoming broadly accepted because it allows the measurement of the free extracellular drug concentrations at the site of action that usually is a tissue. Therefore, the obtained profiles are more predictive of the outcome than the traditional quantification of drugs in plasma (Muller et al. 2004; Brunner and Langer 2006; Sahre et al. 2011).

The principle of MD is based on a concentration difference between two fluid compartments across a semipermeable membrane present in a probe. The probe is placed in the tissue and a pump infuses the perfusion fluid through it, driving the outflow of the "recovered" fluid containing solutes (including drugs) from the tissue, which is collected in a microvial and is called dialysate. The drug concentration in the fluid is usually only a fraction of the extracellular free concentration due to an inability to obtain perfect equilibrium between extracellular fluid and the medium that is pumped through the probe. The fraction is referred to as the "relative recovery" and a calibration procedure is necessary to establish the percentage of the drug that is recovered. There are different methods for calibration, such as method of flow rate variation, no-net-flux, dynamic no-net-flux, retrodialysis, and the use of endogenous reference substance. More information about the technique and calibration procedures can be found elsewhere (Chaurasia 1999; Stenken 1999; Plock and Kloft 2005; Brunner and Derendorf 2006).

MD generates minimal tissue damage during the probe introduction. A certain drawback of the technique is the possibility of infection and the relation of the recovered solute with the flow rate (with larger flow rates the efficiency of the extraction is reduced, however with very low flow rates the time resolution is affected). The technique may be challenging when analyzing non water-soluble substances, since an aqueous perfusate is commonly used (Chaurasia 1999; Brunner and Derendorf 2006) but also because lipophilic compounds may bind to the probe and tubing components (Li et al. 2006b). Additionally, a sensitive analytical technique is necessary to quantify the dialysate that contains low drug concentrations (Chaurasia 1999; Plock and Kloft 2005; Schmidt et al. 2008a). Different analytical methods have been used to quantify antimicrobials in microdalysates, such as high performance liquid chromatography (HPLC) (Hashiguti et al. 1993; Hatfield et al. 1999; Rea et al. 2005; Zhang et al. 2009; Chu et al. 2012), capillary electrophoresis (CE) (Cooley and Lunte 2011; Kaul et al. 2011; Betancourt et al. 2012), mass spectrometry (MS) (Zhang and Beyer 2006; Zhang et al. 2007, 2009), and nuclear magnetic resonance (NMR) (Zhu et al. 2000; Elmeliegy et al. 2011).

6.3 Example Cefpirome, Ertapenem, and Moxifloxacin

MD is considered the method of choice to quantify drugs in the extracellular fluid (Schmidt et al. 2008a). However, all presented techniques have advantages and disadvantages, therefore, in the following tissue PK as evaluated by different techniques will be compared for two hydrophilic drugs, one with low and one with high protein binding, and a lipophilic antibiotic in order to allow for direct comparison of differences.

Cefpirome is a so-called fourth-generation cephalosporin that is characterized by a high stability against beta-lactamases and good activity against Gram-positive and Gram-negative bacteria including *Pseudomonas aeruginosa*, making the antibiotic a frequently used drug in empirical therapy of many infections observed in intensive care including sepsis (Hancock and Bellido 1992; Wiseman and Lamb 1997). Cefpirome is available for parenteral administration only and is given at doses between 1 and 2 g up to three times daily (Strenkoski and Nix 1993; Wiseman and Lamb 1997). Thus, differences in the administered dose must be carefully considered when comparing different PK studies.

Beta-lactam antibiotics are hydrophilic so their ability to pass cell membranes is limited, which in turn results in a low volume of distribution of approximately 20–25 L, i.e., a volume which closely corresponds to the extracellular body water (Wiseman and Lamb 1997; Lipman et al. 2001). Therefore, by mixing low intracellular concentrations with extracellular compartments like done when biopsy specimens are analyzed one would expect that the average concentration is below the interstitial concentration. Indeed, as shown in Table 6.1, with mean C_{max} values of 19.3 μg/ml (after single dose of 1 g) obtained by lung biopsy in non-inflamed lung tissue, concentrations remained considerably below those observed in healthy subjects in interstitium of muscle tissue (C_{max} of 127 μg/ml after single dose of 2 g) (Baldwin et al. 1991; Joukhadar et al. 2002). This PK property of cefpirome explains its in vivo ineffectiveness against intracellular pathogens although susceptibility testing would render some of the bacteria susceptible.

Noteworthy with a C_{max} of 7.2 μg/ml, concentrations of cefpirome in ELF were even lower than in lung biopsy (Baldwin et al. 1991). While one might expect that concentrations in ELF should usually equal those observed in the interstitium, (Schentag 1989) others described than zonulae occludentes in the alveolar epithelium, i.e., the barrier between blood and ELF, heavily inhibit penetration of hydrophilic compounds (Taylor et al. 1965; Effros et al. 1986).

However, also the ability to penetrate into intestinal soft tissues might be impaired in case of certain diseases. While good tissue penetration of cefpirome was consistently detected by skin-blister technique and μD in healthy subjects, (Kavi et al. 1988; Muller et al. 1996; Hollenstein et al. 2000) the plasma-to-tissue equilibration of cefpirome was considerably delayed in septic patients for muscle and subcutaneous adipose tissue (Joukhadar et al. 2002; Sauermann et al. 2005a). This discrepancy between tissue penetration in healthy volunteers and critically ill patients was also observed for other beta-lactams like piperacillin (Brunner et al. 2000), but also other hydrophilic antibiotics like fosfomycin (Frossard et al. 2000;

Table 6.1 Pharmacokinetic properties of cefpirome

Dose g	Health	Site	Method	C_{max} µg/ml	AUC mg h/L	T½ h	Citation
2	Healthy	Muscle	µD	127 ± 15	258.6 ± 24_{0-4}	–	Joukhadar et al. (2002)
2	Healthy	Plasma	Blood draw	210 ± 19	313.2 ± $18.2_{0\text{-}4}$	2.57 ± 0.48	Joukhadar et al. (2002)
1	Healthy	Bronchial mucosa	Biopsy	19.3 ± 1.9	–	–	Baldwin et al. (1991)
1	Healthy	ELF	BAL	7.2 ± 1.1	–	–	Baldwin et al. (1991)
1	Healthy	Blister fluid	Fluid draw	39.2 ± 7.9	200.0 ± $80.5_{0-\infty}$	2.5 ± 0.7	Kavi et al. (1988)

Values are presented as mean ± standard deviation (*SD*)
AUCs of different timeframes, as indicated in the table. To facilitate interpretation, values from literature were calculated to have the same unit
Note BAL bronchial alveolar lavage, *ELF* epithelial lining fluid, *µD* microdialysis

Joukhadar et al. 2003a). Besides a possible shift of fluid and albumin to the extravascular space caused by systemic inflammation and subsequent permeability of the vascular wall, the large amounts of fluids that are administered to maintain sufficient perfusion might substantially change the volume of distribution of hydrophilic compounds (Kinzig et al. 1992). In addition, critically ill patients, especially those suffering from septic shock commonly receive vasopressive catecholamines at large amounts, which was identified as one of the main factors associated with reduced tissue to plasma ratios of several antimicrobial classes (Zeitlinger et al. 2007).

In contrast to soft tissues of healthy subjects penetration of cefpirome into CSF is relatively poor in healthy subjects with ratios between 0.6 and 15 % (Nix et al. 1992). This observation is in line with low CNS penetration of other hydrophilic antibiotics (Kullar et al. 2011). Diffusion barriers at the blood–brain barrier but also active transport systems were discussed as possible underling mechanism of the observed poor penetration (Suzuki et al. 1989).

Like cefpirome, ertapenem belongs to the class of beta-lactams antibiotics. However, in contrast to the cephalosporin the penem has a very high protein binding ranging from 74 to 96 % (Burkhardt et al. 2006; Schmidt et al. 2008b). Clinically, ertapenem has a very broad spectrum of antimicrobial activity and is approved for various soft tissue infections as well as pneumonia and urogenital infections. However, in contrast to other penems it lack an activity against *P. aeruginosa*, thus its use in hospital acquired infections and intensive care is limited (Congeni 2010).

A summary of pharmacokinetic data obtained from a number of compartments is presented in Table 6.2. From the pharmacokinetic point of view, the high protein binding results in markedly different pharmacokinetic properties compared to cefpirome. While for cefpirome the concentration in muscle tissue is comparable to the concentration in plasma (Table 6.1, AUC 258 vs. 313 μg h/ml) the concentration of ertapenem in muscle (AUC 40 μg h/ml) is much lower than in plasma (690 μg h/ml). However, this observation fits perfectly to our knowledge, than only the free fraction of a drug might penetrate in the ISF of tissues.

Interestingly, the concentration in blister fluid behaved completely different for the two antibiotics. While for cefpirome the concentration in blister fluid is much lower (approx. 30 %) of the free concentration in muscle tissue, for ertapenem the concentration in blister fluid exceed that concentration in muscle markedly, probably due to accumulation of protein in the inflamed blister (400 %). This observation once again underlines than skin blister might be a poor model for predicting the free, active extracellular concentration in soft tissues.

Similar to cefpirome the concentration of ertapenem in ELF was much lower than in plasma and was lowest within cells.

Moxifloxacin is a fluoroquinolone with a methoxy-group in position 8 and a bulky side chain in position 7 (Keating and Scott 2004). It is highly active against Gram-positive and Gram-negative bacteria (Blondeau et al. 2000; Maggiolo et al. 2000) and can as well be used to combat infections caused by intracellular pathogens e.g., mycobacteria and Chlamydia (Shandil et al. 2007; Bebear et al. 2008). Additionally, moxifloxacin is more active against mutant bacterial strains

Table 6.2 Pharmacokinetic properties of ertapenem

Dose g	Route	Site	Method	C_{max} μg/ml	AUC μg h/mL	T½ h	Citation
1	i.v.	Muscle	μD	6.7 ± 4.1	$39.7 \pm 24.8_{0-\infty}$	–	Burkhardt et al. (2006)
1 g for 3d	i.v.	Skin blister	Fluid draw	25.0	417.5_{0-24}	–	Laethem et al. (2003)
1 g for 3d	i.v.	Plasma	Blood draw	191.9	688.1_{0-24}	–	Laethem et al. (2003)
1	i.v.	ELF	BAL	4.06	–	–	Burkhardt et al. (2005)
1	i.v.	Alveolar cells	BAL	0.007	–	–	Burkhardt et al. (2005)
1	i.v.	Plasma	Blood draw	103.3 ± 26.3	$359.7 \pm 66.5_{0-24}$	3.8 ± 0.6	Burkhardt et al. (2006)

Values are presented as means (± SD)
AUCs of different time frames as indicated in the table. Values from literature were converted as needed to fit to one unit
Note BAL bronchial alveolar lavage, *ELF* epithelial lining fluid, *i.v.* intravenous, *μD* microdialysis

than other fluoroquinolones and in comparison bears a lower potential to select resistant mutants, probably because it uses at least two mechanisms to inhibit bacterial growth (Talan 2001; Perez-Trallero et al. 2002; Florea et al. 2004). Moxifloxacin is frequently used in the in- and outpatient setting due to its favorable antibacterial spectrum and availability as intravenous (i.v.) and oral formulations. Besides approval for respiratory tract infections, moxifloxacin is also licensed for complicated intra-abdominal infections and complicated skin and skin structure infections, thus making it a drug used in intensive care medicine.

As a lipophilic drug, moxifloxacin penetrates well through cell membranes and therefore it has a high volume of distribution of 143–161 L (Alffenaar et al. 2009). The apparent volume of distribution is larger than the total body volume indicating that moxifloxacin accumulates in certain extravasal compartments of the body. Indeed moxifloxacin was found to be highly concentrated intracellularly (Table 6.3) (Soman et al. 1999). Since moxifloxacin is used to combat infections that occur intracellularly as well as infections in the ISF, information of both intracellular and extracellular concentration is of importance. If one wants to combat intracellular pathogens values from methods that avoid mixture of intracellular and extracellular concentration like measurements of concentrations in isolated macrophages yield substantial information. Indeed, with values of 56.7 ± 1.61 μg/mL the C_{max} of moxifloxacin in AM measured in BAL by far exceeded the concentrations observed in ELF (20.7 ± 1.9 μg/mL) and bronchial mucosa (5.36 ± 1.3 μg/mL) (Soman et al. 1999).

For ISF, with C_{max} values of 1.0 ± 1.0 μg/mL in healthy tissue and 2.3 ± 1.2 μg/mL in inflamed subcutaneous tissue much lower concentrations were obtained by MD (Joukhadar et al. 2003b). In contrast, tissue sampling after surgery, which is a very similar method to tissue biopsy, yielded concentration values between 1.73 and 17.08 μg/mL, (Ober et al. 2009) again underlining that the free extracellular concentrations and thereby the activity of fluoroquinolones might be overestimated by data obtained from tissue homogenization. In case of moxifloxacin, the skin-blister technique with C_{max} value of 1.7 ± 0.3 μg/mL yielded concentrations in the same magnitude to those obtained by MD (Muller et al. 1999b; Joukhadar et al. 2003a, b).

6.4 Impact on Tissue PK by Selected Pathologies

Another major focus of antimicrobial research using MD was to investigate antibiotics tissue distribution in selected patient conditions and pathologies:

6.4.1 Critically Ill Patients

Drug distribution is dependent on different factors that include the properties of the molecule itself. Factors that can modify the tissue distribution in critically ill patients include change in pH, fluid shifts (Gonzalez et al. 2011), and protein

Table 6.3 Pharmacokinetic properties of moxifloxacin

Dose mg	Route	Site	Method	C_{max} µg/ml	AUC µg h/mL	T½ h	Citation
400	i.v.	Muscle	µD	1.2 ± 0.8	$9.5 \pm 5.9_{0-12}$	–	Muller et al. (1999a)
400	i.v.	Skin blister	Fluid draw	1.7 ± 0.3	$16.7 \pm 4.1_{0-12}$	–	Muller et al. (1999b)
400	Orally	ELF	BAL	20.7 ± 1.92	189_{0-24}	–	Soman et al. (1999)
400	Orally	AM	BAL	56.7 ± 1.61	496_{0-24}	–	Soman et al. (1999)
400	Orally	Bronchial mucosa	Biopsy	5.36 ± 1.29	53.8_{0-24}	–	Soman et al. (1999)
400	i.v.	Plasma	Blood draw	4.1 ± 1.2	$26.0 \pm 10.6_{0-24}$	8.6 ± 1.6	Joukhadar et al. (2003)
400	i.v.	Inflamed tissue	µD	2.3 ± 1.2	$12.8 \pm 6.4_{0-\infty}$	4.1 ± 1.3	Joukhadar et al. (2003)
400	i.v.	Healthy tissue	µD	1.0 ± 1.0	$6.9 \pm 7.1_{0-\infty}$	5.2 ± 1.7	Joukhadar et al. (2003)

Values are presented as mean ± SD
AUCs of different time frames as indicated in the table. Values from literature were converted as needed to fit to one unit
Note AM alveolar macrophage, *BAL* bronchial alveolar lavage, *ELF* epithelial lining fluid, *i.v.* intravenous, *µD* microdialysis

binding (Craig and Suh 1978; Gonzalez et al. 2011). Also, a reduction in the cardiac output caused by heart failure can affect the tissue distribution (Gonzalez et al. 2011). For further information on alteration on the pharmacokinetics of anti-infective drugs in critically ill patients, some references are suggested (Roberts and Lipman 2009; Gonzalez et al. 2011). So, considering patients with sepsis and septic shock a higher dose may be necessary to improve anti-infective penetration (Roberts and Lipman 2009).

An Austrian study evaluated if differences in the level of sepsis gravity would generate PK differences in the linezolid profile in plasma and ISF using the MD technique. Eight Patients with severe sepsis and 16 patients with septic shock participated in the study. The healthy controls were obtained from a previous study from the same research group. No substantial difference on the penetration of linezolid into subcutis tissue ($AUC_{0–24}$ for healthy: 129.76 ± 46.18, septic shock: 65.76 ± 32.61 and severe sepsis: 67.70 ± 45.23) and muscle tissue ($AUC_{0–24}$ for healthy: 92.89 ± 23.33, septic shock: 68.74 ± 32.84 and severe sepsis: 52.67 ± 26.00), or in relation to the plasma ($AUC_{0–24}$ for healthy: 78.30 ± 24.97, septic shock: 70.78 ± 28.12 and severe sepsis: 50.91 ± 23.13) was detected for the different levels of disease severity (Thallinger et al. 2008).

Another study investigated the reasons why there is a frequent failure when administering piperacillin to patients with septic shock, even if the isolates are susceptible. Therefore, they studied the penetration of piperacillin (4 g) into the tissues of patients with sepsis and healthy volunteers. The drug concentrations in adipose and muscular tissue were 5–10 fold smaller than in plasma. The penetration ratios were: $AUC_{intersticium}/AUC_{plasma\_free}$ for muscle 0.19 ± 0.03 and 0.55 ± 0.09, for patients and volunteers, respectively; and $AUC_{intersticium}/AUC_{plasma\_free}$ for subcutis 0.10 ± 0.02 and 0.31 ± 0.07, also for patients and volunteers, respectively. The concentration in the adipose tissue remained below the MIC (11 μg/mL) for important pathogens present in septic shock. Even if sufficient concentration were detected in plasma, the concentration in the tissue was insufficient and therefore the absence of success in the treatment may be credited to the lack of sufficient tissue penetration (Joukhadar et al. 2001).

6.4.2 Diabetic Patients

Inflamed tissues and bones are very common in diabetic patients, especially in the lower extremities. Several papers have used MD to evaluate anti-infective concentrations in tissues of diabetic patients.

In a study with nine diabetic patients with chronic limb infections, linezolid was quantified in wound and in healthy thigh. The drug was administered intravenously (600 mg BID) and the MD and plasma samples were obtained after the 3rd to 4th doses. There was no significant difference between the penetration ratios between both tissues, although the variability among patients was high as shown in Table 6.4 (Wiskirchen et al. 2011). Another MD study with three male subjects receiving linezolid was conducted according to the same dosage regimen and evaluated drug penetration during osteomyelitis. However, a rapid equilibrium of the drug from the plasma into the tissue and bone regardless of the inflammation status was detected in the study (Table 6.4) (Traunmuller et al. 2010).

Müller and collaborators used MD to quantify the penetration of ciprofloxacin in inflamed and healthy tissue of six patients with diabetes mellitus. The measured tissue concentrations were lower than the values found for the plasma concentration but no difference in penetration was found for healthy and inflamed tissue (Muller et al. 1999a). A study performed by Bulik and his research group evaluated the concentrations of tigecycline in inflamed and healthy tissue of eight diabetic patients at steady state. The comparison of the concentration in tissues and plasma revealed an absence of difference in drug penetration and also a rapid permeation to the tissue achieving almost the same concentrations in both matrices (Table 6.4) (Bulik et al. 2010).

MD proved to be a powerful tool to quantify the free levels of anti-infective drugs in healthy and inflamed tissues of diabetic patients. In the majority of the studies, no significant difference was detected in tissue penetration. However, when comparing the levels in plasma and tissues, differences in drug concentrations were found in some cases reinforcing the necessity of an evaluation of the drug concentration at the target site.

6.4.3 Obese Patients

The choice of the correct anti-infective agent at the right dosage regiment plays a crucial role in treatment success, safety, and prevention of resistance (Schmidt et al. 2008a). However, dose adjustments according to body weight are not usually performed. Therefore, underdosing in overweight or obese patients may occur (Hollenstein et al. 2001). This supposition is reinforced by the fact that obese patients are more inclined to post-operative infections (Hollenstein et al. 2001; Barbour et al. 2009). Currently, there is no agreement about the correct approach to modify the dosing regimen. One important limitation is that the commonly used methods, as the one based in the ideal body weight (IDW) or based on the absolute body weight (ABW), do not take into consideration the drug distribution into the tissues, where the majority of the infectious diseases are located. So, MD is a useful technique to measure the actual free anti-infective concentrations in the target sites of overweight and obese patients (Hollenstein et al. 2001).

Hollenstein et al. compared the concentrations of ciprofloxacin (2.85 mg/kg) in obese and lean patients at the target site, adjusting the dose by weight. The free drug concentrations were measured with MD in both skeletal muscle and adipose tissue. The concentrations in plasma differed among the groups ($C_{max/obese}$: 9.97 ± 5.64 and $C_{trough/obese}$: 0.44 ± 0.10 μg/ml versus $C_{max/lean}$: 2.59 ± 1.06 and $C_{trough/lean}$: 0.19 ± 0.09 μg/ml), the higher concentrations in the obese group may have safety implications. On the other hand, no significant difference in the

Table 6.4 Tissue penetration ratios in healthy and unhealthy diabetic patients

Drug	Protein Binding	*f*AUC plasma	*f*AUC healthy tissue	*f*AUC unhealthy tissue	Tissue penetration ratio healthy tissue[a]	Tissue penetration ratio unhealthy tissue[a]
Linezolid	14.78 % (3.85–32.03 %)	$fAUC_{0-12} \pm SD$ 51.24 ± 12.72 μg h/ml	$fAUC_{0-12} \pm SD$ 82.76 ± 59.01 μg h/ml	$fAUC_{0-12} \pm SD$ 92.52 ± 60.44 μg h/ml	1.42 (range: 1.08–2.23)	Wound tissue 1.27 (range: 0.86–2.26) Wiskirchen et al. (2011)
	18–26 %	$fAUC_{0-24}$ [median(range)]: 169.1 (162.7–263.2) mg h/L	$fAUC_{0-24}$ [median(range)]: 245.3 (202.3–349.8) mg h/L	$fAUC_{0-24}$ [median(range)]: Inflamed tissue: 210.9 (210.7–212.9) mg h/L Metatarsal bone: 210.4 (165.6–266.0) mg h/L	1.32 ± 0.09	Inflamed tissue: 1.12 ± 0.22 Metatarsal bone: 1.09 ± 0.11 Traunmuller et al. (2010)
Ciprofloxacin		Serum: $fAUC_{0-inf} \pm SD$: 527 ± 318 μg min/ml	$fAUC_{0-inf} \pm SD$: 317 ± 207 μg min/ml	$fAUC_{0-inf} \pm SD$: 278 ± 182 μg min/ml	0.82 ± 0.08 (range, 0.51–0.95	0.78 ± 0.08 (range, 0.56–1.09) Muller et al. (1999)
Tigecycline		$fAUC_{0-24} \pm SD$ 2.65 ± 0.33 μg h/ml	$fAUC_{0-24} \pm SD$ 2.60 ± 1.02 μg h/ml	$fAUC_{0-24} \pm SD$ 2.52 ± 1.15 μg h/ml	98.94 ± 52.75 %	100.00 % ± 44.78 % Bulik et al. (2010)

[a] It is the relation between the tissue fAUC and the plasma fAUC and it is defined as the "penetration ratio"
*f*AUC: area under the curve of the free drug

concentration–time curves was found in the tissues. When evaluating the penetration of the drugs into the tissue, it was significantly lower in the obese group ($AUC_{tissue}/AUC_{plasma}$ for lean subjects 0.45 ± 0.27 versus for obese subjects 0.82 ± 0.36). One possible explanation given by the authors is that the fat cells in obese patients are in higher number and they are in hyperthrophic condition and also the reduced penetration surface area to volume ratio may contribute for this difference. Therefore, the penetration process of ciprofloxacin is diminished in obese patients and the doses have to be adjusted (Hollenstein et al. 2001). Plasma concentrations did not predict the actual target concentrations quantified with MD.

6.5 Protein Binding and Antimicrobial Research

Protein binding is highly significant both in terms of pharmacokinetics and pharmacodynamics for drugs in general and antibiotics in particular (Schmidt et al. 2010; Zeitlinger et al. 2011). It governs antibiotics elimination and distribution at the site of infection since only unbound drug distributes into the extravascular space, and then determine antimicrobial activity (Zeitlinger et al. 2011). However, there is currently no standardization for in vitro or in vivo protein binding determination and although equilibrium dialysis and ultrafiltration are most commonly used, diversity of models and methods precludes between-studies comparisons.

Microdialysis has been proposed for plasma protein binding determination of drugs, including antibiotics such as cephalothin, chloramphenicol, and isoniazid, more than 20 years ago (Herrera et al. 1990). Similar results were obtained when in vitro plasma protein binding experiments were conducted by ultrafiltration or MD. However, the later offers several advantages. In particular, binding equilibrium is not disturbed and the sample temperature can be easily controlled which is not the case with ultrafiltration because of the centrifugation step. Nonspecific adsorption on the ultrafiltration membranes is another problem, especially for compounds with extensive protein binding, but less severe with MD because of the smaller surface area of the membrane. And although these initial experiments were conducted in vitro, the authors indicated that because probes can be readily implanted intravenously, the technique could also be advantageous for in vivo binding determination.

Accordingly, this same group was the first one to our knowledge using intravenous MD in a series of experiments to investigate the pharmacokinetics of acetaminophen in the blood, bile, and liver of an anesthetized rat (Scott et al. 1990; Scott and Lunte 1993). At that time, self made probes were fabricated using regenerated cellulose dialysis fibers. Another paper was published by Evrard and colleagues (Evrard et al. 1996) and commercialized probes for intravenous MD (CMA/20) became available several years later. They were used for direct determination of plasma unbound codeine (Xie and Hammarlund-Udenaes 1998) and morphine (Tunblad et al. 2003) concentrations in rats during brain distribution studies and during investigation of the middle ear distribution of amoxicillin in the awake chinchilla (Huang et al. 2001).

In fact, using intravenous MD for pharmacokinetic studies in small animals presents several advantages over traditional blood sampling for total and/or unbound drug concentrations determinations. The first advantage is that this procedure does not require repeated blood sampling, which in small animals such as rats, may eventually alter hematocrit as well as plasma protein concentrations, and therefore possibly drug pharmacokinetics. However, many antibiotics selected for MD tissue distribution studies in rats are not extensively bound to plasma proteins. Imipenem is an example of such a compound with limited protein binding. But using two CMA/20 probes, one implanted in a jugular vein and the other in a thigh muscle, it was possible to compare the unbound imipenem concentrations versus time profiles in blood and muscle from dialysate samples collected every 10–15 min, without taking blood samples (Marchand et al. 2005b). This was especially useful to investigate the potential effect of hypovolemia on imipenem muscle distribution (Marchand et al. 2005b). Noticeably, intravenous probe recovery was stable over time for the whole study duration (150 min) and virtually similar to muscle probe recovery on average. Microdialysis was also simultaneously used in blood and rats tissue to characterize the distribution of amoxicillin, another antibiotic with limited protein binding (Marchand et al. 2005a).

The second advantage of intravenous MD is to allow direct measurement of unbound drug concentration, and therefore plasma protein binding determination in vivo. This is particularly important since it was recently shown that experimental conditions such as protein source, temperature, pH, electrolytes, and supplements may influence protein binding (Zeitlinger et al. 2011). However, in most studies conducted in rats to compare the unbound tissue and plasma concentrations of antibiotics with extensive protein binding, MD was most often only used for unbound concentrations determinations in tissues, and unbound plasma concentrations were determined by plasma ultrafiltration after blood sampling (Liu et al. 2005).

Yet, plasma protein binding was also occasionally determined by microdialysis but using spiked plasma. As an example the consequences of the extensive and nonlinear protein binding of ceftriaxone on its tissue distribution, have been investigated in rats using MD (Kovar et al. 1997). A microdialysis probe was inserted in a thigh muscle for direct measurement of unbound concentrations but plasma protein binding was determined using MD, in vitro using plasma spiked with ceftriaxone. The nonlinear protein binding was then included in the fitting of the total plasma data, from which it was possible to predict free concentrations in the tissue interstitial space.

The same approach was used more recently in a study conducted with gatifloxacin (Tasso et al. 2008). Microdialysis was used to estimate unbound concentrations in tissues and to determine plasma protein binding in vitro using spiked plasma. Since protein binding (35.5 ± 5.0 %) was constant and independent of the concentration investigated, this average value was used for the prediction of free plasma levels based on measured total concentrations.

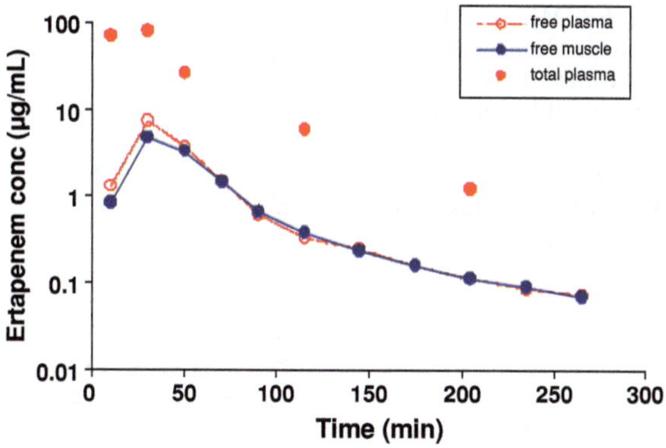

Fig. 6.1 Ertapenem unbound plasma concentrations (*red circle*), unbound muscle concentrations (*filled blue circle*), and total plasma concentrations (*filled red circle*) versus time profiles in a typical rat after a 30 min infusion of ertapenem at a dose equal to 40 mg kg^{-1}

Ertapenem is a carbapenem antibiotic with as opposed to imipenem or meropenem is extensively bound and it was confirmed experimentally that only the unbound compound is responsible for the antimicrobial activity (Nix et al. 2004; Schmidt et al. 2008b). Ertapenem protein binding is quite complex since it is nonlinear and may vary with the disease, as a result of altered protein concentration and/or binding affinity. We have conducted a muscle distribution study of this antibiotic in rats, using MD for the direct determination of unbound concentrations in muscle and blood, and blood sampling for the measurement of total plasma concentrations (unpublished data). Results obtained in a typical rat are presented on Fig. 6.1. It can be observed that free ertapenem plasma concentrations are much lower that corresponding total concentrations but equivalent to muscle interstitial fluid concentrations.

In human, direct in vivo determination of unbound plasma drug concentration was not possible until human intravenous probes became commercially available as happened recently. We have conducted a MD study in six healthy volunteers with ertapenem (1 g) administrated intravenously over 30 min (unpublished data). CMA 60 and CMA 64 MD probes were inserted respectively in the quadriceps and arm vein. Ertapenem was assayed by HPLC at various time points in blood and muscle dialysates, plasma, and plasma ultrafiltrates obtained with the Centrifree® devices. Probes recoveries were determined individually by retrodialysis by drug and varied between 39.2 ± 2.3 and 86.0 ± 3.3 in blood, and between 22.0 ± 8.7 and 77.8 ± 4.9 in muscle. Average unbound fraction assessed from unbound to total plasma AUC ratio was more than twice higher when unbound plasma concentrations were determined by MD ($fu_{\mu D} = 6.4 \pm 4.25$ %) instead of ultrafiltration ($fu_{UF} = 2.3 \pm 1.2$ %) and accordingly unbound plasma concentrations determined by MD were higher than by ultrafiltration as illustrated on Fig. 6.2. Furthermore, unbound muscle to plasma AUC ratio was equal to 1.2 ± 0.4 [0.8–2.0] using unbound plasma concentrations determined by MD and to 2.7 ± 0.9 [1.7–3.9] using unbound plasma concentrations estimated by ultrafiltration. Yet, because higher unbound concentrations in muscle extracellular fluid than in plasma are difficult to explain from a theoretical standpoint (Dahyot et al. 2008), the lower ratio close to unity on average is more likely to reflect the actual value. It

Fig. 6.2 Mean ± SD unbound plasma ertapenem concentrations versus time determined by ultrafiltration (*red circle*) and microdialysis (*blue filled circle*) after 30 min 1 g IV infusion to six healthy volunteers

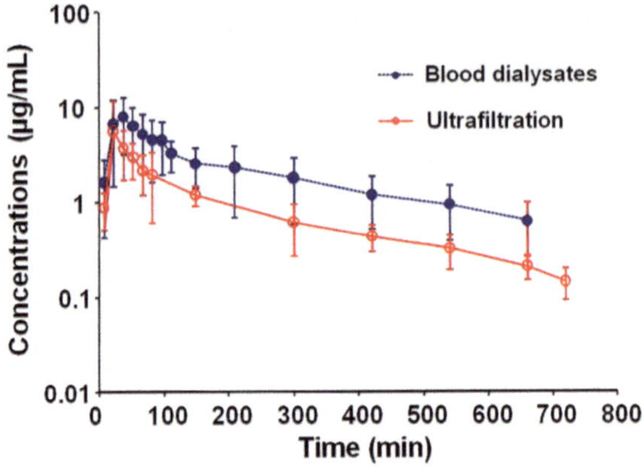

was hypothetized that ertapenem unbound fraction was underestimated using ultrafiltration due to non specific adsorption on the system membranes. This has been described for daptomycin, another antibiotic with extensive protein binding, and the authors made calculations to correct for this problem (Kim et al. 2008). Yet, in this situation intravenous MD appears as an interesting alternative for direct in vivo measurement of the antibiotic unbound concentration.

References

Adembri C, Novelli A (2009) Pharmacokinetic and pharmacodynamic parameters of antimicrobials: potential for providing dosing regimens that are less vulnerable to resistance. Clin Pharmacokinet 48(8):517–528

Alffenaar JW, van Altena R, Bokkerink HJ, Luijckx GJ, van Soolingen D, Aarnoutse RE, van der Werf TS (2009) Pharmacokinetics of moxifloxacin in cerebrospinal fluid and plasma in patients with tuberculous meningitis. Clin Infect Dis 49(7):1080–1082

Andes D, Anon J, Jacobs MR, Craig WA (2004) Application of pharmacokinetics and pharmacodynamics to antimicrobial therapy of respiratory tract infections. Clin Lab Med 24(2):477–502

Baldwin DR, Maxwell SR, Honeybourne D, Andrews JM, Ashby JP, Wise R (1991) The penetration of cefpirome into the potential sites of pulmonary infection. J Antimicrob Chemother 28(1):79–86

Barbour A, Schmidt S, Rout WR, Ben-David K, Burkhardt O, Derendorf H (2009) Soft tissue penetration of cefuroxime determined by clinical microdialysis in morbidly obese patients undergoing abdominal surgery. Int J Antimicrob Agents 34(3):231–235

Bebear CM, de Barbeyrac B, Pereyre S, Renaudin H, Clerc M, Bebear C (2008) Activity of moxifloxacin against the urogenital mycoplasmas Ureaplasma spp., Mycoplasma hominis and Mycoplasma genitalium and Chlamydia trachomatis. Clin Microbiol Infect 14(8):801–805

Betancourt L, Rada P, Paredes D, Hernandez L (2012) In vivo monitoring of cerebral agmatine by microdialysis and capillary electrophoresis. J Chromatogr B Analyt Technol Biomed Life Sci 880(1):58–65

Blondeau JM, Laskowski R, Bjarnason J, Stewart C (2000) Comparative in vitro activity of gatifloxacin, grepafloxacin, levofloxacin, moxifloxacin and trovafloxacin against 4151 Gram-negative and Gram-positive organisms. Int J Antimicrob Agents 14(1):45–50

Brunner M, Derendorf H (2006) Clinical microdialysis: Current applications and potential use in drug development. Trends Anal Chem 25(7):674–680

Brunner M, Langer O (2006) Microdialysis versus other techniques for the clinical assessment of in vivo tissue drug distribution. AAPS J 8(2): E263–E271

Brunner M, Schmiedberger A, Schmid R, Jager D, Piegler E, Eichler HG, Muller M (1998) Direct assessment of peripheral pharmacokinetics in humans: comparison between cantharides blister fluid sampling, in vivo microdialysis and saliva sampling. Br J Clin Pharmacol 46(5):425–431

Brunner M, Pernerstorfer T, Mayer BX, Eichler HG, Muller M (2000) Surgery and intensive care procedures affect the target site distribution of piperacillin. Crit Care Med 28(6):1754–1759

Brunner M, Langer O, Dobrozemsky G, Muller U, Zeitlinger M, Mitterhauser M, Wadsak W, Dudczak R, Kletter K, Muller M (2004) [18F]Ciprofloxacin, a new positron emission tomography tracer for noninvasive assessment of the tissue distribution and pharmacokinetics of ciprofloxacin in humans. Antimicrob Agents Chemother 48(10):3850–3857

Brunner M, Derendorf H, Muller M (2005) Microdialysis for in vivo pharmacokinetic/pharmacodynamic characterization of anti-infective drugs. Curr Opin Pharmacol 5(5):495–499

Bulik CC, Wiskirchen DE, Shepard A, Sutherland CA, Kuti JL, Nicolau DP (2010) Tissue penetration and pharmacokinetics of tigecycline in diabetic patients with chronic wound infections described by using in vivo microdialysis. Antimicrob Agents Chemother 54(12):5209–5213

Burkhardt O, Majcher-Peszynska J, Borner K, Mundkowski R, Drewelow B, Derendorf H, Welte T (2005) Penetration of ertapenem into different pulmonary compartments of patients undergoing lung surgery. J Clin Pharmacol 45(6):659–665

Burkhardt O, Brunner M, Schmidt S, Grant M, Tang Y, Derendorf H (2006) Penetration of ertapenem into skeletal muscle and subcutaneous adipose tissue in healthy volunteers measured by in vivo microdialysis. J Antimicrob Chemother 58(3):632–636

Capitano B, Mattoes HM, Shore E, O'Brien A, Braman S, Sutherland C, Nicolau DP (2004) Steady-state intrapulmonary concentrations of moxifloxacin, levofloxacin, and azithromycin in older adults. Chest 125(3):965–973

Elmeliegy MA, Carcaboso AM, LM LC, Zhang ZM, Calabrese C, Throm SL, Wang F, Baker SJ, Stewart CF (2011) Magnetic resonance imaging-guided microdialysis cannula implantation in a spontaneous high-grade glioma murine model. J Pharm Sci

Chaurasia CS (1999) In vivo microdialysis sampling: theory and applications. Biomed Chromatogr 13(5):317–332

Chu Y, Wang X, Guo J, Li W, Ma X, Zhu Y (2012) Pharmacokinetic study of unbound forsythiaside in rat blood and bile by microdialysis coupled with HPLC method. Eur J Drug Metab Pharmacokinet. Feb 7. [Epub ahead of print]

Congeni BL (2010) Ertapenem. Expert Opin Pharmacother 11(4):669–672

Cooley JC, Lunte CE (2011) Detection of malondialdehyde in vivo using microdialysis sampling with CE-fluorescence. Electrophoresis 32(21):2994–2999

Craig WA, Suh B (1978) Changes in protein binding during disease. Scand J Infect Dis Suppl(14): 239–244

Dahyot C, Marchand S, Bodin M, Debeane B, Mimoz O, Couet W (2008) Application of basic pharmacokinetic concepts to analysis of microdialysis data: illustration with imipenem muscle distribution. Clin Pharmacokinet 47(3):181–189

Effros RM, Mason GR, Silverman P, Reid E, Hukkanen J (1986) Movement of ions and small solutes across endothelium and epithelium of perfused rabbit lungs. J Appl Physiol 60(1):100–107

EMEA (2000) Points to consider on pharmacokinetics and pharmacodynamics in the development of antibacterial medicinal products. (www.emea.eu.int/pdfs/human/ewp/265599en.pdf)

Evrard PA, Deridder G, Verbeeck RK (1996) Intravenous microdialysis in the mouse and the rat: development and pharmacokinetic application of a new probe. Pharm Res 13(1):12–17

FDA (1998) Guidance for Industry. Developing antimicrobial drugs—general considerations for clinical trials. (www.fda.gov/cder/guidance/2580dft.pdf)

Fischman AJ, Alpert NM, Babich JW, Rubin RH (1997) The role of positron emission tomography in pharmacokinetic analysis. Drug Metab Rev 29(4):923–956

Fischman AJ, Babich JW, Bonab AA, Alpert NM, Vincent J, Callahan RJ, Correia JA, Rubin RH (1998) Pharmacokinetics of [18F]trovafloxacin in healthy human subjects studied with positron emission tomography. Antimicrob Agents Chemother 42(8):2048–2054

Florea NR, Tessier PR, Zhang C, Nightingale CH, Nicolau DP (2004) Pharmacodynamics of moxifloxacin and levofloxacin at simulated epithelial lining fluid drug concentrations against Streptococcus pneumoniae. Antimicrob Agents Chemother 48(4):1215–1221

Frossard M, Joukhadar C, Erovic BM, Dittrich P, Mrass PE, Van Houte M, Burgmann H, Georgopoulos A, Muller M (2000) Distribution and antimicrobial activity of fosfomycin in the interstitial fluid of human soft tissues. Antimicrob Agents Chemother 44(10):2728–2732

Gonzalez D, Conrado DJ, Theuretzbacher U, Derendorf H (2011) The effect of critical illness on drug distribution. Curr Pharm Biotechnol 12(12):2030–2036

Gotfried MH, Danziger LH, Rodvold KA (2001) Steady-state plasma and intrapulmonary concentrations of levofloxacin and ciprofloxacin in healthy adult subjects. Chest 119(4):1114–1122

Hancock RE, Bellido F (1992) Factors involved in the enhanced efficacy against gram-negative bacteria of fourth generation cephalosporins. J Antimicrob Chemother 29(Suppl A): 1–6

Hashiguti H, Nakahara D, Maruyama W, Naoi M, Ikeda T (1993) Simultaneous determination of in vivo hydroxylation of tyrosine and tryptophan in rat striatum by microdialysis-HPLC: relationship between dopamine and serotonin biosynthesis. J Neural Transm Gen Sect 93(3):213–223

Hatfield T, Spanis C, McGaugh JL (1999) Response of amygdalar norepinephrine to footshock and GABAergic drugs using in vivo microdialysis and HPLC. Brain Res 835(2):340–345

Herrera AM, Scott DO, Lunte CE (1990) Microdialysis sampling for determination of plasma protein binding of drugs. Pharm Res 7(10):1077–1081

Hollenstein U, Brunner M, Mayer BX, Delacher S, Erovic B, Eichler HG, Muller M (2000) Target site concentrations after continuous infusion and bolus injection of cefpirome to healthy volunteers. Clin Pharmacol Ther 67(3):229–236

Hollenstein UM, Brunner M, Schmid R, Muller M (2001) Soft tissue concentrations of ciprofloxacin in obese and lean subjects following weight-adjusted dosing. Int J Obes Relat Metab Disord 25(3):354–358

Huang Y, Ji P, Inano A, Yang Z, Giebink GS, Sawchuk RJ (2001) Microdialysis studies of the middle ear distribution kinetics of amoxicillin in the awake chinchilla. J Pharm Sci 90(12):2088–2098

Ikawa K, Morikawa N, Ikeda K, Ohge H, Sueda T (2008) Development of breakpoints of carbapenems for intraabdominal infections based on pharmacokinetics and pharmacodynamics in peritoneal fluid. J Infect Chemother 14(4):330–332

Joukhadar C, Frossard M, Mayer BX, Brunner M, Klein N, Siostrzonek P, Eichler HG, Muller M (2001) Impaired target site penetration of beta-lactams may account for therapeutic failure in patients with septic shock. Crit Care Med 29(2):385–391

Joukhadar C, Klein N, Mayer BX, Kreischitz N, Delle-Karth G, Palkovits P, Heinz G, Muller M (2002) Plasma and tissue pharmacokinetics of cefpirome in patients with sepsis. Crit Care Med 30(7):1478–1482

Joukhadar C, Klein N, Dittrich P, Zeitlinger M, Geppert A, Skhirtladze K, Frossard M, Heinz G, Muller M (2003a) Target site penetration of fosfomycin in critically ill patients. J Antimicrob Chemother 51(5):1247–1252

Joukhadar C, Stass H, Muller-Zellenberg U, Lackner E, Kovar F, Minar E, Muller M (2003b) Penetration of Moxifloxacin into Healthy and Inflamed Subcutaneous Adipose Tissues in Humans. Antimicrob Agents Chemother 47(10):3099–3103

Kaul S, Faiman MD, Lunte CE (2011) Determination of GABA, glutamate and carbamathione in brain microdialysis samples by capillary electrophoresis with fluorescence detection. Electrophoresis 32(2):284–291

Kavi J, Andrews JM, Ashby JP, Hillman G, Wise R (1988) Pharmacokinetics and tissue penetration of cefpirome, a new cephalosporin. J Antimicrob Chemother 22(6):911–916

Kearney BP, Aweeka FT (1999) The penetration of anti-infectives into the central nervous system. Neurol Clin 17(4):883–900

Keating GM, Scott LJ (2004) Moxifloxacin: a review of its use in the management of bacterial infections. Drugs 64(20):2347–2377

Kikuchi J, Yamazaki K, Kikuchi E, Ishizaka A, Nishimura M (2007) Pharmacokinetics of gatifloxacin after a single oral dose in healthy young adult subjects and adult patients with chronic bronchitis, with a comparison of drug concentrations obtained by bronchoscopic microsampling and bronchoalveolar lavage. Clin Ther 29(1):123–130

Kikuchi E, Yamazaki K, Kikuchi J, Hasegawa N, Hashimoto S, Ishizaka A, Nishimura M (2008) Pharmacokinetics of clarithromycin in bronchial epithelial lining fluid. Respirology 13(2):221–226

Kim A, Suecof LA, Sutherland CA, Gao L, Kuti JL, Nicolau DP (2008) In vivo microdialysis study of the penetration of daptomycin into soft tissues in diabetic versus healthy volunteers. Antimicrob Agents Chemother 52(11):3941–3946

Kinzig M, Sorgel F, Brismar B, Nord CE (1992) Pharmacokinetics and tissue penetration of tazobactam and piperacillin in patients undergoing colorectal surgery. Antimicrob Agents Chemother 36(9):1997–2004

Korting HC, Schafer-Korting M, Maass L, Klesel N, Mutschler E (1987) Cefodizime in serum and skin blister fluid after single intravenous and intramuscular doses in healthy volunteers. Antimicrob Agents Chemother 31(11):1822–1825

Kovar A, Dalla Costa T, Derendorf H (1997) Comparison of plasma and free tissue levels of ceftriaxone in rats by microdialysis. J Pharm Sci 86(1):52–56

Kullar R, Chin JN, Edwards DJ, Parker D, Coplin WM, Rybak MJ (2011) Pharmacokinetics of single-dose daptomycin in patients with suspected or confirmed neurological infections. Antimicrob Agents Chemother 55(7):3505–3509

Laethem T, De Lepeleire I, McCrea J, Zhang J, Majumdar A, Musson D, Rogers D, Li S, Guillaume M, Parneix-Spake A, Deutsch P (2003) Tissue penetration by ertapenem, a parenteral carbapenem administered once daily, in suction-induced skin blister fluid in healthy young volunteers. Antimicrob Agents Chemother 47(4):1439–1442

Langer O, Muller M (2004) Methods to assess tissue-specific distribution and metabolism of drugs. Curr Drug Metab 5(6):463–481

Li Y, Peris J, Zhong L, Derendorf H (2006a) Microdialysis as a tool in local pharmacodynamics. AAPS J 8(2): E222–E235

Li Y, Peris J, Zhong L, Derendorf H (2006b) Microdialysis as a tool in local pharmacodynamics. AAPS J 08(02):1–33

Lipman J, Wallis SC, Rickard CM, Fraenkel D (2001) Low cefpirome levels during twice daily dosing in critically ill septic patients: pharmacokinetic modelling calls for more frequent dosing. Intensive Care Med 27(2):363–370

Liu P, Fuhrherr R, Webb AI, Obermann B, Derendorf H (2005) Tissue penetration of cefpodoxime into the skeletal muscle and lung in rats. Eur J Pharm Sci 25(4–5):439–444

Maggiolo F, Capra R, Bottura P, Pravettoni G, Migliorino M, Suter F (2000) In vitro activity of moxifloxacin combined with other antibacterials against methicillin-resistant Staphylococcus aureus. J Chemother 12(3):195–198

Marchand S, Chenel M, Lamarche I, Couet W (2005a) Pharmacokinetic modeling of free amoxicillin concentrations in rat muscle extracellular fluids determined by microdialysis. Antimicrob Agents Chemother 49(9):3702–3706

Marchand S, Dahyot C, Lamarche I, Mimoz O, Couet W (2005b) Microdialysis study of imipenem distribution in skeletal muscle and lung extracellular fluids of noninfected rats. Antimicrob Agents Chemother 49(6):2356–2361

Marchand S, Dahyot C, Lamarche I, Plan E, Mimoz O, Couet W (2005c) Lack of effect of experimental hypovolemia on imipenem muscle distribution in rats assessed by microdialysis. Antimicrob Agents Chemother 49(12):4974–4979

Muller M, Haag O, Burgdorff T, Georgopoulos A, Weninger W, Jansen B, Stanek G, Pehamberger H, Agneter E, Eichler HG (1996) Characterization of peripheral-compartment kinetics of antibiotics by in vivo microdialysis in humans. Antimicrob Agents Chemother 40(12):2703–2709

Muller M, Brunner M, Hollenstein U, Joukhadar C, Schmid R, Minar E, Ehringer H, Eichler HG (1999a) Penetration of ciprofloxacin into the interstitial space of inflamed foot lesions in non-insulin-dependent diabetes mellitus patients. Antimicrob Agents Chemother 43(8):2056–2058

Muller M, Stass H, Brunner M, Moller JG, Lackner E, Eichler HG (1999b) Penetration of moxifloxacin into peripheral compartments in humans. Antimicrob Agents Chemother 43(10):2345–2349

Muller M, dela Pena A, Derendorf H (2004). Issues in pharmacokinetics and pharmacodynamics of anti-infective agents: distribution in tissue. Antimicrob Agents Chemother 48(5): 1441–1453

Nix DE, Wilton JH, Velasquez N, Budny JL, Lassman HB, Mitchell P, Divan K, Schentag JJ (1992) Cerebrospinal fluid penetration of cefpirome in patients with non-inflamed meninges. J Antimicrob Chemother 29(Suppl A): 51–57

Nix DE, Matthias KR, Ferguson EC (2004) Effect of ertapenem protein binding on killing of bacteria. Antimicrob Agents Chemother 48(9):3419–3424

Ober MC, Hoppe-Tichy T, Koninger J, Schunter O, Sonntag HG, Weigand MA, Encke J, Gutt C, Swoboda S (2009) Tissue penetration of moxifloxacin into human gallbladder wall in patients with biliary tract infections. J Antimicrob Chemother 64(5):1091–1095

Perez-Trallero E, Garcia-Rey C, Martin-Sanchez AM, Aguilar L, Garcia-de-Lomas J, Ruiz J (2002) Activities of six different quinolones against clinical respiratory isolates of Streptococcus pneumoniae with reduced susceptibility to ciprofloxacin in Spain. Antimicrob Agents Chemother 46(8):2665–2667

Philip-Joet F, Bruguerolle B, Reynaud M, Arnaud A (1992) Correlations between theophylline concentrations in plasma, erythrocytes and cantharides-induced blister fluid and peak expiratory flow in asthma patients. Eur J Clin Pharmacol 43(5):563–565

Pien HH, Fischman AJ, Thrall JH, Sorensen AG (2005) Using imaging biomarkers to accelerate drug development and clinical trials. Drug Discov Today 10(4):259–266

Plock N, Kloft C (2005) Microdialysis–theoretical background and recent implementation in applied life-sciences. Eur J Pharm Sci 25(1):1–24

Rea K, Cremers TI, Westerink BH (2005) HPLC conditions are critical for the detection of GABA by microdialysis. J Neurochem 94(3):672–679

Reckziegel R, Maguilnik I, Goldani LZ (2001) Gentamicin concentration in bile after once-daily versus thrice-daily dosing of 4 mg/kg/day. J Antimicrob Chemother 48(2):327–329

Rennard SI, Basset G, Lecossier D, O'Donnell KM, Pinkston P, Martin PG, Crystal RG (1986) Estimation of volume of epithelial lining fluid recovered by lavage using urea as marker of dilution. J Applied Physiol 60(2):532–538

Roberts JA, Lipman J (2009) Pharmacokinetic issues for antibiotics in the critically ill patient. Crit Care Med 37(3): 840–851; quiz 859

Ryan DM (1993) Pharmacokinetics of antibiotics in natural and experimental superficial compartments in animals and humans. J Antimicrob Chemother 31(Suppl D): 1–16

Sahre M, Naik R, Derendorf H (2011) Applications of microdialysis in skin and soft tissues. In: Tsai T-H (ed) Applications of microdialysis in pharmaceutical sciences. Wiley, London, p 584

Sauermann R, Delle-Karth G, Marsik C, Steiner I, Zeitlinger M, Mayer-Helm BX, Georgopoulos A, Muller M, Joukhadar C (2005a) Pharmacokinetics and pharmacodynamics of cefpirome in subcutaneous adipose tissue of septic patients. Antimicrob Agents Chemother 49(2):650–655

Sauermann R, Karch R, Langenberger H, Kettenbach J, Mayer-Helm B, Petsch M, Wagner C, Sautner T, Gattringer R, Karanikas G, Joukhadar C (2005b) Antibiotic abscess penetration: fosfomycin levels measured in pus and simulated concentration-time profiles. Antimicrob Agents Chemother 49(11):4448–4454

Schentag JJ (1989) Clinical significance of antibiotic tissue penetration. Clin Pharmacokinet 16(Suppl 1):25–31

Schmidt S, Banks R, Kumar V, Rand KH, Derendorf H (2008a) Clinical microdialysis in skin and soft tissues: an update. J Clin Pharmacol 48(3):351–364

Schmidt S, Rock K, Sahre M, Burkhardt O, Brunner M, Lobmeyer MT, Derendorf H (2008b) Effect of protein binding on the pharmacological activity of highly bound antibiotics. Antimicrob Agents Chemother 52(11):3994–4000

Schmidt S, Gonzalez D, Derendorf H (2010) Significance of protein binding in pharmacokinetics and pharmacodynamics. J Pharm Sci 99(3):1107–1122

Scott DO, Lunte CE (1993) In vivo microdialysis sampling in the bile, blood, and liver of rats to study the disposition of phenol. Pharm Res 10(3):335–342

Scott DO, Sorensen LR, Lunte CE (1990) In vivo microdialysis sampling coupled to liquid chromatography for the study of acetaminophen metabolism. J Chromatogr 506:461–469

Shandil RK, Jayaram R, Kaur P, Gaonkar S, Suresh BL, Mahesh BN, Jayashree R, Nandi V, Bharath S, Balasubramanian V (2007) Moxifloxacin, ofloxacin, sparfloxacin, and ciprofloxacin against Mycobacterium tuberculosis: evaluation of in vitro and pharmacodynamic indices that best predict in vivo efficacy. Antimicrob Agents Chemother 51(2):576–582

Soman A, Honeybourne D, Andrews J, Jevons G, Wise R (1999) Concentrations of moxifloxacin in serum and pulmonary compartments following a single 400 mg oral dose in patients undergoing fibre-optic bronchoscopy. J Antimicrob Chemother 44(6):835–838

Stenken JA (1999) Methods and issues in microdialysis calibration. Anal Chim Acta 379:337–358

Strenkoski LC, Nix DE (1993) Cefpirome clinical pharmacokinetics. Clin Pharmacokinet 25(4):263–273

Suzuki H, Sawada Y, Sugiyama Y, Iga T, Hanano M (1989) Facilitated transport of benzylpenicillin through the blood-brain barrier in rats. J Pharmacobiodyn 12(3):182–185

Talan DA (2001) Clinical perspectives on new antimicrobials: focus on fluoroquinolones. Clin Infect Dis 32(Suppl 1): S64–S71

Tasso L, Bettoni CC, Oliveira LK, Dalla Costa T (2008) Evaluation of gatifloxacin penetration into skeletal muscle and lung by microdialysis in rats. Int J Pharm 358(1–2):96–101

Taylor AE, Guyton AC, Bishop VS (1965) Permeability of the alveolar membrane to solutes. Circ Res 16:353–362

Thallinger C, Buerger C, Plock N, Kljucar S, Wuenscher S, Sauermann R, Kloft C, Joukhadar C (2008) Effect of severity of sepsis on tissue concentrations of linezolid. J Antimicrob Chemother 61(1):173–176

Thys JP, Vanderhoeft P, Herchuelz A, Bergmann P, Yourassowsky E (1988) Penetration of aminoglycosides in uninfected pleural exudates and in pleural empyemas. Chest 93(3):530–532

Traunmuller F, Schintler MV, Spendel S, Popovic M, Mauric O, Scharnagl E, Joukhadar C (2010) Linezolid concentrations in infected soft tissue and bone following repetitive doses in diabetic patients with bacterial foot infections. Int J Antimicrob Agents 36(1):84–86

Tunblad K, Jonsson EN, Hammarlund-Udenaes M (2003) Morphine blood-brain barrier transport is influenced by probenecid co-administration. Pharm Res 20(4):618–623

Ungerstedt U, Pycock C (1974) Functional correlates of dopamine neurotransmission. Bull Schweiz Akad Med Wiss 30(1–3):44–55

Van Bambeke F, Barcia-Macay M, Lemaire S, Tulkens PM (2006) Cellular pharmacodynamics and pharmacokinetics of antibiotics: current views and perspectives. Curr Opin Drug Discov Devel 9(2):218–230

Wiseman LR, Lamb HM (1997) Cefpirome. A review of its antibacterial activity, pharmacokinetic properties and clinical efficacy in the treatment of severe nosocomial infections and febrile neutropenia. Drugs 54(1):117–140

Wiskirchen DE, Shepard A, Kuti JL, Nicolau DP (2011) Determination of tissue penetration and pharmacokinetics of linezolid in patients with diabetic foot infections using in vivo microdialysis. Antimicrob Agents Chemother 55(9):4170–4175

Xie R, Hammarlund-Udenaes M (1998) Blood-brain barrier equilibration of codeine in rats studied with microdialysis. Pharm Res 15(4):570–575

Zeitlinger M, Muller M, Joukhadar C (2005) Lung microdialysis—a powerful tool for the determination of exogenous and endogenous compounds in the lower respiratory tract (mini-review). Aaps J 7(3): E600–E608

Zeitlinger BS, Zeitlinger M, Leitner I, Muller M, Joukhadar C (2007) Clinical scoring system for the prediction of target site penetration of antimicrobials in patients with sepsis. Clin Pharmacokinet 46(1):75–83

Zeitlinger MA, Derendorf H, Mouton JW, Cars O, Craig WA, Andes D, Theuretzbacher U (2011) Protein binding: do we ever learn? Antimicrob Agents Chemother 55(7):3067–3074
Zhanel GG, Dueck M, Hoban DJ, Vercaigne LM, Embil JM, Gin AS, Karlowsky JA (2001) Review of macrolides and ketolides: focus on respiratory tract infections. Drugs 61(4):443–498
Zhang MY, Beyer CE (2006) Measurement of neurotransmitters from extracellular fluid in brain by in vivo microdialysis and chromatography-mass spectrometry. J Pharm Biomed Anal 40(3):492–499
Zhang MY, Hughes ZA, Kerns EH, Lin Q, Beyer CE (2007) Development of a liquid chromatography/tandem mass spectrometry method for the quantitation of acetylcholine and related neurotransmitters in brain microdialysis samples. J Pharm Biomed Anal 44(2):586–593
Zhang QL, Hu JH, Zhu QG, Li FQ, Liu JY, Wang D (2009) Development of a novel HPLC-MS/MS method for the determination of aconitine and its application to in vitro and rat microdialysis samples. Biomed Chromatogr 23(7):692–699
Zhu Y, Wong PS, Cregor M, Gitzen JF, Coury LA, Kissinger PT (2000) In vivo microdialysis and reverse phase ion pair liquid chromatography/tandem mass spectrometry for the determination and identification of acetylcholine and related compounds in rat brain. Rapid Commun Mass Spectrom 14(18):1695–1700

Chapter 7
Skin and Transdermal Drug Delivery

Advantages and Challenges

Eva Benfeldt

Abstract This chapter will give a review of microdialysis sampling in the skin with emphasis on dermal sampling and topical application of drugs and other substances in the form of transdermal patches, solutions, or topical formulations. Microdialysis as a research technique is suitable for a number of different study designs for sampling in the skin or subcutaneous tissues. This versatility includes the option of preclinical studies in animals as well as ex vivo human skin set ups. Alterations in the barrier properties of the skin, introduced intentionally or associated with skin diseases, will alter the penetration of drugs and other substances considerably. Furthermore, the effect of even small alterations in a topical formulation can be detected by simultaneous sampling in several areas in the same individual, enabling studies of bioavailability and bioequivalence in study groups of relatively few volunteers or patients. Whenever the dermis is the target tissue, dermal microdialysis can be considered to be the method of choice for acquisition of human in vivo data. The current status with regulatory authorities as well as the study planning for successful sampling will be described.

7.1 Introduction

In this chapter, the focus will be on topically applied drugs and substances and on studies performed in human volunteers or patients. Some introductory comments concerning the technical aspect of dermal and subcutaneous microdialysis sampling are given below.

E. Benfeldt (✉)
Department of Dermatology, Faculty of Health Sciences, University of Copenhagen, Roskilde Hospital, Koegevej 7-13, DK 4000 Roskilde, Denmark
e-mail: benfeldt@post5.tele.dk

M. Müller (ed.), *Microdialysis in Drug Development*, AAPS Advances in the Pharmaceutical Sciences Series, DOI: 10.1007/978-1-4614-4815-0_7,

7.2 Technical Issues in Dermal Microdialysis

There are many choices during the preparation of in vivo experimentation. The most important one is probably the study design itself, as the between-subjects variability will, for some topical applications, be much larger than the intersubject variability (see later Sects. 7.5 and 7.6). Whenever possible a design using the opportunity for several test and reference areas in the same person should be chosen.

7.2.1 Choice of Probe Type and Perfusate

Different types of probes for dermal microdialysis are available (Holmgaard et al. 2010) and the majority of these probes are inserted in the skin through a small guide cannula as the probes are made of fragile material. The linear probe has a unidirectional flow and compared to the other probe types which have a very small diameter—down to approximately 200 μm for some structures (Benfeldt et al. 1999). Linear probes will, however, need to penetrate the skin twice when inserted, since they have both an inlet and an outlet. The typical probe used in clinical studies relating to systemic drugs is the concentric probe which, like the side-by-side and loop design probes, has an inlet and outlet placed in parallel with each other. The latter probe types require guide cannulas of larger dimensions (e.g. 500 μm) than those necessary for insertion of the linear probe, which leads to a larger insertion trauma (see later)—and this is a drawback.

In cutaneous microdialysis the perfusate is most often an isotonic saline solution or a Ringer solution. Depending on the lipophilicity of the drug, the perfusate medium may have to be modified to allow more lipophilic substances to enter the probe. Substances such as albumin, Intralipid®, and Encapsin® have been used (Carneheim and Stahle 1991; Ward et al. 2003; Holmgaard et al. 2012; Au et al. 2012). The enhancing effect of adding binding agents such as α, β, γ-cyclodextrins to the perfusate has been evaluated by studying the recovery of several eicosanoids in vitro. Similarly, the effect of adding small organic molecules such as ethanol, propylene glycol, and dimethylsulfoxide has been studied, and the inclusion of arachidonic acid in the perfusate has been shown to increase in vitro relative recovery for hydrophilic analytes (Sun and Stenken 2003).

7.2.2 Application Site

In cutaneous microdialysis the most frequently used area of application is the dermis of the volar forearm (Benfeldt et al. 1999; Au et al. 2012; Benfeldt and Groth 1998; Groth and Serup 1998; Benfeldt 1999, 2007; Ortiz et al. 2008, 2009, 2011;

Tettey-Amlalo et al. 2009). The reason for this is twofold—this is the "standard area" for investigations into non-invasive measurements of skin blood flow and barrier perturbation. Second, the area is easily accessible, usually not hairy and not very convex. If there is a need for shaving this should be done preferably three days before the experiment to prevent an artifact by skin barrier perturbation.

Only one arm is used for cutaneous microdialysis at a time, since that arm must be at rest and tied up during the experiment. It is uncomfortable to have both arms tied up! Blood drawn from the other arm can provide measurements of plasma drug concentrations and thus the systemic drug delivery, if relevant.

7.2.3 Insertion Procedure, Trauma, and Exclusion Criteria

The probes are inserted under clean or sterile procedures after gentle wash of the skin area. To minimize the discomfort during insertion, local anesthesia (Groth and Serup 1998; Anderson et al. 1994; Petersen 1998) as well as application of ice packs onto the skin can be used (Au et al. 2012; Tettey-Amlalo et al. 2009; Kellogg et al. 2008). The probes are inserted using a 19–23 G cannula as a guide, which is inserted horizontally in the dermis or subcutis. The probe is then inserted in the opposite direction through the open tip of the hollow cannula and tested to secure the functionality. Subsequently, the guide cannula is withdrawn with the probe still in place. The accuracy of placing the probe at the intended depth in the skin depends on the training and experience of the laboratory personnel, and the depth should be measured routinely by ultrasound scanning as part of the experimental protocol, see Sect. 7.6.3.

Microdialysis is a minimally invasive technique, but the skin trauma and histamine release evoked by the insertion will cause a reversible increase in the local blood flow, increased skin thickness, and hyperemia in both animals and humans. The tissue trauma needs to subside before sampling can begin and an appropriate equilibration period of a minimum of 60–90 min in human skin is advisable (Groth and Serup 1998; Anderson et al. 1992, 1994). The hyperemia reaction lasts between 90 and 135 min (Petersen et al. 1992), but complete normalization of the skin perfusion may not occur (Krogstad et al. 1996). The use of local anesthesia has been found to reduce the trauma reaction in man in several studies (Groth and Serup 1998; Anderson et al. 1994; Petersen 1998).

The presence of a microdialysis probe will elicit an inflammatory response after approximately 12 h; infiltration of lymphocytes will begin and after 32 h scar tissue may appear (Krogstad et al. 1996; Ault et al. 1994). Following insertion of concentric probes more extensive tissue disruption compared to a linear probe has been reported (Ault et al. 1992, 1994). The larger reaction could be caused by the larger diameter of the guide cannula used for implantation of concentric probes (e.g. ~500 μm).

The tissue reactions are generally reversible and lasting tissue damage is negligible. However, all the cytokines and cells associated with, e.g.,inflammatory disease are involved in the trauma reaction (Stenken et al. 2010). In my experience the development of skin changes, visible at clinical examination after, e.g., 3 months is very rare. One exception is if the volunteer or patient is prone to keloid formation, and for this reason we recommend screening for this as an exclusion criterion of the study protocol, thus avoiding dermal microdialysis in keloid-prone individuals.

7.3 Pharmacokinetics Obtained in the Skin

The skin is an excellent organ for microdialysis sampling since the tissue is fairly homogeneous and the extracellular fluid is equilibrated with the systemic circulation. The equilibration also applies for small water-soluble compounds, e.g., glucose or histamine whereas for the larger molecules, applied as topical formulations, this is not always the case.

The first human dermal microdialysis studies examined the cutaneous penetration of ethanol, the release of histamine in response to skin manipulation, and the vascular changes in the skin after insertion of dermally placed microdialysis probes (Groth and Serup 1998; Anderson et al. 1991, 1994).

7.3.1 Clinical Studies in the Skin

The distribution of orally or intravenously administered antimicrobial substances has been sampled in the skin (Borg et al. 1999; Joukhadar et al. 2001; Klimowicz et al. 2002; Bielecka-Grzela and Klimowicz 2003). Dermal microdialysis allows sampling of the drug of interest as well as its metabolites in the dermis. Dermal sampling of drugs which exercise their action in the skin or can be followed by a biomarker gives the opportunity for pharmacokinetic-pharmacodynamic studies (Kreilgaard 2002).

7.3.2 Topical Drug Application

Dermal microdialysis has been used for in vivo studies of pharmacokinetics of topically applied drugs, pesticides, and other substances; for an overview please see Table 3 in review by Holmgaard et al. (2010). The substances have been administered as creams, ointments, gels, or in transdermal therapeutic systems (patches, so-called TTS). Not all substances have been successfully sampled, whenever attempted, and difficulties have in particular been encountered with very lipophilic substances such as the corticosteroids and the sex hormones.

Cross et al. have studied the active ingredients in an anti-inflammatory formulation containing methyl salicylate and found indication of direct (regional) tissue penetration and not solely redistribution by systemic circulation (Cross et al. 2007). Very recently the effect of convective transport, which can assist the penetration of, e.g., NSAID following topical application, has been shown to be associated with the drug being highly plasma protein bound (Dancik et al. 2012).

Studies of transdermally administered medication have explored the kinetics of drug release from different patches containing, e.g., nicotine and estradiol, (Hegemann et al. 1995; Muller et al. 1995). Also, felbinac (TTS—transdermal) and diclofenac (spray gel), two topical drugs for treatment of muscle inflammation, have been examined using microdialysis. Based on the high drug concentrations found in the skin and subcutaneous tissues, respectively, both administration modalities were shown to be possible alternatives to orally administered NSAID for superficial muscle inflammation (Brunner et al. 2005; Shinkai et al. 2008). Ketoprofen penetration kinetics following application as a TTS (with the target being intra-articular inflammation) has been measured in the skin as well as in the knee joint by microdialysis. Sufficient drug concentration for inhibition of prostaglandin E2 production were found in the joint (Shinkai et al. 2008).

The consequences of influencing the skin barrier, whether it is a physical or chemical disruption, will for most substances mean an increase in bioavailability and drug penetration, and an overview of the effect of skin barrier changes is found in Sect. 7.6.1.

Cutaneous microdialysis has also been used in studies of topical drug penetration in cutaneous tumor tissue, involved in the so-called photodynamic therapy for basal cell carcinomas (Wennberg et al. 2000; Sandberg et al. 2008).

The effect of iontophoresis on topical drug delivery (acyclovir, propranolol, methotrexate and flurbiprofen) to/through the skin and the subsequent systemic delivery has also been studied by microdialysis in the dermis. The conclusion is that iontophoretic drug delivery could be demonstrated to be efficient in delivering a high and fast drug penetration to the dermis and underlying tissue while maintaining low systemic exposure (Stagni et al. 2000, 2004; Stagni and Shukla 2003; Mathy et al. 2005).

Recommended reviews of dermal microdialysis in studies of topically applied drugs and substances are Schmidt et al. (2008) and Holmgaard et al. (2010).

7.4 Bioavailability

The bioavailability of a topically applied drug is defined as "the rate and extent to which the active ingredient or active moiety is absorbed and becomes available at the target site. For drugs that are not intended to be absorbed into the bloodstream, bioavailability reflects the rate and extent to which the active ingredient becomes available at the relevant site within the skin" as per FDA "Guidance for Industry" (US-FDA 2002). A recent study demonstrates the use of microdialysis to determine the bioavailability in a study comparing orally administered diclofenac with topically administered diclofenac. Thus the bioavailability at the target site was measured by placing the microdialysis probe in the subcutaneous tissue as well as the muscular tissue. The relative bioavailability in both muscular and subcutaneous tissue after topical application was significantly higher than after oral administration and the measured plasma levels were much lower when the drug was topically administered compared to oral administration (Benfeldt 1999; Tegeder et al. 1999, 2002). Similarly, the diffusion of ketoprofen from transdermal patch application to the knee joint fluid has been demonstrated in rats and pigs (Shinkai et al. 2011). A very recent study explores, by microdialysis sampling in subcutaneous tissues and skeletal muscle, the effect of alterations in the galenic composition of four novel topical diclofenac formulations under development. An improved drug delivery to the tissue (2.7-fold improved) over the reference product, a commercially available diclofenac gel, was demonstrated (Brunner et al. 2011).

7.5 Bioequivalence

FDA describes bioequivalence as a comparative test between two products using specified criteria. Bioequivalence is defined as "the absence of a significant difference in the rate and extent to which the active ingredient or active moiety in pharmaceutical equivalents or pharmaceutical alternatives becomes available at the site of drug action when administered at the same molar dose under similar conditions in an appropriately designed study" (US-FDA 2002). However, the assessment of bioequivalence for locally acting and targeted delivery drugs has presented great challenges to science in the approval of generic drugs.

Kreilgaard et al. published the first human study demonstrating the potential of cutaneous microdialysis for bioequivalence studies of topically applied drugs in 2001 (Kreilgaard et al. 2001). The aim of that study was to evaluate the cutaneous bioequivalence of a lipophilic drug (lidocaine) applied in a novel topical microemulsion vehicle, compared to a conventional oil-in-water emulsion. Earlier Kreilgaard had published an experimental study in rats documenting that dermal drug delivery of hydrophilic and lipophilic drugs was improved by microemulsion vehicles (Kreilgaard 2001). Subsequently Kreilgaard proved that microemulsion vehicles can increase dermal drug delivery of lipophilic drugs in humans, and that the microdialysis technique combined with an appropriate pharmacokinetic model provides high sensitivity in this bioequivalence study of a topically applied formulation (Kreilgaard 2002).

When addressing the issue of dermal microdialysis variability the source of variability in these studies can be extracted from data obtained in hairless rats (which are genetically inbred and of the same sex and age and thus have minimum variability). The increased variability observed in dermal sampling of the topically applied drugs can be seen to arise from interindividual variability in skin penetration kinetics/barrier function and or microcirculation, unrelated to the microdialysis sampling methodology, which is followed by an internal calibrator simultaneously with the ongoing penetration process (Simonsen et al. 2004).

The comparability of dermato-pharmacokinetics, so-called tape-stripping methodology, and dermal microdialysis was demonstrated in a recent study evaluating bioequivalence of lidocaine in ointment and cream (Benfeldt et al. 2007). The study showed agreement between the two methods, finding a 3–5-fold higher lidocaine absorption from the cream formulation over the ointment formulation. Statistical calculations from this study indicated that bioequivalence evaluation of topical formulations based on dermal microdialysis sampling may be conducted using 27 subjects and applying two probes in each application site or 18 subjects using three probes in each application site with 90 % confidence interval and 80–125 % bioequivalence limits (Benfeldt et al. 2007), which are the limits applicable to topical bioequivalence evaluations as per FDA protocols (Shah et al. 1998).

A theoretical–statistical microdialysis paper found, based on data in the literature, that an evaluation of topical bioequivalence, conducted by duplicate sampling of both formulations in the same human volunteer, could be expected to be conclusive within 80–125 % confidence limits when a population of approximately 20 subjects was to be used (McCleverty et al. 2006). This number of participants was similar to what was calculated in the study on topical lidocaine formulations (Benfeldt et al. 2007). In the latter study, 61 % of the experimental variability could be ascribed to intersubject variability—an important finding for planning of future studies. This result was corroborated by Tettey-Amlalo et al. who demonstrated a very low variability using the exact same probe structure for sampling drug penetration from a topical ketoprofen gel formulation (Tettey-Amlalo et al. 2009). However, different formulations may demonstrate different variabilities when the cutaneous penetration is studied by dermal microdialysis; in a study of topical metronidazole creams the variability was higher and the number of participants for a conclusive bioequivalence evaluation would have been 34 (Ortiz et al. 2011). Nevertheless, in comparison with the 6–700 subjects needed for a clinical comparative study, these numbers are small (Lionberger 2008).

Another recent study has compared topical penetration of two commercially available tetracycline formulations and also found high variability (Incecayir et al. 2011), whereas a study of an aggressive ethanol solution of clobetasol propionate—a drug which it has previously not been feasible to sample in the dermis—showed that using Intralipid® as the perfusate the drug could now be reproducibly sampled (Au et al. 2012).

Dermal microdialysis sampling may also enable bioequivalence studies in diseased skin (Benfeldt et al. 1999; Ortiz et al. 2008, 2009), which can be argued to be closer to the clinical study than bioequivalence evaluations conducted in healthy subjects. However, variability is likely to be increased and possibly too problematic, see Sect. 7.6.1 below.

7.6 Sources of Variability

7.6.1 Skin Barrier Function

Damage of the skin barrier implies an increase in transepidermal water loss, TEWL, which can be quantified by measurements above the skin surface. In a number of in vivo dermal microdialysis studies the impact of experimentally induced skin barrier perturbation on the cutaneous penetration of different substances has been demonstrated (Benfeldt et al. 1999; Ortiz et al. 2008; Tsai et al. 2003; Morgan et al. 2003; Klimowicz et al. 2007). Benfeldt et al. have studied the effect of different barrier disruption methods on cutaneous penetration in humans and hairless rats (Benfeldt and Serup 1999). The studies demonstrated highly increased drug penetration in tape stripped skin (157- and 170-fold, respectively, in comparison to the penetration in unmodified skin) and in skin with irritant dermatitis (46- and 80-fold increased penetration). Since the same probe type, perfusate, flow rate, and topical drug solution were used in these studies, a direct comparison between drug penetration in hairless rats and human volunteers could be made. This showed a 46-fold increase in penetration across rat skin when compared with human skin, while increases in penetration, induced by barrier perturbation were of the same order of magnitude (Benfeldt 1999).

Other more recent studies have compared, e.g., the penetration of acyclovir and salicylic acid on disrupted skin barriers using microdialysis in the dermis and tape stripping (Klimowicz et al. 2007), and the penetration of a metronidazole cream formulation (1 %) applied to the forearm skin in areas of both irritant dermatitis and normal skin (Ortiz et al. 2008). Furthermore, dermal microdialysis sampling showed a significant threefold increased penetration of topically applied metronidazole in skin with atopic dermatitis compared with unaffected skin (Ortiz et al. 2009).

7.6.2 Microcirculation

The bioavailability of topically applied products in skin and underlying tissues is not only dependent on the integrity of the skin barrier, but also on the local blood flow. Vasodilatation as well as vasoconstriction can be physiologically or pharmacologically induced and will have a large influence on the local blood flow. The skin concentration of a topically applied drug will increase if the blood flow is diminished, whereas an increased blood flow enhances the uptake and subsequent systemic distribution and elimination of the drug from the skin (Ault et al. 1994). Experimental studies which induce vasoconstriction or vasodilatation have demonstrated that the bioavailability of topically as well as systemically applied test substances is highly influenced by changes in the microcirculation of the skin (Borg et al. 1999; Mathy et al. 2005; Seki et al. 2004; Chaturvedula et al. 2005)—an influence much stronger than the influence of variations in probe depth.

Topically administered substances have been studied, with added noradrenaline for vasoconstriction and the nitric oxide donor glyceryltrinitrate for vasodilatation, delivered by the microdialysis probe. The changes seen in the dialysate concentration reflected the changes in the microcirculation (Mathy et al. 2005).

7.6.3 Probe Depth in the Skin

An influence of probe depth on the drug levels sampled following topical drug application is likely from a theoretical point of view. The recommendation is to measure (normally by 20 MHz ultrasound scanning) the probe depth in three separate scans along the length of the probe in situ. With experience, probes can be inserted with great a accuracy and low variability (Benfeldt et al. 2007; Tettey-Amlalo et al. 2009), e.g., 0.7 ± 0.15 mm mean ± SD (Ortiz et al. 2008). The preferable insertion depth is 0.6–1.0 mm in the dermis.

One study regarding transdermal delivery of nicotine showed a correlation between depth and drug concentration, but only when the analysis included different skin layers (both dermal and subcutaneous probe placement) (Muller et al. 1995). Following this, several studies have been unable to show a correlation between the depth of the probe and the drug concentration (Benfeldt et al. 2007; Ortiz et al. 2008; Tettey-Amlalo et al. 2009; Hegemann et al. 1995; Muller et al. 1997). However, in a very recent microdialysis study, conducted in ex vivo human donor skin, an inverse relationship between the depth of the probe in the dermis and the amount of drug sampled following topical penetration is demonstrated. The result is of relevance to the in vivo situation, and it can be predicted that the differences in sampling at different probe depths will have a more significant impact in the beginning of a study or in studies of short duration.

Based on this study it can be recommended that studies of topical drug penetration using dermal microdialysis sampling should include measurements of probe depth and that efforts should be made to minimize probe depth variability (e.g., to have few and similarly trained persons undertaking insertion as well as scanning for feedback of probe depth achieved) (Holmgaard et al. 2012).

7.7 Advantages and Limitations

The methodological challenges that may influence in vivo experiments may be typically identified through well-planned in vitro experiments as a part of the planning phase. Founded on pre-experimental troubleshooting, conducted previous to in vivo experimentation, reproducible results with acceptable variability and validated analysis can be achieved for most drugs.

7.7.1 Advantages

- Microdialysis captures the pharmacological events where they take place in the tissues, providing high-resolution real-time details.
- There is no fluid extraction from the tissue.
- Dermal microdialysis sampling allows testing of both test and reference product at the same time in the same individuals.
- Both the drug of interest and the metabolites.
- The method can provide protein-free samples, which is often an analytical advantage.
- Cessation of enzyme degradation in the samples.
- Dermal microdialysis is a relatively inexpensive method to use once the microdialysis pumps have been acquired.
- The probes allow sampling as well as delivery of substances.
- Multiple application sites.
- Good reproducibility.
- Dermal microdialysis sampling of topical drug formulation in the bioavailability/bioequivalence setting does not depend on drug concentrations in the formulations being the same.
- Dermal microdialysis sampling can be used in the presence of barrier perturbation or skin disease (unlike other methods for skin penetration assessment).

7.7.2 Limitations

- Drugs with a very high lipophilicity are excellent for topical application but less favorable for sampling by microdialysis (the tape-stripping method will often be more suitable).
- An in vitro relative recovery of less than 4 % will most often characterize a compound as unsuitable for microdialysis studies due to an expected even lower in vivo recovery.
- Some topical drug formulations are of very low drug concentrations—the analysis of the dialysate will unavoidably be very challenging.
- For protein-bound substances it is often necessary to add a protein to the perfusate to increase recovery, which will often result in more complex analytical procedures.
- Low variability in probe insertion and probe manufacturing depends on experienced personnel.
- The dialysate concentration will decrease with increasing flow rate and vice versa since the relative recovery of substances is flow dependent. However, if very low flow rates are used, the time resolution can be compromised.

- Analytical procedures may require extensive modifications prior to in vivo experiments.
- Both drug concentrations in the tissue as well as recovery by microdialysis sampling are known to be influenced by blood flow and this must be considered in the planning stage of an experiment.
- The duration of a cutaneous microdialysis experiment is in many instances limited for practical reasons, and this creates a limitation for slowly penetrating substances (this can be overcome by the use of portable pumps, which in turn typically have relatively few syringe spaces).
- As a consequence of the above-mentioned relationships between topical dose, dermal concentration, and the ensuing concentration in the microdialysates, microdialysis sampling in the skin will hardly ever be the right method for toxicological studies of low dose/real life skin exposures—other methods will be more relevant (Cross et al. 2007).

7.8 Regulatory Perspective

Present FDA requirements for bioequivalence determination of topical drug products depend on the pharmacological class of the dosage form. The blanching assay using a chromameter has been recommended for glucocorticoid dosage forms. All other topical drug products were first suggested to be determined using the dermatopharmacokinetic (so-called tape-stripping) methodology (DPK) (US-FDA 1998). The FDA subsequently withdrew the draft guidance and the current determination of bioequivalence relies on comparative clinical trials between the generic drug and the innovator drug. As clinical trials often require several hundreds of participants this is therefore a very time-consuming and expensive procedure (Shah et al. 1998; Lionberger 2008).

The FDA launched the "Critical Path Initiative" that focused on the challenges involved in the development of new drugs in 2004. Concern was raised regarding the stagnating and declining innovation of biomedical products due to the continuing increase in costs of medical product development and called upon new knowledge to improve predictability and efficiency along the critical path from laboratory concept to commercial product (US-FDA 2004).

The FDA has emphasized the value and importance of obtaining drug concentrations in the human living tissue and has recently been supportive of the use of microdialysis in humans to achieve this kind of information. The FDA has been receptive to microdialysis data and has used microdialysis data in both pre- and post approval situations and as part of an overall preclinical and clinical pharmacology setting, as an addition to in vivo bioavailability experiments. Furthermore, microdialysis has been recommended to pharmaceutical companies by the FDA as a way to investigate specific safety topics pertaining to systemic drug delivery of drugs intended for topical application and action (Chaurasia et al. 2007).

Microdialysis sampling is expected to become central in the development of new drugs as well as line extensions, and is expected to gain even more recognition among the different regulatory authorities in the years to come.

7.9 Future Research

A key issue in the development of a standardized protocol is the reproducible insertion of the probe at a consistent depth within the skin. Reproducible probe insertion is a skill which is only improved with practice. It is evident that implantation in either the superficial dermis or the subcutaneous tissue will affect the data collected. A thorough evaluation of this methodological issue, preferably studied under human in vivo conditions, using drugs of varying MW and lipophilicity, is needed.

Most microdialysis studies are of limited time duration, typically less than 8 h, and histological studies of the skin response to probe implantation have not showed signs of tissue inflammation (Sandberg et al. 2008). Studies with more extended observation and sampling periods have, however, demonstrated that infiltration of lymphocytes and even development of scar tissue may be observed over time (Wennberg et al. 2000; Sandberg et al. 2008).

Another concern, particularly relevant to cutaneous microdialysis, is the use of microdialysis in studies of inflammatory and immune-mediated diseases. Here, the molecules of interest will very often be the same as those generated during the insertion trauma and subsequent wound repair. Care should be taken to incorporate the proper controls in order to confirm the correct relation between molecule, implantation trauma, and disease process (Stenken et al. 2010). Studies of the exact relation between probe implantation (depth, diameter, time since implantation) and the ensuing tissue damage will be important in the further development of the microdialysis methodology, also from an ethical point of view.

In vitro drug penetration studies using static or flow-through diffusion cells have been extensively used in the past, and there is a pertinent need to establish a correlation between results from in vitro penetration studies of topical drugs and in vivo data obtained by microdialysis methodology.

Only few microdialysis studies in diseased skin have been performed. An improved knowledge of the impact of the structural changes in diseased or otherwise impaired skin on topical drug penetration in vivo will improve the development of topical therapies. Furthermore, the relevance is supported by the fact that a substantial fraction of the general population suffer from different skin conditions that make the skin barrier potentially more permeable—an aspect relevant also to the occupational setting (Fulzele et al. 2007).

There is a significant potential for microdialysis methodology in drug research, preclinical test phases, and clinical studies as well as for use in industry for regulatory approval purposes. Whether the method is employed for sampling of pharmacokinetic or pharmacodynamic information, studies of inflammation, ischemia or homeostasis, microdialysis in the skin as a technique offers a unique opportunity for real-time chronological sampling in the target tissue.

References

Anderson C, Andersson T, Molander M (1991) Ethanol absorption across human skin measured by in vivo microdialysis technique. Acta Derm Venereol 71(5):389–393

Anderson C, Andersson T, Andersson RG (1992) In vivo microdialysis estimation of histamine in human skin. Skin Pharmacol 5(3):177–183

Anderson C, Andersson T, Wardell K (1994) Changes in skin circulation after insertion of a microdialysis probe visualized by laser Doppler perfusion imaging. J Invest Dermatol 102(5):807–811

Au WL, Skinner MF, Benfeldt E, Verbeeck RK, Kanfer I (2012) Application of dermal microdialysis for the determination of bioavailability of clobetasol propionate applied to the skin of human subjects. Skin Pharmacol Physiol 25(1):17–24

Ault JM, Lunte CE, Meltzer NM, Riley CM (1992) Microdialysis sampling for the investigation of dermal drug transport. Pharm Res 9(10):1256–1261

Ault JM, Riley CM, Meltzer NM, Lunte CE (1994) Dermal microdialysis sampling in vivo. Pharm Res 11(11):1631–1639

Benfeldt E (1999) In vivo microdialysis for the investigation of drug levels in the dermis and the effect of barrier perturbation on cutaneous drug penetration. Studies in hairless rats and human subjects. Acta Derm Venereol Suppl (Stockh) 206:1–59

Benfeldt E, Groth L (1998) Feasibility of measuring lipophilic or protein-bound drugs in the dermis by in vivo microdialysis after topical or systemic drug administration. Acta Derm Venereol 78(4):274–278

Benfeldt E, Serup J (1999) Effect of barrier perturbation on cutaneous penetration of salicylic acid in hairless rats: in vivo pharmacokinetics using microdialysis and non-invasive quantification of barrier function. Arch Dermatol Res 291(9):517–526

Benfeldt E, Serup J, Menne T (1999) Effect of barrier perturbation on cutaneous salicylic acid penetration in human skis: in vivo pharmacokinetics using microdialysis and non-invasive quantification of barrier function. Br J Dermatol 140(4):739–748

Benfeldt E, Hansen SH, Volund A, Menne T, Shah VP (2007) Bioequivalence of topical formulations in humans: evaluation by dermal microdialysis sampling and the dermatopharmacokinetic method. J Invest Dermatol 127(1):170–178

Bielecka-Grzela S, Klimowicz A (2003) Evaluation of ofloxacin penetration into the skin after a single oral dose assessed by cutaneous microdialysis. Pol J Pharmacol 55(4):613–618

Borg N, Gotharson E, Benfeldt E, Groth L, Stahle L (1999) Distribution to the skin of penciclovir after oral famciclovir administration in healthy volunteers: comparison of the suction blister technique and cutaneous microdialysis. Acta Derm Venereol 79(4):274–277

Brunner M, Dehghanyar P, Seigfried B, Martin W, Menke G, Muller M (2005) Favourable dermal penetration of diclofenac after administration to the skin using a novel spray gel formulation. Br J Clin Pharmacol 60(5):573–577

Brunner M, Davies D, Martin W, Leuratti C, Lackner E, Muller M (2011) A new topical formulation enhances relative diclofenac bioavailability in healthy male subjects. Br J Clin Pharmacol 71(6):852–859. doi:10.1111/j.1365-2125.2011.03914.x

Carneheim C, Stahle L (1991) Microdialysis of lipophilic compounds—a methodological study. Pharmacol Toxicol 69(5):378–380

Chaturvedula A, Joshi DP, Anderson C, Morris R, Sembrowich WL, Banga AK (2005) Dermal, subdermal, and systemic concentrations of granisetron by iontophoretic delivery. Pharm Res 22(8):1313–1319

Chaurasia CS, Muller M, Bashaw ED, Benfeldt E, Bolinder J, Bullock R, Bungay PM, DeLange EC, Derendorf H, Elmquist WF, Hammarlund-Udenaes M, Joukhadar C, Kellogg DL Jr, Lunte CE, Nordstrom CH, Rollema H, Sawchuk RJ, Cheung BW, Shah VP, Stahle L, Ungerstedt U, Welty DF, Yeo H (2007) AAPS-FDA workshop white paper: microdialysis principles, application and regulatory perspectives. Pharm Res 24(5):1014–1025

Cross SE, Innes B, Roberts MS, Tsuzuki T, Robertson TA, McCormick P (2007) Human skin penetration of sunscreen nanoparticles: in vitro assessment of a novel micronized zinc oxide formulation. Skin Pharmacol Physiol 20(3):148–154

Dancik Y, Anissimov YG, Jepps OG, Roberts MS (2012) Convective transport of highly plasma protein bound drugs facilitates direct penetration into deep tissues after topical application. Br J Clin Pharmacol 73(4):564–578

Fulzele SV, Babu RJ, Ahaghotu E, Singh M (2007) Estimation of proinflammatory biomarkers of skin irritation by dermal microdialysis following exposure with irritant chemicals. Toxicology 237(1–3):77–88

Groth L, Serup J (1998) Cutaneous microdialysis in man: effects of needle insertion trauma and anaesthesia on skin perfusion, erythema and skin thickness. Acta Derm Venereol 78(1):5–9

Hegemann L, Forstinger C, Partsch B, Lagler I, Krotz S, Wolff K (1995) Microdialysis in cutaneous pharmacology: kinetic analysis of transdermally delivered nicotine. J Invest Dermatol 104(5):839–843

Holmgaard R, Nielsen JB, Benfeldt E (2010) Microdialysis sampling for investigations of bioavailability and bioequivalence of topically administered drugs: current state and future perspectives. Skin Pharmacol Physiol 23(5):225–243

Holmgaard R, Benfeldt E, Nielsen JB, Gatschelhofer C, Sorensen JA, Höfferer C, Bodenlenz M, Pieber TR, Sinner F (2012) Comparison of open-flow microperfusion and microdialysis methodologies when sampling topically applied fentanyl and benzoic acid in human dermis ex vivo. Pharm Res 29(7):1808–1820

Holmgaard R, Benfeldt E, Bangsgaard N, Sorensen JA, Brosen K, Nielsen F, Nielsen JB (2012b) Probe depth matters in dermal microdialysis sampling of topical penetration. An ex vivo study in human skin. Skin Pharmacol Physiol 25(1):9–16

Incecayir T, Agabeyoglu I, Derici U, Sindel S (2011) Assessment of topical bioequivalence using dermal microdialysis and tape stripping methods. Pharm Res 28(9):2165–2175. doi:10.1007/s11095-011-0444-3

Joukhadar C, Frossard M, Mayer BX, Brunner M, Klein N, Siostrzonek P, Eichler HG, Muller M (2001) Impaired target site penetration of beta-lactams may account for therapeutic failure in patients with septic shock. Crit Care Med 29(2):385–391

Kellogg DL, Zhao JL, Wu Y (2008) Neuronal nitric oxide synthase control mechanisms in the cutaneous vasculature of humans in vivo. J Physiol (London) 586(3):847–857

Klimowicz A, Bielecka-Grzela S, Tomaszewska U (2002) A simple and rapid liquid chromatographic method for the determination of metronidazole and its hydroxymetabolite in plasma and cutaneous microdialysates. Acta Pol Pharm 59(5):327–331

Klimowicz A, Farfal S, Bielecka-Grzela S (2007) Evaluation of skin penetration of topically applied drugs in humans by cutaneous microdialysis: acyclovir vs. salicylic acid. J Clin Pharm Ther 32(2):143–148

Kreilgaard M (2001) Dermal pharmacokinetics of microemulsion formulations determined by in vivo microdialysis. Pharm Res 18(3):367–373

Kreilgaard M (2002) Influence of microemulsions on cutaneous drug delivery. Adv Drug Deliv Rev 54(Suppl 1):S77–S98

Kreilgaard M, Kemme MJ, Burggraaf J, Schoemaker RC, Cohen AF (2001) Influence of a microemulsion vehicle on cutaneous bioequivalence of a lipophilic model drug assessed by microdialysis and pharmacodynamics. Pharm Res 18(5):593–599

Krogstad AL, Jansson PA, Gisslen P, Lonnroth P (1996) Microdialysis methodology for the measurement of dermal interstitial fluid in humans. Br J Dermatol 134(6):1005–1012

Lionberger RA (2008) FDA critical path initiatives: opportunities for generic drug development. AAPS J 10(1):103–109

Mathy FX, Lombry C, Verbeeck RK, Preat V (2005) Study of the percutaneous penetration of flurbiprofen by cutaneous and subcutaneous microdialysis after iontophoretic delivery in rat. J Pharm Sci 94(1):144–152

McCleverty D, Lyons R, Henry B (2006) Microdialysis sampling and the clinical determination of topical dermal bioequivalence. Int J Pharm 308(1–2):1–7

Morgan CJ, Renwick AG, Friedmann PS (2003) The role of stratum corneum and dermal microvascular perfusion in penetration and tissue levels of water-soluble drugs investigated by microdialysis. Br J Dermatol 148(3):434–443

Muller M, Schmid R, Wagner O, Vonosten B, Shayganfar H, Eichler HG (1995) In vivo characterization of transdermal drug transport by microdialysis. J Controlled Release 37(1–2):49–57

Muller M, Mascher H, Kikuta C, Schafer S, Brunner M, Dorner G, Eichler HG (1997) Diclofenac concentrations in defined tissue layers after topical administration. Clin Pharmacol Ther 62(3):293–299

Ortiz PG, Hansen SH, Shah VP, Menne T, Benfeldt E (2008) The effect of irritant dermatitis on cutaneous bioavailability of a metronidazole formulation, investigated by microdialysis and dermatopharmacokinetic method. Contact Dermatitis 59(1):23–30

Ortiz PG, Hansen SH, Shah VP, Menne T, Benfeldt E (2009) Impact of adult atopic dermatitis on topical drug penetration: assessment by cutaneous microdialysis and tape stripping. Acta Derm Venereol 89(1):33–38

Ortiz PG, Hansen SH, Shah VP, Sonne J, Benfeldt E (2011) Are marketed topical metronidazole creams bioequivalent? Evaluation by in vivo microdialysis sampling and tape stripping methodology. Skin Pharmacol Physiol 24(1):44–53

Petersen LJ (1998) Measurement of histamine release in intact human skin by microdialysis technique. Clinical and experimental findings. Dan Med Bull 45(4):383–401

Petersen LJ, Skov PS, Bindslev-Jensen C, Sondergaard J (1992) Histamine release in immediate-type hypersensitivity reactions in intact human skin measured by microdialysis. A preliminary study. Allergy 47(6):635–637

Sandberg C, Halldin CB, Ericson MB, Larko O, Krogstad AL, Wennberg AM (2008) Bioavailability of aminolaevulinic acid and methylaminolaevulinate in basal cell carcinomas: a perfusion study using microdialysis in vivo. Br J Dermatol 159(5):1170–1176

Schmidt S, Banks R, Kumar V, Rand KH, Derendorf H (2008) Clinical microdialysis in skin and soft tissues: an update. J Clin Pharmacol 48(3):351–364

Seki T, Wang A, Yuan D, Saso Y, Hosoya O, Chono S, Morimoto K (2004) Excised porcine skin experimental systems to validate quantitative microdialysis methods for determination of drugs in skin after topical application. J Controlled Release 100(2):181–189

Shah VP, Flynn GL, Yacobi A, Maibach HI, Bon C, Fleischer NM, Franz TJ, Kaplan SA, Kawamoto J, Lesko LJ, Marty JP, Pershing LK, Schaefer H, Sequeira JA, Shrivastava SP, Wilkin J, Williams RL (1998) Bioequivalence of topical dermatological dosage forms—methods of evaluation of bioequivalence. Pharm Res 15(2):167–171

Shinkai N, Korenaga K, Takizawa H, Mizu H, Yamauchi H (2008a) Percutaneous penetration of felbinac after application of transdermal patches: relationship with pharmacological effects in rats. J Pharm Pharmacol 60(1):71–76

Shinkai N, Korenaga K, Mizu H, Yamauchi H (2008b) Intra-articular penetration of ketoprofen and analgesic effects after topical patch application in rats. J Controlled Release 131(2):107–112

Shinkai N, Korenaga K, Okumura Y, Mizu H, Yamauchi H (2011) Microdialysis assessment of percutaneous penetration of ketoprofen after transdermal administration to hairless rats and domestic pigs. Eur J Pharm Biopharm 78(3):415–421. doi:10.1016/j.ejpb.2011.03.005

Simonsen L, Jorgensen A, Benfeldt E, Groth L (2004) Differentiated in vivo skin penetration of salicylic compounds in hairless rats measured by cutaneous microdialysis. Eur J Pharm Sci 21(2–3):379–388

Stagni G, Shukla C (2003) Pharmacokinetics of methotrexate in rabbit skin and plasma after IV-bolus and iontophoretic administrations. J Controlled Release 93(3):283–292

Stagni G, O'Donnell D, Liu YJ, Kellogg DL, Morgan T, Shepherd AM (2000) Intradermal microdialysis: kinetics of iontophoretically delivered propranolol in forearm dermis. J Controlled Release 63(3):331–339

Stagni G, Ali ME, Weng D (2004) Pharmacokinetics of acyclovir in rabbit skin after IV-bolus, ointment, and iontophoretic administrations. Int J Pharm 274(1–2):201–211

Stenken JA, Church MK, Gill CA, Clough GF (2010) How minimally invasive is microdialysis sampling? A cautionary note for cytokine collection in human skin and other clinical studies. AAPS J 12(1):73–78. doi:10.1208/s12248-009-9163-7

Sun L, Stenken JA (2003) Improving microdialysis extraction efficiency of lipophilic eicosanoids. J Pharmaceut Biomed Anal 33(5):1059–1071

Tegeder I, Muth-Selbach U, Lotsch J, Rusing G, Oelkers R, Brune K, Meller S, Kelm GR, Sorgel F, Geisslinger G (1999) Application of microdialysis for the determination of muscle and subcutaneous tissue concentrations after oral and topical ibuprofen administration. Clin Pharmacol Ther 65(4):357–368

Tegeder I, Brautigam L, Podda M, Meier S, Kaufmann R, Geisslinger G, Grundmann-Kollmann M (2002) Time course of 8-methoxypsoralen concentrations in skin and plasma after topical (bath and cream) and oral administration of 8-methoxypsoralen. Clin Pharmacol Ther 71(3):153–161

Tettey-Amlalo RN, Kanfer I, Skinner MF, Benfeldt E, Verbeeck RK (2009) Application of dermal microdialysis for the evaluation of bioequivalence of a ketoprofen topical gel. Eur J Pharm Sci 36(2–3):219–225

Tsai JC, Shen LC, Sheu HM, Lu CC (2003) Tape stripping and sodium dodecyl sulfate treatment increase the molecular weight cutoff of polyethylene glycol penetration across murine skin. Arch Dermatol Res 295(4):169–174

US-FDA (1998) Guidance for industry: topical dermatological drug product NDAs and ANDAs-in vivo bioavilability, bioequivalence, in vitro release, and associated studies. Draft Guidance, June 1998, U.S. Department of Health and Human Services, Food and Drug Administration, Center for Drug Evaluation and Research (CDER)

US-FDA (2002) Guidance for industry bioavailability and bioequivalence studies for orally administered drug products—general considerations. http://www.fda.gov/cder/guidance/index.htm

US-FDA (2004) Challenge and opportunity on the critical path to new medicinal products

Ward KW, Medina SJ, Portelli ST, Doan KMM, Spengler MD, Ben MM, Lundberg D, Levy MA, Chen EP (2003) Enhancement of in vitro and in vivo microdialysis recovery of SB-265123 using intralipid (R) and encapsin (R) as perfusates. Biopharm Drug Dispos 24(1):17–25

Wennberg AM, Larko O, Lonnroth P, Larson G, Krogstad AL (2000) Delta-aminolevulinic acid in superficial basal cell carcinomas and normal skin—a microdialysis and perfusion study. Clin Exp Dermatol 25(4):317–322

Chapter 8
Translational Approaches for Predicting CNS Drug Effects Using Microdialysis

Elizabeth de Lange

Abstract CNS drug efficacy is largely dependent on target site kinetics of the unbound drug that may be strongly influenced by transport across the blood–brain barrier (BBB) and intracerebral distribution. It is, therefore, essential to have information on the unbound CNS target site pharmacokinetics, as this may distinctively differ from (unbound) plasma pharmacokinetics. Microdialysis studies on morphine, M3G, and M6G have shown the impact of different conditions on BBB transport and/or PK–PD relationships. Such heterogeneity in BBB transport and PK–PD relationships should be encompassed in the development of translational models that include specific expressions of the processes on the causal path between drug administration and CNS drug effects. An example on the development of a preclinical data-based translational model on dopaminergic inhibition is presented showing that microdialysis may provide critical and quantitative information on rate and extent of mechanisms between drug administration and CNS effects of a drug in different settings, which, combined with PK–PD modeling approaches, serves as the basis for generic translational models for prediction of CNS effects in varying conditions.

8.1 Introduction

Knowledge on transport into the brain is important for drugs that directly act on targets in the central nervous system (CNS) (Westerhout et al. 2011) such as anticonvulsants, antidepressants, anesthetics, antibiotics, and antinociceptive, and

E. de Lange (✉)
Division of Pharmacology, Leiden/Amsterdam Center for Drug Research,
Leiden University, Leiden, The Netherlands
e-mail: lange_l@lacdr.leidenuniv.nl

M. Müller (ed.), *Microdialysis in Drug Development*, AAPS Advances in the Pharmaceutical Sciences Series, DOI: 10.1007/978-1-4614-4815-0_8,

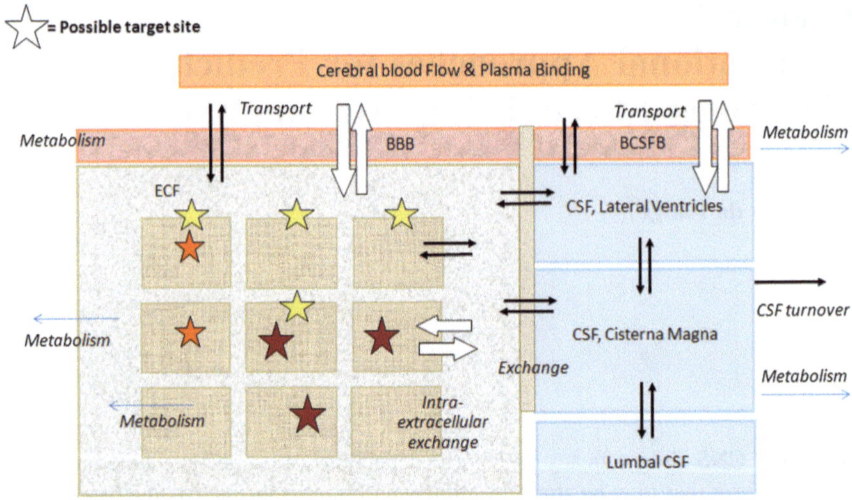

Fig. 8.1 Schematic representation of blood and brain compartments and routes for exchange of drugs (passive transport, *black arrows*; active transport (*big white arrows*). The *stars* represent possible target sites

anticancer agents. The intensity and duration of the effects of a drug depend on the unbound drug kinetics at the target site (Breimer and Danhof 1997; Welty et al. 1993). Many processes after drug administration play a role in the final CNS target site kinetics (Fig. 8.1), such as plasma pharmacokinetics, the rate and extent of blood–brain barrier (BBB) transport and intracerebral distribution, and brain elimination (Hammarlund-Udenaes et al. 1997; de Lange and Danhof 2002; Liu et al. 2005; Collins and Dedrick 1983; Wang and Welty 1996; Syvänen et al. 2012).

- **Plasma pharmacokinetics**—After administration a drug will distribute into and out of body compartments. It may reside in the unbound form or associated with binding sites, in plasma (e.g. plasma proteins) and binding sites in tissues. These binding processes vary in rate and extent, and together with elimination processes determine the time-dependent unbound drug concentration in plasma available for brain distribution and target interaction ("free-drug hypothesis") (Jolliet et al. 1997; de Rick et al. 1987; Lin et al. 1987; van Steeg et al. 2009).
- **BBB transport**—The BBB is located at the brain capillary endothelium and is characterized by the presence of narrow tight junctions, a low rate of vesicular transport, lack of fenestrations and intercellular clefts, and a continuous basal lamina. The BBB prevents direct contact between brain and the blood by tight junctions that connect the endothelial cells of the brain vessels. In addition, a barrier function is created by specific metabolizing enzymes and efflux transporters, such as P-glycoprotein located within the endothelial cells, that may actively remove drugs from the brain. Therefore, the BBB can play a major role in limiting the delivery of systemically administered drugs to the CNS

(Abott and Revest 1991; Rubin and Staddon 1999; Cornford 1985; Brosman and Claudio 1998; Begley 2004; Tamai et al. 2000; Cordon-Cardo et al. 1989; Schinkel et al. 1995; de Boer et al. 2003; Zhang et al. 2004).

- **Intracerebral distribution**—The brain has many anatomical structures with different characteristics. The main compartments are the brain extracellular fluid (ECF), the brain intracellular space, and the cerebrospinal fluid (CSF). After passage of the BBB, a drug enters the brain ECF. Then, extra-intracellular exchange of the drug will be determined by passive diffusion as well as by functionality of transporters in the brain parenchyma. (Lee et al. 2001; Scism et al. 1997; Suzuki et al. 1997). Also, the kinetics of drug exchange between brain ECF and CSF should also be considered (Cserr 1971, 1984; Williams et al. 1995; Fenstermacher et al. 1974; Del Bigio 1995; Malhotra et al. 1994; Aird 1984), as well as the contribution of drug transport across the blood–CSF barrier (BCSFB) (Wijnholds et al. 2000; de Lange 2004; Segal 1998).

Finally, it should be realized that all these pharmacokinetic processes result from interaction between the drug and the biological system (Danhof et al. 2007), and the rate and extent of these processes may differ substantially between different drugs having different physico-chemical properties (Oldendorf 1974). This indicates that it is important to strictly distinguish between pharmacokinetic (PK) and pharmacodynamic (PD) processes by having information on CNS target site kinetics, and that we therefore should have information on the unbound drug concentration in brain (de Lange et al. 2005; Westerhout et al. 2011). The distinction between biological system and drug characteristics should be made clear by studying different drugs in the same setting, or to use the same drug in different settings (e.g., by drug transporter inhibition or disease condition).

In many cases, CNS targets are membrane bound receptors facing the brain ECF, or enzymes within the brain ECF. This makes information on brain ECF concentrations highly valuable (de Lange and Danhof 2002; Watson et al. 2009; Jeffrey and Summerfield 2010; Westerhout et al. 2011). For intracellular targets, obtaining in vivo information is more complicated. There are no means to directly monitor brain intracellular concentration–time profiles. At best it can be derived by combining different experimental approaches (Hammarlund-Udenaes et al. 2008; Fridén et al. 2007). Moreover, the intracellular space is highly heterogeneous. Though not being the ideal information, data on brain cell exposure of the unbound drug in the brain ECF seems to remain relevant to date.

Intracerebral microdialysis is widely considered to be the best technique to monitor the concentrations of compounds in brain ECF over time (Cremers et al. 2009). Combination with simultaneous serial blood sampling from the same animal is a powerful approach to study PK properties related to BBB transport and intracerebral distribution mechanisms (de Lange et al. 1994, 1997). It has been applied in numerous brain distribution studies, and below examples of studies are given to demonstrate its added value to unravel processes between plasma PK and target site concentrations.

In this chapter, a number of microdialysis studies on a selection of opioids are presented for those being very illustrative on the impact of BBB transport and intracerebral distribution and/or on the CNS effects (Letrent et al. 1998, 1999; Stain-Texier et al. 1999; Bouw et al. 2000, 2001a, b; Tunblad et al. 2003, 2004, 2005; Groenendaal et al. 2007a, b; Bengtsson et al. 2009; Ederoth et al. 2004; Xie et al. 2000). Moreover, these studies have been performed for different drugs and dosages, in different conditions, and in different subjects, clearly showing that BBB transport differs in rate and extent in the different settings. When it comes to prediction of CNS drug effects, this heterogeneity/variability is exactly what needs to be addressed. To that end, experimental and modeling approaches for the development of predictive PKPD translational models is needed (Danhof et al. 2007). Finally, the contribution of rat intracerebral microdialysis data in the development of a mechanism-based translational model on dopaminergic inhibition with prolactin concentrations in blood as a PD marker, able to predict human CNS effects, is demonstrated (Stevens et al. 2011, 2012).

8.2 Heterogeneity and Variability in CNS Target Site PK (and PD)

Living mammals share mostly the same biological mechanisms, which is the basis for the use of interspecies extrapolation in drug development. However, the problem lies within the heterogeneity of their rates and extents in the causal path between drug administration and CNS effects. This explains why the same dose in different subjects (/under different conditions) may result in different effects. Heterogeneity is caused by differences in genetic background, species, gender, tissue, age, diet, (pathologic) conditions, drug treatment, etc. (Karssen et al. 2001; Kooij et al. 2010; de Lange et al. 2005; Mulder et al. 2001; Ravenstijn et al. 2007, 2012). It is therefore surprising that, in most cases, still, the dose-effect or at best the plasma-effect relationships are used for extrapolation. In this section, a number of microdialysis studies on morphine and its main metabolites are selected because those are very illustrative for showing the impact of conditions that underlie the differences in BBB transport and intracerebral distribution and/or CNS effects, by differences in transporter functionality, intracerebral distribution, species, age, and disease conditions (Tables 8.1, 8.2, and 8.3).

Morphine (Table 8.1). Because data were available to indicate that the anesthetic effects of morphine could be modulated by Pgp inhibition, Letrent et al. (1999) studied the brain distribution and antinociceptive effects of morphine with microdialysis in the rat. Upon co-administration of the Pgp inhibitor GF120918, the half-life of morphine in the brain increased about threefold, while also the antinociceptive effect by morphine was increased. These effects appeared to be associated with modulation of the BBB transport of morphine. Stain-Texier et al. (1999) studied the brain distribution of morphine by microdialysis and also used

Table 8.1 Investigations in different conditions on morphine distribution into and within the brain and/or its effects

Opioid	Parameter	Condition	Mice	Rat	Pig	Sheep	Human
Morphine	BBB transport	Control	Xie et al. (1999)	Bouw et al. (2000); Groenendaal et al. (2007a, b)		Bengtsson et al. (2009)	
		Active transport inhibition	Xie et al. (1999)	Letrent et al. (1999); Tunblad et al. (2003); Groenendaal et al. (2007a, b)			
		Disease			Tunblad et al. (2004)	Bengtsson et al. (2009)	Ederoth et al. (2004)
		Age				Bengtsson et al. (2009)	
	Extra-intracellular distribution	Control		Stain-Texier et al. (1999)			
	PKPD	Control		Bouw et al. (2000); Groenendaal et al. (2007a, b)			
		Active transport inhibition		Letrent et al. (1999); Groenendaal et al. (2007a, b)			

Table 8.2 Investigations in different conditions on morphine-3-glucuronide (M3G) distribution into the brain

Opioid	Parameter	Condition	Mice	Rat	Pig	Sheep	Human
M3G	BBB transport	Control		Xie et al. (2000)		Bengtsson et al. (2009)	
		Active transport inhibition		Xie et al. (2000)			
		Disease				Bengtsson et al. (2009)	
		Age				Bengtsson et al. (2009)	

Table 8.3 Investigations in different conditions on morphine-6-glucuronide (M6G) distribution into and within the brain

Opioid	Parameter	Condition	Mice	Rat	Pig	Sheep	Human
M6G	BBB transport	Control		Bouw et al. (2001a, b); Tunblad et al. (2005)			
		Active transport inhibition		Tunblad et al. (2005)			
	Extra-intracellular distribution			Stain-Texier et al. (1999); Bouw et al. (2001a, b)			

brain samples for determination of total brain concentrations to have information on brain ECF versus intracellular distribution. It was found that morphine brain intracellular levels were approximately four times higher than brain ECF levels.

Effect of active transport mechanisms on BBB transport of morphine was further studied by Tunblad et al. (2003), using co-administration of probenecid. This increased the AUCBrainECF/AUCblood, unbound ratios (Kp,uu; Hammarlund-Udenaes et al. 2008) from ~0.3 to ~0.4, while the elimination half-life of brain ECF concentrations increased from ~60 to ~120 min. This clearly indicated that morphine is a substrate for probenecid-sensitive transporter(s) at the BBB.

The contribution of BBB transport of morphine to the delay in antinociceptive effect was assessed in a quantitative manner first by Bouw et al. (2000, 2001a, b) by compartmental PK–PD modeling. Morphine was administered intravenously in the rat, in medium and high dose. The Kp,uu of morphine was ~0.25, and showed some dose dependency. This ratio being lower than unity indicates the involvement of active efflux of morphine across the BBB. The PK of morphine was related to the anesthetic effect-time profiles that displayed a clear delay relative to the blood concentration-time profile. The delay had a half-life of about 30 min. Microdialysis data revealed that this delay was to a large extent determined by slow BBB transport, leaving a delay of only 5 min between brain ECF concentrations and the anesthetic effect. This means that the BBB transport of morphine has a significant impact on the effect delay toward the antinociceptive effect.

Groenendaal et al. (2007a, b) further characterized the nonlinear distribution kinetics of morphine in rat brain in a strict quantitative and dose-independent manner. Using three different dosages of morphine, a broad range on plasma concentrations as well as the resulting brain ECF concentrations was obtained. The Pgp inhibitor GF120918 (elacridar) was used to investigate the contribution of Pgp-mediated efflux from the brain. A PK population distribution model was developed. Non-linear distribution kinetics in brain ECF was observed and separate expressions were identified for passive diffusion, active saturable BBB influx as well as BBB active efflux in terms of capacity and velocity. Thus, it was found that brain distribution of morphine is determined by three factors (1) limited passive diffusion (2) active efflux, reduced by ~50 % by Pgp inhibition, and (3) low capacity active influx. This implies that morphine is prone to blood concentration-dependency and sensitivity to drug–drug interactions.

For morphine, further, the influence of target site distribution including P-glycoprotein (Pgp) function on the PK–PD correlations of morphine's actions in rat brain was investigated, using the EEG effects as the PD endpoint. (Groenendaal et al. 2007a, b) A low, medium, and high dose of morphine was studied, with the low dose in absence and presence of the Pgp inhibitor elacridar. The delay between morphine blood concentrations and its effects on the EEG was profound, and dose dependent. The effects of elacridar on the distribution kinetics of morphine in the "effect compartment" were consistent with the distribution in brain ECF as estimated by intracerebral microdialysis Groenendaal et al. (2007a, b). However, the time course of morphine concentrations at the site of action in the brain, as deduced from the mathematical EEG target site model, was distinctly different from the brain ECF concentrations, indicating that the PK–PD relationship of morphine cannot yet be fully understood and therefore not predicted.

BBB transport of morphine was also studied in diseased brain following experimentally induced meningitis in piglets (Tunblad et al. 2004). After a control period, meningitis was induced by lipopolysaccharide injection in Cisterma Magna. When meningitis was established, the morphine experiment was repeated. The The Kp,uu of morphine was ~0.45 during the control period, and ~0.95 during meningitis. The increase in the Kp,uu of morphine during meningitis implied a decreased active efflux and an increased passive diffusion of morphine over the BBB. The elimination half-life of morphine in brain was longer than in blood during both periods, and was unaffected by meningitis. These data showed that the morphine distribution into the brain is significantly increased during meningitis as compared with the control situation.

The effect of age on the distribution of morphine across the BBB was studied in a sheep model in which also the effect of neonatal asphyxia on brain drug distribution was studied Bengtsson et al. (2009). Premature lambs and ewes were compared. Morphine Kp,uu was 1.2 and 1.9 for the sheep and premature lambs, respectively, indicating that active influx into the brain decreases with age. Induced asphyxia did not affect transport of morphine or M3G across the BBB. Morphine unbound volume of distribution of unbound (Vu,brain) measurements were higher in sheep than in premature lambs. The morphine Kp,uu was above unity, indicating active transport into the brain; influx was significantly higher in premature lambs than in adult sheep. These results in sheep differ from those in humans, rats, mice, and pigs where a net efflux of morphine from the brain is observed.

Ederoth et al. (2004) hypothesized that the active efflux of morphine from brain to blood would also occur in the human brain, and that brain injury would alter BBB transport and intracerebral distribution of morphine. BBB transport characteristics of morphine were studied in patients with traumatic brain injury with microdialysis probes inserted in "better brain tissue" and in "worse brain tissue", referring to the degree of traumatic injury. It was found that the Kp,uu was $\sim$0.65 in the "better brain tissue", while being $\sim$80 % in the "worse brain tissue". Also, increase in the rate of BBB transport from "better brain tissue" to "worse brain tissue" was indicated.

Thus, clear differences in brain distribution of morphine were demonstrated to be seen upon transport inhibition, by disease conditions, and between species. Also, CNS effects of morphine were condition dependent.

Morphine-3-glucuronide (M3G, Table 8.2). Xie et al. (2000) studied the BBB transport characteristics of M3G in the rat, absence and presence of probenecid, using a crossover experimental design. Normal Kp,uu value of M3G was about $\sim$0.08, and increased about twofold upon co-administration of probenecid. Bengtsson et al. (2009) further investigated for M3G the effect of age and neonatal asphyxia on brain distribution in a sheep model. Induced asphyxia in the premature lambs did not affect transport of M3G across the BBB. The M3G Kp,uu values were 0.27 and 0.17 in sheep and premature lambs, indicating a net efflux from the brain in both groups. All together these studies indicate that M3G differs in brain distribution between rat and sheep, and that in sheep there is a net increase in brain distribution with age, without an effect of neonatal asphyxia.

Morphine-6-glucuronide (M6G, Table 8.3) The other morphine metabolite, morphine-6-glucuronide (M6G) is considered more potent than morphine. Stain-Texier et al. (1999) studied the brain distribution of M6G by microdialysis in the rat and also used brain samples for determination of total brain concentrations to have information on brain ECF versus intracellular distribution. The M6G brain ECF concentrations were about 125-fold higher than the calculated intracellular levels. Thus, it was shown that M6G is almost exclusively distributed into brain ECF, which is highly favorable for exposing the opioid receptors.

The quantitative PKPD relationship of M6G for its antinociceptive action was further investigated by Bouw et al. (2001a, b). The half-life of the BBB transport rate for M6G in the rat was about 25 min in arterial and in venous blood and about 60 min in brain ECF. The Kp,uu value was $\sim$0.25, being below unity, and indicate active efflux being involved in BBB transport of M6G. The unbound volume of distribution was $\sim$20 (V/W) % of the brain that actually corresponds with the ECF space. For M6G about 50 % of the half-life in delay between blood and antinociceptive effect could be attributed to slow transport across the BBB.

Also, for M6G the impact of probenecid on BBB transport was investigated (Tunblad et al. 2005). The Kp,uu value for controls was about 0.3. This value being lower than unity indicates the presence of efflux transport of M6G at the level of the BBB. However, BBB efflux did not appear to be probenecid sensitive, as probenecid only decreased the systemic elimination of M6G.

Taken together these studies show considerable differences in CNS distribution and/or effects depending on condition and/or species. In order to be predictive there is a need for the development of translational PK–PD models.

8.3 Approaches for Development of Predictive Translational PK–PD Models

For a proper CNS effect the drug should have the ability to access the relevant target site within the CNS (de Lange et al. 2005). As already indicated, apart from BBB permeability and functionality, this also depends on other factors such as plasma PKs and intrabrain distribution. These factors are controlled by many mechanisms. Each mechanism has its particular influence by its specific rate and extent, and thereby plays a more or less important role in having the drug at the *right place, right time, and at the right concentration.*

PK–PD modeling typically utilizes the time course of the drug concentration in plasma as a measure of exposure. Commonly, a compartmental model is used to describe the time course of the drug concentration in plasma. This is important since drug concentration versus time profiles can differ widely between drugs and for the same drug, between species and individuals (Danhof et al. 2007). In a compartmental model, drug disposition is characterized as the transfer of drug between interconnected hypothetical compartments, which serves to mimic the drug absorption, distribution, and elimination processes. However, most drugs have their target site in a tissue, rather than plasma. For those drugs, distribution to the site of action may represent a rate-limiting step for producing the biological effect. This is reflected in a delay of the pharmacological effect relative to the drug concentration in plasma (hysteresis) (Welty et al. 1993) and can be incorporated in the model if solely based upon diffusion. However, while useful for descriptive purposes, such an approach bears its limits in that it is not truly mechanistic, and therefore lacks extrapolative and predictive power for more “complex” target site distribution processes, as is the case for many CNS drugs. To improve especially the interspecies extrapolation of PKs, a more physiologically based PK modeling approach should be used for more translational properties. This infers that more advanced experimental designs are needed (Garrido et al. 2000; Gabrielsson and Green 2009; Danhof et al. 2007, 2008; Ploeger et al. 2009).

As biological system mechanisms are *concurrently working*, there is a need for *integrated* in vivo experiments, e.g., that the experiments *address multiple mechanisms* (/biomarkers thereof) *at the same time*. Using animals, we can learn more on the interrelationship of the different PK processes, by performing integrative studies (van der Graaf and Danhof 1997; Gabrielsson and Green 2009; van Steeg et al. 2009; Westerhout et al. 2011) in which variables are systematically varied (e.g., inhibition of an efflux transporter or induction of pathological state, or using a different drug or route of administration). These are so-called integrative cross-compare designed studies. We can thereby dissect contributions of individual mechanisms in animals. This provides the links to the human situation.

In translational models specific expressions are needed that quantitatively characterize processes on the causal path between drug administration and effect. These includes target site distribution, target binding and activation, transduction, PD interactions, and homeostatic feedback mechanisms (Mandema et al. 1991; Cox et al. 1998; van der Graaf and Danhof 1997; van Steeg et al. 2009). Ultimately, also the effects on disease processes and disease progression have to be considered. These can be characterized by biomarkers, and to that end the following classification of biomarkers (Danhof et al. 2005) is very useful.

- **Type 0 biomarkers** refer to the genotype—or phenotype as determinant of the drug response that influences target site exposure or response due to variation in the expression of, e.g., enzymes or receptors. They are commonly used as covariates in PK–PD models.
- **Type 1 biomarkers** refer to drug concentrations in general and at the target site in particular. As previously pointed out, quantitative biomarkers that represent the target site distribution of drugs and metabolites for compounds that act in the CNS are difficult to obtain in man, but readily available in vivo in animals (de Lange et al. 2005).
- **Type 2 biomarkers** refer to the degree of target occupancy. In theory, effects may occur at different degrees of target occupancy and may be species dependent. The relationship between target occupancy and effect is therefore important for the understanding of inter- and intra-individual variability. Information on target occupancy is available by bioassays in vitro and can also be noninvasively measured in humans by positron emission tomography (Kapur et al. 2000; Kvernmo et al. 2008).
- **Type 3 biomarkers** refer to quantification of the target site activation. By means of in vitro bioassays information can be obtained on receptor activation in animal and man. Techniques such as electro-encephalo-grams (Kropf and Kuschinsky 1993; Vorobyov et al. 2003) and functional-magnetic resonance imaging can obtain specific receptor activation in preclinical and clinical in vivo setting.
- **Type 4 biomarkers** refer to physiological measures in the integral biological system, which are often controlled by homeostatic feedback mechanisms. Such measures can, for example, be on pituitary hormones that play a very important role in communication between CNS and periphery (Freeman et al. 2000).
- **Type 5 biomarkers** characterize disease processes and are particularly useful in clinical settings. (However, an important question is whether type 5 biomarkers can be identified in animal models of disease).
- **Type 6 biomarkers** refer to clinical endpoints, such as occurrence of a disease, symptom, sign, or laboratory abnormality that links to target outcomes.

Obtaining information on multiple types of mechanisms (/biomarkers) allows the development of better models and increases its accuracy. It is, however, not always necessary to obtain information on each step/type of biomarker, as the parsimony of the biological system can be applied in mechanism-based PK–PD modeling. But, still, to obtain enough information from the clinical setting is extremely costly, and for clear reasons limited. This makes that we have to make use of in vivo animal studies and apply translational pharmacology approaches. To that end, quantitative preclinical PK data should be obtained in refined animal models and for biomarkers of the effect one should focus on the ones that can be measured in both animal and human.

8.4 Development of a Preclinical Translational PK–PD Model for the D2 Receptor System

8.4.1 Preclinical Translation; Focus on Different Sites of Administration

In this study, a preclinical translational PK–PD model was developed in the rat following intravenous administration of remoxpride as a model drug for dopaminergic D2 receptor inhibition, using the levels of the pituitary hormone prolactine as a PD readout, and compared to intranasal administration (Stevens et al. 2011, 2012). Since dopamine is an important neurotransmitter in hypothalamic control, pituitary hormones have high potential as (translational) type 4 biomarkers for dopaminergic activity in the brain (Freeman et al. 2000). Prolactin is synthesized in the lactotrophs of the pituitary and its release into plasma will occur upon dopaminergic inhibition.

It was an interesting question whether intranasal administration could offer benefit for therapy of dopaminergic drugs, as many of these drugs have limited use as they are subjected to substantial and/or highly variable first pass effects, or are limited in brain distribution due to the BBB (Illum 2000, 2004; Costantino et al. 2007; Dhuria et al. 2009).

Intranasal administration could be an attractive mode of delivery for these drugs, as it could provide a higher bioavailability due to avoidance of a hepatic first-pass effect and might also offer a more rapid onset of CNS action. However, a direct transport route from the nasal cavity into the brain was questioned. To quantify PKs in plasma and brain ECF following intranasal administration and to possibly identify direct nose-to-brain transport in a quantitative manner, a comparison with intravenous and intranasal administration of remoxipride (REM) obtained within the same subjects and experimental setting had to be made. Moreover, serially obtained plasma- and brain ECF samples were needed.

As anesthesia and stress significantly influence nose physiology, a more advanced animal model for intranasal administration in freely moving animals

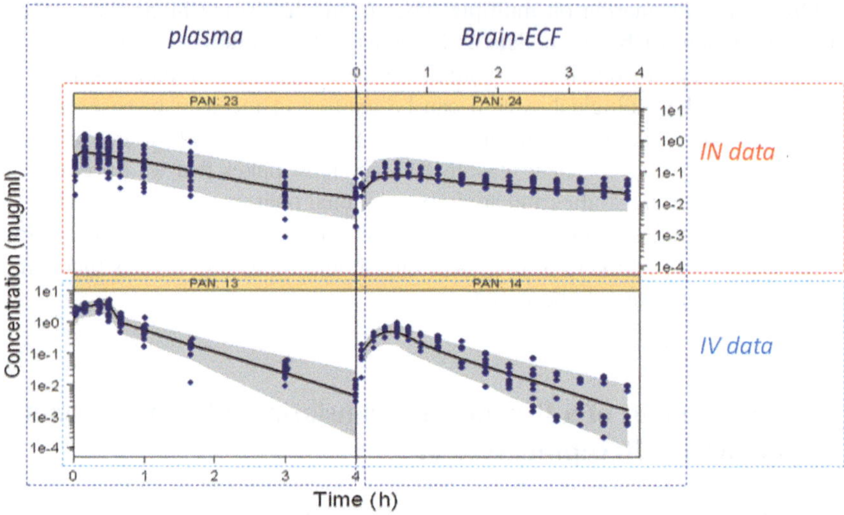

Fig. 8.2 Visual predictive check for 1,000 simulation of the REM PK model. Depicted are the observed data for intranasal (IN) and intravenous (IV) administration of REM in the rat, for plasma and brain ECF compartment (*filled circle*), and the medium of of the simulations (*black line*) with the *grey area's* representing the 95 % confidence intervals

under minimum stress conditions was developed (Stevens et al. 2009). REM was administered at different dosages, in freely moving rats, by the intranasal and intravenous route. Plasma- and brain ECF concentration-time profiles were obtained using microdialysis. The plasma and brain ECF data were simultaneously analyzed using nonlinear mixed effects modeling to identify the existence of direct nose-to-brain transport in a quantitative manner. Restricted brain distribution was found, indicated by Kp,uu (0–4 h) values of 0.3 and 0.2 after intranasal and intravenous administration, respectively. (It should be noted that this ratio is an underestimate for the intranasal administration, as, unanticipated, the elimination half-life was increased such that AUC values (0–∞) could not be adequately determined).

A multi compartment PK model with two absorption compartments (nose-to-systemic and nose-to-brain) was found to best describe the observed PK data. Absorption was described in terms of bioavailability and rate. Total bioavailability following intranasal administration was 89 %, of which 75 % was attributed to direct nose-to brain transport. Direct nose-to-brain absorption rate was slow, explaining prolonged brain ECF exposure after intranasal compared to intravenous administration. By this approach explicit separation and quantitation of systemic- and direct nose-to-brain transport after intranasal administration of REM in the rat could be made. (Fig. 8.2)

Thus, having characterized REM PK at the target site (brain ECF) in a semi-physiology based manner, it was anticipated to allow for better prediction of PD

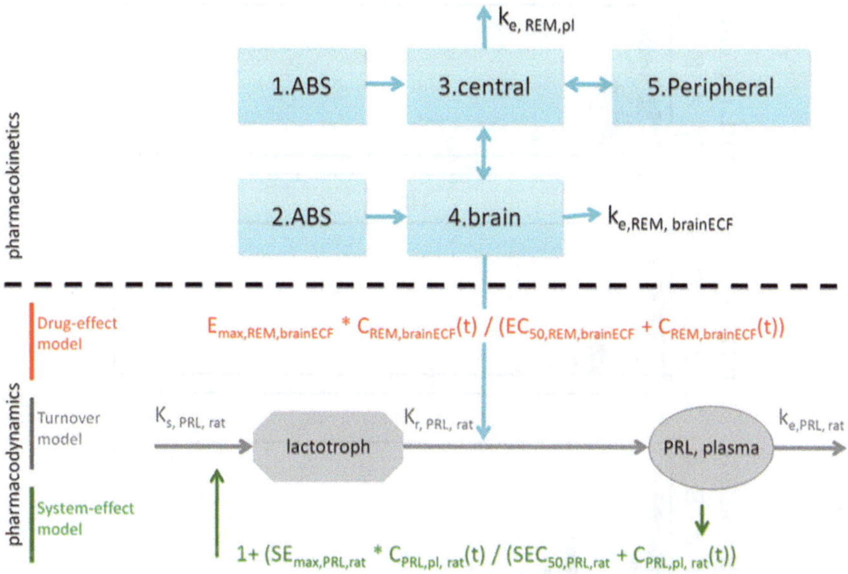

Fig. 8.3 The translational REM PK–PD model. This model encompasses three parts: REM PK-model: Following IN administration, total absorption of REM is defined by systemic uptake (1.ABS) and direct nose-to-brain transport (2.ABS). REM is distributed (*blue arrows*) in a central-, peripheral-, and brain compartment, and eliminated from plasma (ke, REM, pl) and brain ECF (ke, REM, brainECF). PKPD model: An Emax model for relation between brain ECF REM concentrations and PRL release. A turnover model described PRL synthesis rate (ks, PRL, rat), release of PRL from lactotrophs into plasma (kr, PRL, rat), and PRL plasma elimination (ke, PRL, rat) system-feedback model. An Emax model for the relation between PRL plasma concentration and rate of synthesis

effects in a PK–PD model. After assessment of baseline variation in prolactin plasma concentrations, the prolactin response (plasma concentrations) of REM was also measured following different single intravenous doses of REM. Also, the prolactin response was measured following double low dosing of REM with different time intervals between the dosing, to get information on the synthesis of prolactin in the lactotrophs of the pituitary (Fig. 8.3). The mechanistic PK–PD model consisted of:

- A PK model for REM concentrations in brain ECF.
- A pool model incorporating prolactin synthesis, storage in lactotrophs, release into and elimination from plasma.
- A positive feedback component interconnecting prolactin plasma concentrations and prolactin synthesis.
- A dopamine antagonism component interconnecting REM brain ECF concentrations and stimulation of prolactin release.

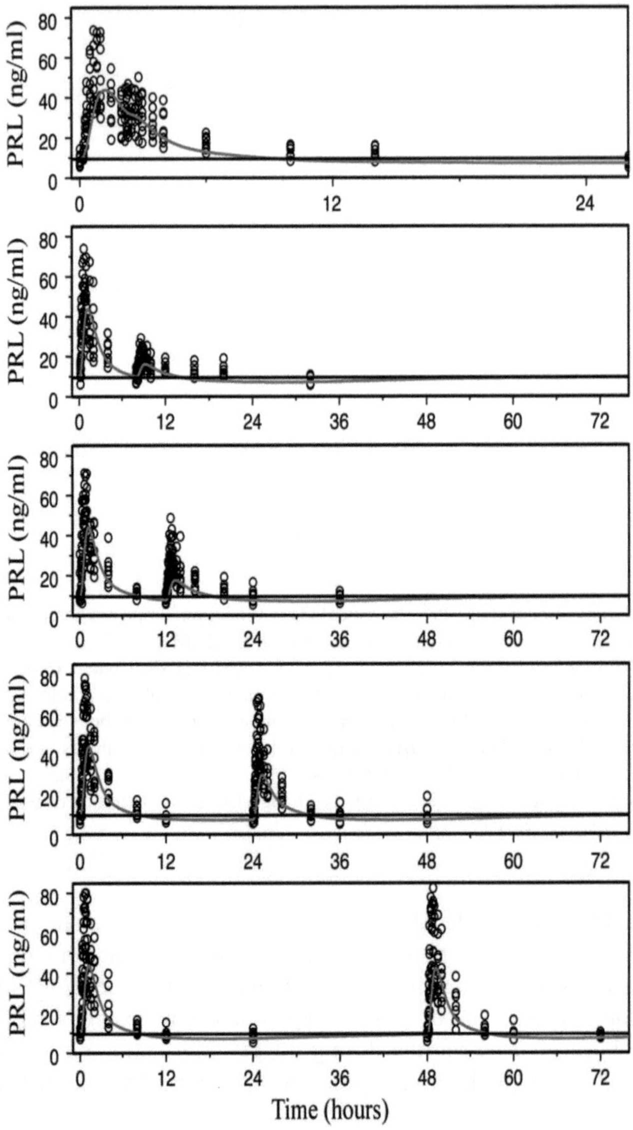

Fig. 8.4 Translation to human PD. Based on estimated human brain ECF PK the predicted PRL PK (*grey line*) is in accordance with the clinical data (O)

The most important finding was the positive feedback on prolactin synthesis in the lactotrophs, in contrast to the negative feedback in the previous models on the PK–PD correlation of REM. Then, the translational value of this PK–PD relationship was tested using the dataset obtained in rats following intranasal administration of the different single doses of REM.

8.4.2 Translation From Animal to Human

Quantification of drug-specific and biological system-specific parameters in translational PK–PD models provides the opportunity to scale the animal model to man. Allometric scaling of drug PK properties and biological system-specific parameters has been used in translational investigations, with reasonable degree of success, to predict drug effects in humans (Yassen et al. 2007; Zuideveld et al. 2007). PD properties are more difficult to scale compared to PK properties, since PD parameters are often not related to bodyweight (e.g., receptor occupancy, transduction, maximal effect, etc.). However, this information can be obtained by in vitro bioassays, and, for many drugs and endogenous compounds, clinical information is often readily available in the literature (e.g. target binding characteristics of dopaminergic compounds, (Kvernmo et al. 2006, 2008).

This provides the opportunity to replace the drug and biological system parameters estimated in rat by the human values, and thus provide an extrapolation to man. Subsequent simulation studies can provide early insight on the clinical applicability of a drug, at an early stage in drug development. Also, as the preclinical-derived translational mechanism-based PK–PD model describes the drug-effect relationship and the biological system, clinical studies is anticipated to suffice with fewer individuals and less samples per individual, for proof of concept in man. Ultimately, simultaneous modeling of relatively large animal and small clinical datasets, allows further investigations on critical factors of animal-to-human extrapolation in a strict quantitative manner.

As previously pointed out, quantitative biomarkers that represent the target site distribution of drugs and metabolites for compounds that act in the CNS are difficult to obtain in man, but more readily available in vivo in animals.

8.4.3 Prediction of Human PKPD Based on a Mechanism-Based PKPD Model Developed in Rats

Following simulation of human REM brain ECF concentrations, PD extrapolation from rat to humans was performed, using allometric scaling in combination with independent information on the values of biological system-specific (in vitro) parameters in human as prior knowledge (Ben Jonathan et al. 2008). The PK–PD model successfully predicted the system prolactin response in humans (Fig. 8.4), indicating that positive feedback on prolactin synthesis and allometric scaling there of could be a new feature in describing complex homeostatic mechanisms.

8.5 Conclusions

To develop treatments with improved safety and efficacy, one of the scientific challenges is to understand the biological mechanisms underlying the PKPD relationships of CNS drugs. PK–PD modeling is the golden standard to investigate such complex mechanisms. Often, these models include plasma drug concentration-effect relationships. However, when the target site is the brain ECF that surrounds membrane-bound receptors, a more mechanistic approach should be aimed at understanding the drug concentrations at the target site. To that end, microdialysis may provide critical and quantitative information on rate and extent of mechanisms between drug administration and CNS effects of a drug in different settings, which, combined with PK–PD modeling approaches, serves as the basis for generic translational models for prediction of CNS effects in varying conditions.

References

Abott NJ, Revest PA (1991) Control of brain endothelial permeability. Cerebrovasc Brain Met Rev 3:39–72

Aird RB (1984) A study of intrathecal, cerebrospinal fluid-to-brain exchange. Exp Neurol 86:342–358

Begley DJ (2004) ABC transporters and the blood–brain barrier. Curr Pharm Des 10:1295–1312

Ben Jonathan N, LaPensee CR, LaPensee EW (2008) What can we learn from rodents about prolactin in humans? Endocrin Rev 29:1–41

Bengtsson J, Ederoth P, Ley D, Hansson S, Amer-Wåhlin I, Hellström-Westas L, Marsál K, Nordström CH, Hammarlund-Udenaes M (2009) The influence of age on the distribution of morphine and morphine-3-glucuronide across the blood–brain barrier in sheep. Br J Pharmacol 157(6):1085–1096

Bouw MR, Gardmark M, Hammarlund-Udenaes M (2000) Pharmacokinetic–pharmacodynamic modelling of morphine transport across the blood–brain barrier as a cause of the antinociceptive effect delay in rats—a microdialysis study. Pharm Res 17:1220–1227

Bouw R, Ederoth P, Lundberg J, Ungerstedt U, Nordstrom CH, Hammarlund-Udenaes M (2001a) Increased blood–brain barrier permeability of morphine in a patient with severe brain lesions as determined by microdialysis. Acta Anest Scand 45:390–392

Bouw MR, Xie R, Tunblad K, Hammarlund-Udenaes M (2001b) Blood–brain barrier transport and brain distribution of morphine-6-glucuronide in relation to the antinociceptive effect in rats—pharmacokinetic/pharmacodynamic modelling. Br J Pharmacol 134:1796–1804

Breimer DD, Danhof M (1997) Prediction of the time course of drug effects in vivo in health and disease (intensity and duration). Clin Pharmacokinet 32:259–267

Brosman CF, Claudio L (1998) Brain microvasculature in multiple sclerosis. In: Pardridge WM (ed) Introduction to the blood–brain barrier; methodology, biology and pathology, Cambridge University Press, pp 386–400

Collins JM, Dedrick LD (1983) Distributed model for drug delivery to CSF and brain tissue. J Am Physiol 14:R303–R310

Cordon-Cardo B, O'Brien JP, Casals D, Rittman-Grauer L, Biedler JL, Melamed MR, Bertino JR (1989) Multidrug-resistance gene (P-glycoprotein) is expressed by endothelial cells at blood–brain barrier sites. Proc Natl Acad Sci U S A 86:689

Cornford EM (1985) The blood–brain barrier, a dynamic regulatory interface. Mol Physiol 7:219–260

Costantino HR, Illum L, Brandt G, Johnson PH, Quay SC (2007) Intranasal delivery: physicochemical and therapeutic aspects. Int J Pharm 337:1–24

Cox EH, Kerbusch T, van der Graaf PH, Danhof M (1998) Pharmacokinetic–pharmacodynamic modeling of the electroencephalogram effect of synthetic opioids in the rat. Correlation with binding at the μ-opioid receptor. J Pharmacol Exp Ther 284:1095–1103

Cremers TI, de Vries MG, Huinink KD, Loon JP, Hart MV, Ebert B, Westerink BH, de Lange EC (2009) Quantitative microdialysis using modified ultraslow microdialysis: direct rapid and reliable determination of free brain concentrations with the metaquant technique. J Neurosci Methods 178(2):249–254

Cserr HF (1971) Physiology of the choroid plexus. Physiol Rev 51:273–311

Cserr HF (1984) Convection of brain interstitial fluid. In: Shapiro K, Marmarou A, Portnoy H (eds) *Hydrocephalus*. Raven Press, New York, pp 59–68

Danhof M, Alvan G, Dahl SG, Kuhlmann J, Paintaud G (2005) Mechanism-based pharmacokinetic–pharmacodynamic modeling-a new classification of biomarkers. Pharm Res 22: 1432–1437

Danhof M, de Jongh J, de Lange EC, Della Pasqua O, Ploeger BA, Voskuyl RA (2007) Mechanism-based pharmacokinetic–pharmacodynamic modeling: biophase distribution, receptor theory, and dynamical systems analysis. Annu Rev Pharmacol Toxicol 47:357–400

Danhof M, de Lange EC, Della Pasqua OE, Ploeger BA, Voskuyl RA (2008) Mechanism-based pharmacokinetic–pharmacodynamic (PKPD) modeling in translational drug research. Trends Pharmacol Sci 29:186–191

de Boer AG, van der Sandt I, Gaillard PJ (2003) The role of drug transporters at the blood–brain barrier. Annu Rev Pharmacol Toxicol 43:629–656

de Lange EC, Danhof M (2002) Considerations in the use of cerebrospinal fluid pharmacokinetics to predict brain target concentrations in the clinical setting: implications of the barriers between blood and brain. Clin Pharmacokinet 41:691–703

de Lange EC, Danhof M, de Boer AG, Breimer DD (1994) Critical factors of intracerebral microdialysis as a technique to determine the pharmacokinetics of drugs in rat brain. Brain Res 666:1–8

de Lange EC, Ravenstijn PGM, Groenendaal D, van Steeg TS (2005) Toward the prediction of CNS drug effect profiles in physiological and pathological conditions using microdialysis and mechanism-based pharmacokinetic–pharmacodynamic modeling. AAPS J 7(3), Article 54

de Lange EC (2004) Potential role of ABC transporters as a detoxification system at the blood–cerebrospinal fluid-barrier. Adv Drug Del Rev 56(12):1793–1809

de Lange ECM, Danhof M, de Boer AG, Breimer DD (1997) Methodological considerations of intracerebral microdialysis in pharmacokinetic studies on blood–brain barrier transport of drugs. Brain Res Rev 25:27–49

de Rick AF, Belpaire FM, Dello C, Bogaert MG (1987) Influence of enhanced AAG concentration on protein binding, pharmacokinetics and anti-arrhythmic effect of lidocaine in the dog. J Pharmacol Exp Ther 241:289–293

Del Bigio MR (1995) The ependyma: a protective barrier between brain and cerebrospinal fluid. Glia 14:1–13

Dhuria SV, Hanson LR, Frey WH (2009) Intranasal delivery to the central nervous system: mechanisms and experimental considerations. J Pharm Sci 99:1654–1673

Ederoth P, Tunblad K, Bouw R, Lundberg CJ, Ungerstedt U, Nordström CH, Hammarlund-Udenaes M (2004) Blood–brain barrier transport of morphine in patients with severe brain trauma. Br J Clin Pharmacol 57(4):427–435

Fenstermacher JD, Patlak CS, Blasberg RG (1974) Transport of material between brain extracellular fluid, brain cells and blood. Fed Proc 33:2070–2074

Freeman ME, Kanyicska B, Lerant A, Nagy G (2000) Prolactin: structure, function, and regulation of secretion. Physiol Rev 80:1523–1631

Fridén M, Gupta A, Antonsson M, Bredberg U, Hammarlund-Udenaes M (2007) In vitro methods for estimating unbound drug concentrations in the brain interstitial and intracellular fluids. Drug Metab Dispos 35:1711–1719

Gabrielsson J, Green AR (2009) Quantitative pharmacology or pharmacokinetic pharmacodynamic integration should be a vital component in integrative pharmacology. J Pharmacol Exp Ther 331:767–774

Garrido M, Gubbens-Stibbe J, Tukker E et al (2000) Pharmacokinetic–pharmacodynamic analysis of the EEG effect of alfentanil in rats following beta-funaltrexamine-induced mu-opioid receptor “knockdown” in vivo. Pharm Res 17:653–659

Groenendaal D, Freijer J, de Mik D, Bouw MR, Danhof M, de Lange EC (2007a) Pharmacokinetic–pharmacodynamic modelling of the electroencephalogram effects of morphine: the influence of biophase equilibration and P-glycoprotein interaction. Br J Pharmacol 151(5):713–20. Epub Apr 30 2007

Groenendaal D, Freijer J, de Mik D, Bouw MR, Danhof M, de Lange EC (2007) Population pharmacokinetic modelling of non-linear brain distribution of morphine: influence of active saturable influx and P-glycoprotein mediated efflux. Br J Pharmacol 151(5):701–12. Epub Apr 30 2007

Hammarlund-Udenaes M, Fridén M, Syvänen S, Gupta A (2008) On the rate and extent of drug delivery to the brain. Pharm Res 25(8):1737–1750

Hammarlund-Udenaes M, Paalzow LN, de Lange ECM (1997) Drug equilibration across the blood–brain-barrier—pharmacokinetic considerations based on the microdialysis method. Pharm Res 14:128–134

Illum L (2000) Transport of drugs from the nasal cavity to the central nervous system. Eur J Pharm Sci 11:1–18

Illum L (2004) Is nose-to-brain transport of drugs in man a reality? J Pharm Pharmacol 56:3–17

Jeffrey P, Summerfield S (2010) Assessment of the blood–brain barrier in CNS drug discovery. Neurobiol Dis 37:33–37

Jolliet P, Simon N, Bree F et al (1997) Blood-to-brain transfer of various oxicams: effects of plasma binding on their brain delivery. Pharm Res 14:650–656

Kapur S, Zipursky R, Jones C, Remington G, Houle S (2000) Relationship between dopamine D2 occupancy, clinical response, and side effects: a double-blind PET study of first-episode schizophrenia. Am J Psychiatry 157:514–520

Karssen AM, Meijer OC, van der Sandt ICJ, Lucassen PJ, de Lange ECM, de Boer AG, de Kloet ER (2001) Multidrug resistance P-glycoprotein hampers the access of cortisol but not of corticosterone to mouse and human brain. Endocrinology 142:2686–2694

Kooij G, van Horssen J, de Lange EC, Reijerkerk A, van der Pol SM, van Het Hof B, Drexhage J, Vennegoor A, Killestein J, Scheffer G, Oerlemans R, Scheper R, van der Valk P, Dijkstra CD, de Vries HE. (2010). T lymphocytes impair P-glycoprotein function during neuroinflammation. J Autoimmun 34(4):416–425

Kropf W, Kuschinsky K (1993) Effects of stimulation of dopamine D1 receptors on the cortical EEG in rats: different influences by a blockade of D2 receptors and by an activation of putative dopamine autoreceptors. Neuropharmacol 32:493–500

Kvernmo T, Hartter S and Burger E (2006) A review of the receptor-binding and pharmacokinetic properties of dopamine agonists. Clin Therapeutics 28:1065–1078

Kvernmo T, Houben J, Sylte I (2008) Receptor-binding and pharmacokinetic properties of dopaminergic agonists. Curr Top Med Chem 8(12):1049–1067

Lee G, Dallas S, Hong M, Bendayan R (2001) Drug transporters in the central nervous system: brain barriers and brain parenchyma considerations. Pharmacol Rev 53:569–596

Letrent SP, Pollack GM, Brouwer KR, Brouwer KL (1998) Effect of GF120918, a potent P-glycoprotein inhibitor, on morphine pharmacokinetics and pharmacodynamics in the rat. Pharm Res 15:599–605

Letrent SP, Pollack GM, Brouwer KR, Brouwer KL (1999) Effects of a potent and specific P-glycoprotein inhibitor on the blood-brain barrier distribution and antinociceptive effect of morphine in the rat. Drug Metab Dispos 27:827–834

Lin TH, Sawada Y, Sugiyama Y, Iga T, Hanano M (1987) Effects of albumin and alpha 1-acid glycoprotein on the transport of imipramine and desipramine through the blood-brain barrier in rats. Chem Pharm Bull (Tokyo) 35:294–301

Liu X, Smith BJ, Chen C, Callegari E, Becker SL, Chen X, Cianfrogna J, Doran AC, Doran SD, Gibbs JPN, Hosea J, Liu Nelson FR, Szewc MA, van Deusen J (2005) Use of a physiologically based pharmacokinetic model to study the time to reach brain equilibrium: an experimental analysis of the role of blood–brain barrier permeability, plasma protein binding, and brain tissue binding. J Pharmacol Exp Ther 313:1254–1262

Malhotra BK, Lemaire M, Sawchuk RJ (1994) Investigation of the distribution of EAB 515 to cortical ECF and CSF in freely moving rats utilizing microdialysis. Pharm Res 11:1223–1231

Mandema JW, Sansom LN, Dios-Viéitez MC, Hollander-Jansen M, Danhof M (1991) Pharmacokinetic–pharmacodynamic modelling of the EEG effects of benzodiazepines. Correlation with receptor binding and anticonvulsant activity. J Pharmacol Exp Ther 257:472–478

Mulder M, Blokland A, van den Berg DJ, Schulten H, Bakker AHF, Terwel D, Honig W, de Kloet ER, Havekes LM, Steinbusch HWM, de Lange ECM (2001) Apolipoprotein E protects against neuropathology induced by a high-fat diet and maintains the integrity of the blood–brain barrier during aging. Lab Invest 81:953–960

Oldendorf WH (1974) Lipid solubility and drug penetration of the blood–brain barrier. Proc Exp Biol Med 14:813–816

Ploeger BA, van der Graaf PH, Danhof M (2009) Incorporating receptor theory in mechanism-based pharmacokinetic–pharmacodynamic (PKPD) modeling. Drug Metab Pharmacokinet 24:3–15

Ravenstijn PG, Merlini M, Hameetman M, Murray TK, Ward MA, Lewis H, Ball G, Mottart C, de Ville de Goyet C, Lemarchand T, van Belle K, O'Neill MJ, Danhof M, de Lange EC (2007) The exploration of rotenone as a toxin for inducing parkinson's disease in rats, for application in BBB transport and pkpd experiments. J Pharmacol Toxicol Meth 57(2): 114–130

Ravenstijn PGM, Drenth H, O'Neill MJ, Danhof M, de Lange (2012) ECM evaluation of BBB transport and CNS drug metabolism in diseased and control brain side: application for L-DOPA in a unilateral rat model of Parkinson's disease. FBCNS, accepted

Rubin LL, Staddon JM (1999) The cell biology of the blood–brain barrier. Annu Rev Neurosci 22:11–28

Schinkel AH, Wagenaar E, van Deemter L, Mol CAAM, Borst P (1995) Absence of the mdr1a p-glycoprotein in mice affects tissue distribution and pharmacokinetics of dexamethasone, digoxin, and cyclosporin A. J Clin Invest 96:1698–1705

Scism JL, Powers KM, Artru AA, Chambers AC, Lewis L, Adkison KK, Kalhorn TF, Shen DD (1997) Effects of probenecid on brain–cerebrospinal fluid–blood distribution kinetics of E-Delta(2)-valproic acid in rabbits. Drug Metab Disp 25:1337–1346

Segal MB (1998) The blood–CSF barrier and the choroid plexus. In: Pardridge MD (ed) Introduction to the blood–brain barrier, methodology, biology and pathology. Cambridge University Press, pp 251–258

Stain-Texier F, Boschi G, Sandouk P, Scherrmann JM (1999) Elevated concentrations of morphine 6-beta-D-glucuronide in brain extracellular fluid despite low blood–brain barrier permeability. Br J Pharmacol 128:917–924

Stevens J, Suidgeest J, van der Graaf PH, Danhof M, de Lange EC (2009) A new minimal-stress freely-moving rat model for preclinical studies on intranasal administration of CNS drugs. Pharm Res 26(8):1911–1917

Stevens J, Ploeger B, van der Graaf PH, Danhof M, de Lange ECM (2011a) Systemic- and direct nose-to-brain transport in the rat; a mechanistic pharmacokinetic model for remoxipride after intravenous and intranasal administration. Drug Metab Disp 39(12):2275–2282

Stevens J, Ploeger B, vd Graaf PH, Danhof M, and de lange ECM (2012). Translational pharmacology of intranasal administration of dopamine receptor agonists and antagonists. J Pharmacokin Pharmacodyn (in press)

Suzuki H, Terasaki T, Sugiyama Y (1997) Role of efflux transport across the blood–brain barrier and blood–cerebrospinal fluid barrier on the disposition of xenobiotics in the central nervous system. Adv Drug del rev 25:257–285

Syvänen S, Schenke M, van den Berg DJ, Voskuyl RA and de Lange ECM (2012) Alteration in P-glycoprotein functionality affects intrabrain distribution of drugs more than brain entry—A study in rats subjected to status epilepticus by kainate. AAPS J 14(1): 87–96

Tamai I, Yamashita J, Kido Y et al (2000) Limited distribution of new quinolone antibacterial agents into brain caused by multiple efflux transporters at the blood–brain barrier. J Pharmacol Exp Ther 295:146–152

Tunblad K, Hammarlund-Udenaes M, Jonsson EN (2005) Influence of probenecid on the delivery of morphine-6-glucuronide to the brain. Eur J Pharm Sci 24:49–57

Tunblad K, Ederoth P, Gardenfors A, Hammarlund-Udenaes M, Nordstrom CH (2004) Altered brain exposure of morphine in experimental meningitis studied with microdialysis. Acta Anaesthesiol Scand 48:294–301

Tunblad K, Jonsson EN, Hammarlund-Udenaes M (2003) Morphine blood–brain barrier transport is influenced by probenecid co-administration. Pharm Res 20:618–623

van der Graaf PH, Danhof M (1997) Analysis of drug-receptor interactions in vivo: a new approach in pharmacokinetic–pharmacodynamic modelling. Int J Clin Pharmacol Ther 35:442–446

van Steeg TJ, Boralli VB, Krekels EHJ, Slijkerman P, Freijer J, Danhof M, de Lange EC (2009) Influence of plasma protein binding on pharmacodynamics: estimation of in vivo receptor affinities of b blockers using a new mechanism-based PK–PD modelling approach. J Pharm Sci 98(10):3816–3828

Vorobyov VV, Schibaev NV, Morelli M, Carta AR (2003) EEG modifications in the cortex and striatum after dopaminergic priming in the 6-hydroxydopamine rat model of Parkinson's disease. Brain Res 972:177–185

Wang YF, Welty DF (1996) The simultaneous estimation of the influx and efflux blood–brain barrier permeabilities of gabapentin using a microdialysis–pharmacokinetic approach. Pharm Res 13:398–403

Watson J, Wright S, Lucas A, Clarke KL, Viggers J, Cheetham S, Jeffrey P, Porter R, Read KD (2009) Receptor occupancy and brain free fraction. Drug Metab Dispos 37:753–760

Welty DF, Schielke GP, Vartanian MG, Taylor CP (1993) Gabapentin anticonvulsant action in rats: disequilibrium with peak drug concentrations in plasma and brain microdialysate. Epilepsy Res 16:175–181

Westerhout J, Danhof M, de Lange EC (2011) Preclinical prediction of human brain target site concentrations: considerations in extrapolating to the clinical setting. J Pharm Sci 100(9):3577–3593

Wijnholds J, de Lange ECM, Scheffer GL, van den Berg DJ, Mol CAAM, van der Valk M, Schinkel AH, Scheper RJ, Breimer DD, Borst P (2000) Multidrug resistance protein 1 protects the choroid plexus epithelium and contributes to the blood–cerebrospinal fluid barrier. Clin Invest 105:279–285

Williams SA, Davson H, Segal MB (1995) Transport of the nucleoside thymidine, in the central nervous system: the blood–cerebrospinal fluid and blood–brain barriers. In: Greenwood J, Begley DJ, Segal MB (eds) New concepts of a blood–brain barrier. Plenum Press, New York

Xie R, Hammarlund-Udenaes M, de Boer AG, de Lange ECM (1999) The role of P-glycoprotein in blood-brain barrier transport of morphine: transcortical microdialysis studies in mdr1a(−/−) and mdr1a(+/+) mice. Br J pharmacol 131(8):563–568

Xie R, Bouw MR and Hammarlund-Udenaes M (2000). Modelling of the blood-brain barrier transport of morphine-3-glucuronide studied using microdialysis in the rat: involvement of probenecid-sensitive transport. Br J Pharmacol 131(8):1784–1792

Yassen A, Olofsen E, Kan J, Dahan A, Danhof M (2007) Animal-to-human extrapolation of the pharmacokinetic and pharmacodynamic properties of buprenorphine. Clin Pharmacokinet 46:433–447

Zhang Y, Schuetz JD, Elmquist WF, Miller DW (2004) Plasma membrane localization of multidrug resistance-associated protein homologs in brain capillary endothelial cells. J Pharmacol Exp Ther 311:449–455

Zuideveld KP, van der Graaf PH, Peletier LA, Danhof M (2007) Allometric scaling of pharmacodynamic responses: application to 5-Ht1A receptor mediated responses from rat to man. Pharm Res 24:2031–2039

Chapter 9
Extracellular Amyloid-β Protein Dynamics in Alzheimer's Disease

John R. Cirrito

Abstract Alzheimer's disease (AD) is caused by the accumulation of the amyloid-β (Aβ) peptide within the brain extracellular space. Aβ is produced within neurons then secreted into the extracellular fluid. Normally, Aβ is found as a soluble, monomeric peptide within the brain interstitial fluid (ISF) and cerebrospinal fluid. During the pathogenesis of AD, however, Aβ changes conformation then aggregates into toxic higher ordered species, including soluble oligomers and insoluble plaques, which appear to underlie the etiology of AD. Conversion into these toxic species of Aβ appears to be concentration dependent and, at least partially, occurs within the ISF. Consequently, understanding the metabolic processes that regulate Aβ generation, particularly within the ISF, has implications for understanding pathogenesis as well as treatment. In order to study ISF Aβ metabolism in a physiological setting, several groups have utilized in vivo microdialysis both in animal models of disease as well as human patients. Synaptic transmission, the events that underlie brain function, directly leads to Aβ generation in vivo; suppressing synaptic activity rapidly reduces ISF Aβ levels whereas increasing activity increases Aβ levels. In mouse models of AD, behaviors such as stress and sleep/wake cycles also dynamically modulate ISF Aβ generation. Following traumatic brain injury(TBI), cortical ISF Aβ levels in humans is correlated with neurological status and synaptic transmission; comatose patients with initially low synaptic activity have low ISF Aβ levels which rise as neurological status improve. Conversely, in humans 10 % of individuals with temporal lobe epilepsy (TLE) develop Aβ plaques. The absolute concentration of Aβ within the ISF appears

J. R. Cirrito (✉)
Department of Neurology, Hope Center for Neurological Disorders, and the Knight Alzheimer's Disease Research Center, Washington University School of Medicine, St. Louis, MO, USA
e-mail: cirritoj@neuro.wustl.edu

M. Müller (ed.), *Microdialysis in Drug Development*, AAPS Advances in the Pharmaceutical Sciences Series, DOI: 10.1007/978-1-4614-4815-0_9,

directly related to whether a brain region will develop plaques or not. Microdialysis provides a powerful tool to assess brain ISF Aβ levels longitudinally through a wide variety of experimental and behavioral paradigms.

9.1 Alzheimer's Disease and Amyloid-β

Alzheimer's disease (AD) is characterized by two hallmark brain pathologies: extracellular plaques consisting primarily of the amyloid-β (Aβ) peptide and intracellular neurofibrillary tangles consisting primarily of hyperphosphorylated tau. While tau inclusions are likely to contribute to neuronal damage and dysfunction in AD, genetic, biochemical, and animal model studies strongly suggest that Aβ plays a central role in the pathogenesis of the disease (Selkoe 2001; Holtzman et al. 2011). Aβ is a 38–43 amino acid peptide derived from the amyloid precursor protein (APP) through sequential proteolytic cleavage by β-secretase then γ-secretase. Aβ is generated intracellularly within endosomes then secreted into the brain extracellular fluid as a normal process. Since Aβ is found within the extracellular space, also where plaques form, elucidating mechanisms that influence extracellular Aβ is likely to be crucial for understanding and treating AD.

In humans, Aβ is produced throughout the body and can be found in all bodily fluids (Seubert et al. 1992). Normally, this peptide is in a soluble, non-toxic conformation and, to date, has no identified function. A growing literature suggests Aβ may inhibit certain forms of synaptic transmission; however, whether this inhibition may occur normally or only within a pathologic state, if at all, is still unclear (Hartley et al. 1999; Klyubin et al. 2004; Shankar et al. 2008). At AD pathogenesis, this normally soluble peptide changes conformation, becomes insoluble, and deposits within the extracellular space of the neocortex and hippocampus as toxic Aβ plaques. Conversion of soluble Aβ into toxic species such as plaques or oligomers is concentration dependent, meaning that elevated levels of ISF Aβ are likely to hasten the formation of toxic structures. This process begins 10–20 years prior to the onset of disease symptoms (Morris and Price 2001; Mintun et al. 2006). Appropriately, brain regions such as the cortex and hippocampus that develop Aβ pathology are also the regions responsible for the cognitive faculties that are most affected in AD.

AD was first described by Alois Alzheimer in 1906 and for decades Aβ plaques appeared to be permanent structures. In part, this is attributable to the static methodology used to evaluate Aβ pathology, typically postmortem. With the advent of dynamic assessment techniques, such as ex vivo tissue slice preparations and in vivo two-photon microscopy imaging (Bacskai and Hyman 2002; Bacskai et al. 2002; Klunk et al. 2002; Brendza et al. 2003), Aβ pathology in living tissues could be observed which demonstrated how dynamic these structures can be under the appropriate conditions. For instance, glial cellular processes can engulf and degrade Aβ within plaques (Rogers et al. 2002; Wyss-Coray et al. 2003). The current impression is that plaque development is race between formation and degradation; when formation is more rapid than degradation then the number and size of plaques increase. Unfortunately, it appears that once plaques begin to develop, degradation is a much slower process which allows plaques to accumulate. Optical approaches can be used to study insoluble forms of Aβ, however cannot be used to assess soluble forms of the peptide. To assess Aβ within the soluble phase, biochemical techniques are necessary.

9.2 Brain Interstitial Fluid Aβ

When Aβ is secreted from neurons within the brain, it directly enters the brain extracellular fluid, or interstitial fluid (ISF). It is within the brain extracellular space where Aβ is found aggregated as plaques. Also, if Aβ has any function, then it is likely within this compartment. It remains controversial where plaques originate. Several groups have suggested that Aβ aggregates intracellularly, possibly in lipid rafts, and then the already aggregated Aβ is released into the extracellular space as the seed for a plaque (Mizuno et al. 1999). Aβ within the ISF then attaches to the seed causing the plaque to expand. In contrast, using brain tissue transplants from APP transgenic mice into wild-type mice Meyer-Leuhmann et al. demonstrated in vivo that extracellular Aβ is capable of forming the original seed for a plaque as well (Meyer-Luehmann et al. 2003). The extent to which each of these pathways contributes to plaque formation is unknown. Regardless of the where the seed, or nidus, a plaque polymerization occurs, it is clear that Aβ that builds onto existing Aβ aggregate scaffolding comes from the ISF, as this is the only source in direct contact with plaques.

9.3 Using Microdialysis to Measure Brain ISF Aβ

In vivo microdialysis has been used to study brain ISF Aβ to the exclusion of all other brain compartments that can also contain Aβ (e.g., intracellular or insoluble extracellular) (Cirrito et al. 2003). Standard biochemical extractions of protein from tissue result in mixing of Aβ from various brain compartments, whereas microdialysis permits specific sampling of only extracellular molecules. In many cases, biochemical extraction of Aβ of brain tissue does not correlate with Aβ levels within the ISF (Hong et al. 2011). Though microdialysis is a common technique to sample small molecules within the brain ISF, sampling larger peptides, such as Aβ, requires several technical modifications.

A general complication for protein microdialysis is that many proteins exist in association with other proteins or molecules. As such, while a peptide may theoretically be able to diffuse across a semi-permeable membrane into a microdialysis probe, if associated to other molecules the effective size of the peptide may be too large for recovery. Aβ within the ISF appears to bind to larger molecules

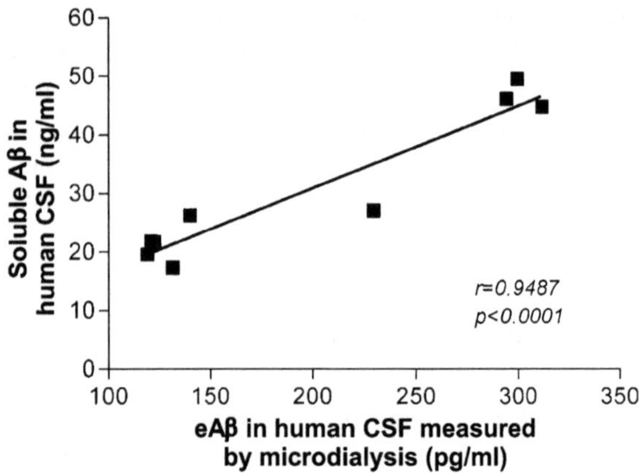

Fig. 9.1 Exchangeable Aβ. The concentration of exchangeable Aβ (eAβ) measured by microdialysis and total soluble Aβ sampled directly are highly correlated within a sample of human CSF in vitro (Pearson's $r = 0.9487$; $p < 0.0001$)

such as apolipoprotein E or clusterin (DeMattos et al. 2004). These complexes would be too large to enter most microdialysis probes. Not necessarily surprisingly, only a fraction of Aβ within a physiological fluid is directly measurable by microdialysis, called "exchangeable Aβ" (eAβ) (Cirrito et al. 2003). Importantly, however, the exchangeable pool of eAβ is proportional to the total amount of soluble Aβ present (Fig. 9.1). Another complicating factor to Aβ microdialysis is that Aβ is an unusually "sticky" molecule in which it inherently sticks to itself as well as plastics and glassware. Including a carrier protein, such as albumin, within the microdialysis perfusion buffer is necessary to recover any measurable amount of Aβ at all.

One power of microdialysis is that it measures molecules within the ISF over time while animals are awake and behaving. This enables the study of dynamic changes in a molecule in near-physiological conditions as opposed to a single static, postmortem measurement. In the case of APP transgenic mouse models of AD, Aβ can be sampled every 30 min over several days. Microdialysis studies have demonstrated that Aβ levels change very rapidly within the brain.

Steady-state Aβ levels are a balance between production and elimination; each of which can be modulated to alter Aβ levels over a time span of minutes to hours. Elimination pathways such as blood–brain barrier transport out of the brain (Shibata et al. 2000; Cirrito et al. 2005b) or proteolytic degradation (Yin et al. 2006; Farris et al. 2007) reduces ISF Aβ levels with a half-life of 1–2 h (Cirrito et al. 2003). Aβ production is regulated by neuronal activity; elevations in activity can increase ISF Aβ levels within minutes (Cirrito et al. 2005a). In contrast to insoluble Aβ within plaques which is long lived and difficult to eliminate, the soluble peptide changes dynamically in response to a variety of conditions.

9.4 Regulation of Aβ Production

One major pathway for Aβ generation is endocytosis of full length APP from the plasma membrane into the endocytic compartment, where β-secretase and γ-secretase act to produce Aβ (Vassar et al. 1999; Lah and Levey 2000). In the brain this occurs within neurons. Synaptic and neuronal activities regulate how much Aβ is generated via two distinct, but related, mechanisms. Synaptic transmission within the presynaptic terminal of a neuron increases endocytosis which directly drives APP cleavage into Aβ. Independently, neurotransmitter receptors can be activated on neurons to induce cell signaling pathways to alter secretase activity and subsequent Aβ generation. Interestingly, neurons can have cell signaling events that suppress Aβ generation while simultaneously increasing synaptic transmission and Aβ production; the net change in ISF Aβ generation would be a balance of these mechanisms.

9.5 High Levels of Neuronal Activity Correlate with Aβ Deposition in Humans

Several studies in humans suggest that neuronal activity may be linked to Aβ deposition as plaques. For instance, 10 % of individuals with temporal lobe epilepsy (TLE) develop Aβ plaques throughout the temporal lobe at ages when AD pathology would otherwise be rare (Mackenzie and Miller 1994; Gouras et al. 1997). Although the epileptic tissue where the plaques occur is not necessarily normal, these individuals share dramatically elevated neuronal activity compared to activity in the same brain regions of normal individuals. Individuals as young as 30 years old can exhibit diffuse Aβ plaques at autopsy. Interestingly, even though Aβ plaques are present, TLE does not confer a greater risk of developing AD possibly.

Other studies linking neuronal activity with Aβ deposition come from recent brain imaging work. Studies by Buckner and colleagues demonstrate that brain areas that have a high level of basal metabolic activity are the same brain areas that are most vulnerable to Aβ deposition in human AD patients (Buckner et al. 2005, 2009). Studies in animal models of AD also demonstrate the amount of neuronal activity within a brain region tightly correlates with ISF Aβ concentration and the amount of plaque formation (Bero et al. 2011). This suggests that areas that have the most overall neuronal activity are particularly vulnerable to Aβ deposition.

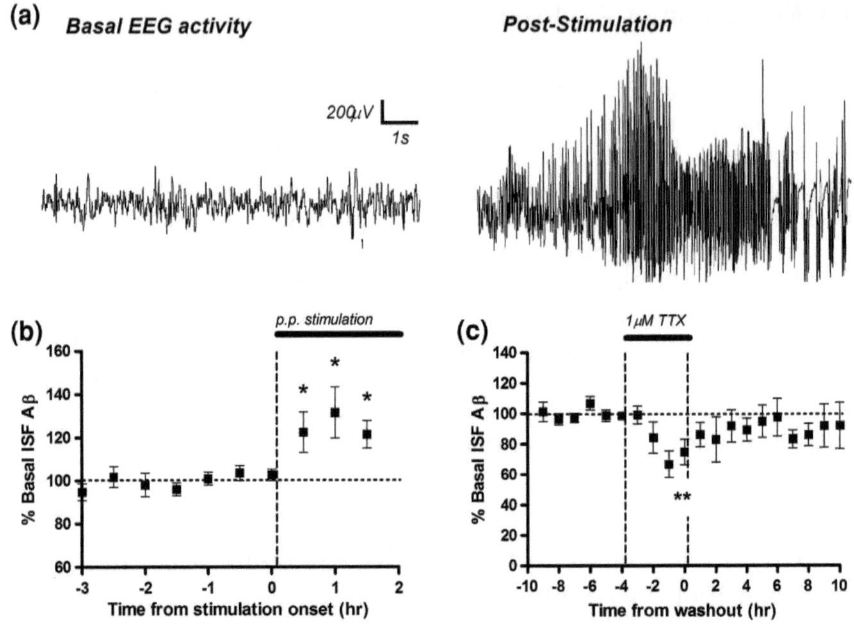

Fig. 9.2 Synaptic activity dynamically modulates ISF Aβ levels in vivo. **a** Electrical stimulation of the perforant pathway induces hippocampal seizures and **b** increases ISF Aβ levels by 25 % within 30 min. **c** Direct administration of TTX to the hippocampus gradually lowers ISF Aβ levels over 4 h which is completely reversible within hours once TTX is removed from the microdialysis probe

9.6 Synaptic Transmission Dynamically Regulates ISF Aβ Levels In Vivo

To determine if increased synaptic activity alters ISF Aβ levels acutely, our group electrically stimulated the hippocampus to increase neuronal activity while measuring ISF Aβ every 30 min using microdialysis (Cirrito et al. 2005a). These probes also had recording electrodes attached in order to measure neuronal activity using electroencephalography (EEG). High frequency, electrical stimulation of the perforant pathway generates seizures within the hippocampal formation (Sloviter et al. 1996). These seizures are evident in EEG, or extracellular field potential, recordings within the hippocampus (Fig. 9.2a). Hippocampal seizures increased ISF Aβ levels by 25 % within the first 30 min of stimulation compared to prestimulation Aβ levels (Fig. 9.2b). In fact, as long as seizures continued, ISF Aβ levels remained elevated.

To determine if a more subtle increase in synaptic activity can also modulate ISF Aβ levels, 25 μM picrotoxin was administered directly into the hippocampus via reverse microdialysis. While high doses of picrotoxin can cause seizures (>200 μM), this low dose of the $GABA_A$ receptor antagonist causes occasional synchronous spikes in EEG activity, but does not cause seizures. Picrotoxin treatment increases ISF Aβ levels by 20 % within the first 30 min of treatment with levels reaching a maximum increase of 45 % by 4 h (Fig. 9.2b). Interestingly, a low dose of picrotoxin elevates ISF Aβ more than electrically stimulated seizures did. Stimulation-induced seizures cause brief bursts of very high levels of EEG activity (fast frequency and high amplitude) followed by interictal suppression of activity whereas picrotoxin caused a sustained, low level of elevated activity (spikes of spontaneous activity a 2–4 Hz). It is possible that the distinct kinetics and overall change in synaptic activity in these two paradigms account for the varying degrees of ISF Aβ alteration.

Blocking synaptic transmission also reduces ISF Aβ levels. Tetrodotoxin (TTX) is a sodium channel blocker produced in puffer fish. If ingested in food or administered peripherally, TTX is lethal. If TTX is administered directly to the brain, however, it blocks action potential propagation and locally inhibits neuronal activity. Mice were continuously administered TTX via reverse microdialysis directly into the hippocampus for 4 h. During this treatment, EEG activity within the hippocampus gradually declined to zero with a concurrent decrease in ISF Aβ levels (Cirrito et al. 2005a). After 4 h of treatment, ISF Aβ levels decreased to 70 % of basal levels (Fig. 9.2c). The effect of TTX was reversible; TTX treatment reduced EEG activity and Aβ levels; however, both of these measures gradually returned to baseline levels over 4 h following removal of the drug from the microdialysis perfusion buffer.

TTX reduces, but does not completely prevent, neuronal activity. While evoked synaptic transmission is blocked by TTX, spontaneous activity is not. In order to block synaptic transmission more completely, a bolus injection of tetanus toxin can be given directly into the hippocampus surrounding a microdialysis probe. Tetanus toxin is taken into the presynaptic terminal and cleaves VAMP2, which is necessary for synaptic vesicle exocytosis. This does not interfere with upstream action potential formation and propagation but does prevent exocytosis, thereby blocking all types of local synaptic activity. Tetanus toxin reduces the levels of ISF Aβ to approximately 25 % of burden (Cirrito et al. 2005a). This dramatic reduction in Aβ levels strongly suggests that synaptic vesicle exocytosis is a critical event involved in Aβ release from neurons.

9.7 Endocytosis Within the Presynaptic Terminal Causes Aβ Generation and Secretion

Synaptic transmission involves fusion of the synaptic vesicle with the presynaptic terminal, release of neurotransmitter into the synaptic cleft, and subsequent recycling of the vesicle membrane and protein by clathrin-mediated endocytosis (Newton et al. 2006). Separately, CME is also responsible for APP internalization

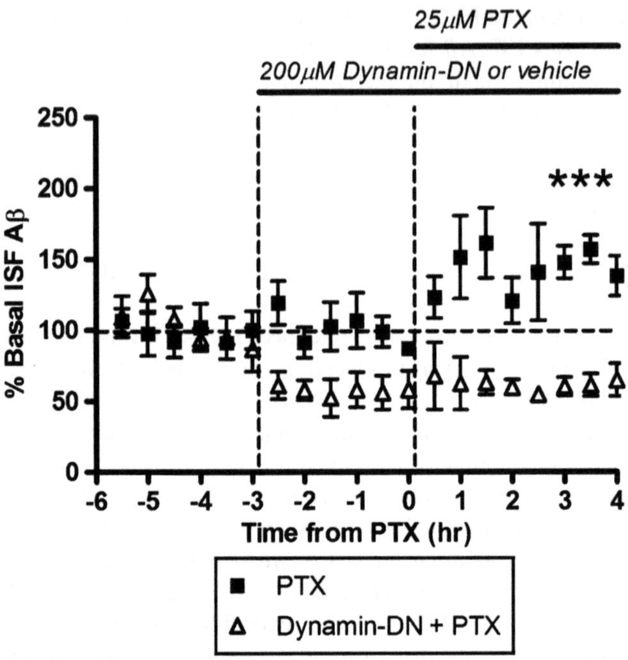

Fig. 9.3 Endocytosis is required for synaptic-dependent Aβ generation. Inhibition of endocytosis using a dominant negative peptide against dynamin (Dynamin-DN) reduces ISF Aβ levels and blocks the synaptic-dependent increase in ISF Aβ caused by infusion of picrotoxin (PTX), a $GABA_A$ receptor antagonist

and subsequent Aβ generation. We hypothesized that synaptic transmission and Aβ generation were linked via endocytosis (Koo and Squazzo 1994).

When synaptic vesicles release their neurotransmitter, the synaptic vesicle membrane typically fuses with the plasma membrane. The vesicular membrane and the associated proteins are then recycled from the plasma membrane by CME and used to replenish the pool of synaptic vesicles. As synaptic activity increases, more synaptic vesicle membrane will be endocytosed from the plasma membrane. As presynaptic endocytosis increases, so does APP internalization and subsequent Aβ generation. Within clathrin-coated vesicles, APP co-localizes with several synaptic vesicle markers (Marquez-Sterling et al. 1997). Increasing synaptic activity, but blocking endocytosis completely blocks Aβ generation as assessed by in vivo microdialysis (Cirrito et al. 2008) (Fig. 9.3). It appears that synaptic activity and Aβ generation are only linked by circumstance; APP is coincidently endocytosed with synaptic vesicle membrane as opposed to playing a functional role in synaptic recycling itself.

9.8 Neurotransmitter Second Messenger Signaling Can Suppress Aβ Generation

Synaptic transmission within the presynaptic part of a neuron directly leads to increased Aβ levels; however, activation of neurotransmitter receptors can have separate, and sometimes opposite, effects on Aβ levels. Activation of postsynaptic receptors, such as M1 muscarinic acetylcholine receptors and serotonin receptors, induce signaling pathways which result in altered processing of APP, thereby affecting Aβ production (Nitsch et al. 2000; Beach et al. 2001; Nelson et al. 2007). Both of these neurotransmitter receptors appear to increase α-secretase cleavage of APP; this protease cleaves within the Aβ sequence which precludes formation of the peptide. Deletion of M1 receptors increases amyloid plaques in animal models of AD (Davis et al. 2010). Similarly, an increase in serotonin signaling using selective serotonin reuptake inhibitor (SSRI) antidepressants significantly reduces ISF Aβ levels and plaques (Cirrito et al. 2011).

9.9 Synaptic Activity Can Simultaneously Increase and Decrease Aβ Levels

Regulation of synaptic Aβ generation and regulation of APP processing function independently to control Aβ generation. In unique cases, however, both of these mechanisms can be activated by the same trigger. Several studies have assessed the role of glutamatergic NMDA receptors (NMDA-Rs) in Aβ metabolism using neuronal culture. Many of these studies demonstrate that NMDA upregulates α-secretase cleavage of APP thus reducing Aβ generation (Fazeli et al. 1994; Gordon-Krajcer et al. 2002; Marcello et al. 2007). Two papers in particular however show in neuronal culture that NMDA modulates Aβ generation in opposite directions; one shows NMDA elevates Aβ (Lesne et al. 2005) whereas the other shows NMDA reduces Aβ. The discrepancy between these in vitro findings is not immediately obvious based on the methods used. Ironically, in vivo studies using microdialysis demonstrate that NMDA-R activation may actually both increase and decrease Aβ generation depending on the level of receptor activation (Verges et al. 2011). Infusion of low doses of NMDA into the brain activates these receptors and elevates presynaptic Aβ generation in vivo. In contrast, high doses of NMDA reduce Aβ by altering processing of APP into Aβ (Fig. 9.4a). Activating a relatively small number of NMDA-Rs depolarizes neurons which increase synaptic transmission and presynaptic Aβ generation. Blocking synaptic activity with TTX completely blocks this increase in Aβ. In contrast, when a large number of receptors are activated, calcium influx into neurons activates second messenger pathways that culminate in extracellular regulated kinase (ERK) activation. ERK increases α-secretase cleavage of APP so that although synaptic transmission causes endocytosis, APP has already been cleaved in a way that no longer produces Aβ at the

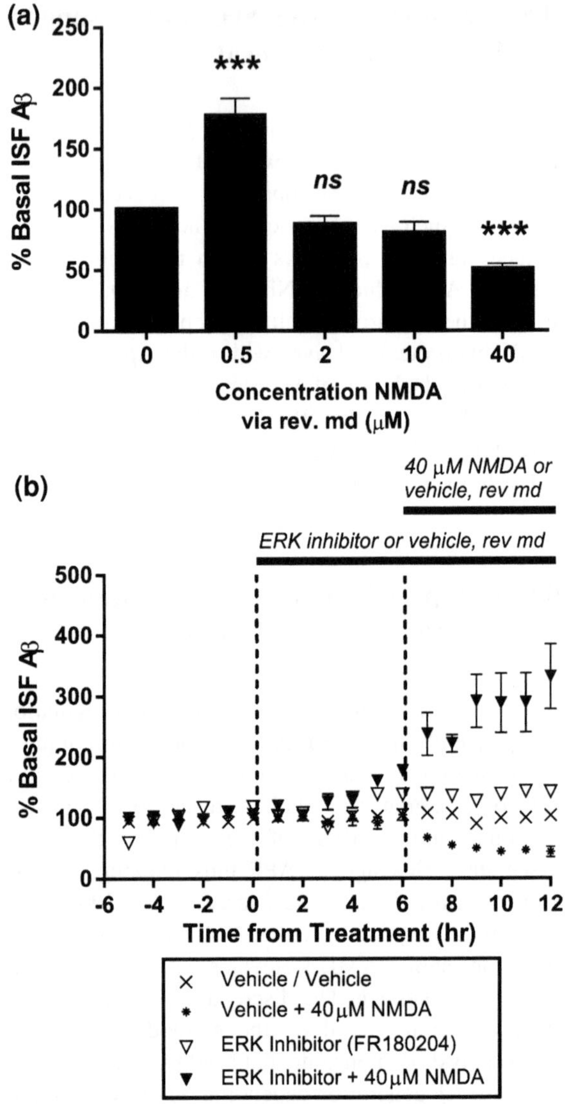

Fig. 9.4 NMDA causes a differential dose-dependent change in ISF Aβ levels in vivo. **a** Infusion of low doses on NMDA (0.5 μM) into the brain by reverse microdialysis increases ISF Aβ levels whereas high doses of NMDA infusion decrease Aβ levels. **b** Inhibition of ERK (FR180204 by reverse microdialysis) increases baseline ISF Aβ levels and blocks the increase in Aβ caused by high doses of NMDA

presynaptic terminal. Blocking ERK activation then applying high doses of NMDA prevents the signaling-dependent change in Aβ and unmasks presynaptic Aβ generation (Fig. 9.4b). The difference between the Lesne and Hoey findings in neuronal culture may be the degree of NMDA-R expression or activity which would shift the balance between the presynaptic Aβ generation and APP processing mechanisms depending on the conditions in each model.

9.10 Model of Synaptic Aβ Metabolism

Postsynaptic receptor activation can lead to presynaptic vesicle exocytosis, meaning inverse mechanisms that regulate Aβ can work concurrently within the same neuron. Low doses of exogenous NMDA depolarize neurons, elevate synaptic transmission, and increase Aβ generation and secretion (Fig. 9.5a). In contrast, high doses of NMDA activate the ERK signaling pathway which downregulates APP processing into Aβ despite increased synaptic activity (Fig. 9.5b). This depression in Aβ appears to be in part due to enhanced α-secretase enzymatic activity. We propose that at low levels of NMDA-R activation, very little ERK is activated, thus the balance is shifted so that synaptic transmission drives presynaptic Aβ generation. Strong NMDA-R activation however recruits ERK which reduces APP processing into Aβ. The balance between enhancing Aβ generation and depressing APP processing determines the total amount of Aβ that a neuron will produce at a given time. Interestingly, mid-range doses of NMDA have no net change in ISF Aβ levels; this likely represents an equal balance of these competing pathways.

Neuronal depolarization and synaptic transmission lead to APP endocytosis, Aβ generation, then secretion (Cirrito et al. 2008). This appears to be a normal metabolic pathway for Aβ generation. Increasing synaptic activity in a cell may simply drive more APP through this pathway and therefore generate more Aβ. In contrast, changing APP processing via secretase activity alters the ability of a cell to produce Aβ. APP is primarily cleaved by α-secretase within the secretory pathway (Lammich et al. 1999) which is upstream of endocytic events that lead to Aβ generation (Koo and Squazzo 1994). Thus, in the presence of high NMDA, elevated α-secretase activity cleaves APP which reduces the substrate available for later Aβ generation within endosomes. When ERK signaling is inhibited, however, high doses of NMDA have no effect on the APP substrate, thus a dramatic increase in synaptic Aβ generation is unmasked.

9.11 Physiologic Fluctuations in ISF Aβ Levels

Pharmacological manipulations, such as described above, can rapidly alter ISF Aβ generation which has provided important details about mechanisms that regulate peptide levels in the brain. These mechanisms appear to be at work normally as well as in the setting of Aβ pathologies similar to AD. Without any external manipulation, however, Aβ levels are still dynamically regulated. ISF Aβ fluctuates in relation to the sleep-wake cycle (Kang et al. 2009). In a mouse model of AD, a microdialysis probe was implanted into the hippocampus as well as electrodes to carefully monitor the sleep state of the mouse. When animals were awake ISF Aβ levels were high, whereas levels were reduced when animals slept. Over the course of 24 h, ISF Aβ levels were cyclic with almost a twofold difference from peak to trough. The regulation appears directly related to signaling by the

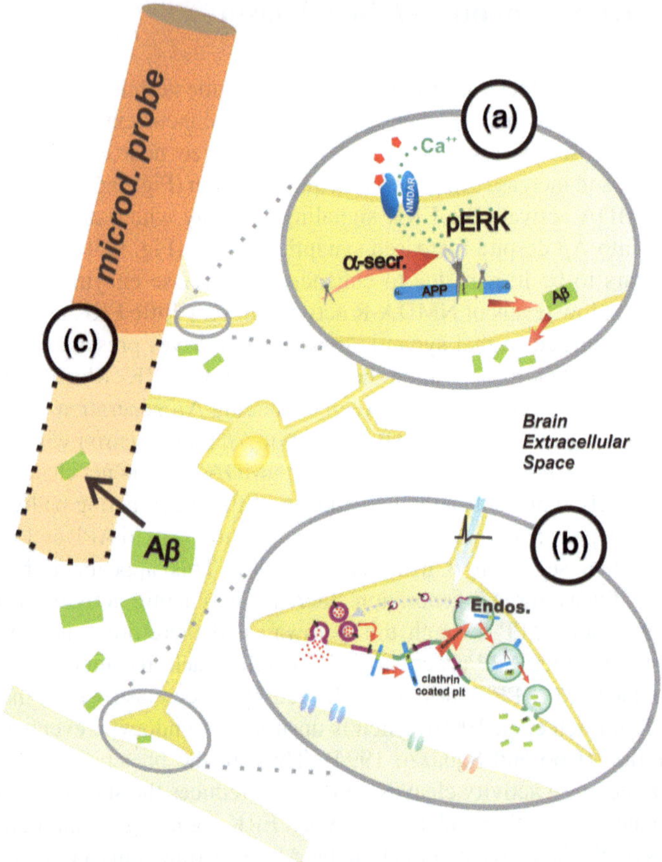

Fig. 9.5 Model of synaptic-dependent Aβ regulation. **a** Neurotransmitter receptors, such as NMDA-Rs, can activate second messenger signaling cascades that culminate in altered APP processing into Aβ. **b** Within the presynaptic terminal, synaptic transmission drives endocytosis of APP. Internalized APP is cleaved to produce Aβ within endosomes which is then secreted into the brain ISF. **c** Microdialysis probes can recover ISF Aβ and measure dynamic changes in peptide levels over time as a consequence of both pharmacological and physiological alterations in Aβ metabolism

neuropeptide orexin which is responsible for promoting the waking state. In humans CSF Aβ levels also fluctuate over the course of 24 h; while it is difficult to determine the extent that CSF Aβ is linked to neuronal activity, those changes are consistent with the synaptic-dependent changes observed in mouse ISF.

Aβ levels can also be altered by physiologic state. Stress has been associated to increased risk of developing AD (Wilson et al. 2003; Wilson et al. 2005; Csernansky et al. 2006). Similarly, mouse models of AD that are chronically stressed develop more Aβ plaques and have impaired memory (Dong et al. 2004).

Stress acutely increases brain ISF Aβ levels within hours which appears directly related to synaptic mechanisms (Kang et al. 2009).

Traumatic brain injury (TBI) can suppress cortical ISF Aβ levels in humans (Brody et al. 2008; Magnoni et al. 2011). In cases involving severe TBI including when patients are in a comatose state, neuronal activity is dramatically suppressed and ISF Aβ levels are low. As a patient's neurological and cognitive status improves, ISF Aβ levels increase as well. In fact the change in neurological status and level of Aβ within the ISF are very tightly correlated.

9.12 Conclusion

ISF Aβ levels are dynamically regulated by a variety of processes and mechanisms within the brain. The concentration of Aβ within the brain is a key determinant of whether and when the peptide will aggregate into toxic species which eventually precipitate AD symptoms. A multitude of factors could alter Aβ levels, including neurological status, psychiatric drugs, as well as the amount of sleep someone receives. The extent that each of these factors regulates Aβ generation over a lifetime likely contributes to an individual's risk of developing AD. Understanding the processes that regulate Aβ levels will be crucial to find and test therapeutic strategies in AD. Microdialysis has provided one useful and unique approach to study such Aβ dynamics.

References

Bacskai BJ, Hyman BT (2002) Alzheimer's disease: what multiphoton microscopy teaches us. Neuroscientist 8:386–390

Bacskai BJ, Klunk WE, Mathis CA, Hyman BT (2002) Imaging amyloid-beta deposits in vivo. J Cereb Blood Flow Metab 22:1035–1041

Beach TG, Walker DG, Potter PE, Sue LI, Fisher A (2001) Reduction of cerebrospinal fluid amyloid beta after systemic administration of M1 muscarinic agonists. Brain Res 905:220–223

Bero AW, Yan P, Roh JH, Cirrito JR, Stewart FR, Raichle ME, Lee JM, Holtzman DM (2011) Neuronal activity regulates the regional vulnerability to amyloid-beta deposition. Nat Neurosci 14:750–756

Brendza RP, Simmons K, Bales KR, Paul SM, Goldberg MP, Holtzman DM (2003) Use of YFP to study amyloid-beta associated neurite alterations in live brain slices. Neurobiol Aging 24:1071–1077

Brody DL, Magnoni S, Schwetye KE, Spinner ML, Esparza TJ, Stocchetti N, Zipfel GJ, Holtzman DM (2008) Amyloid-beta dynamics correlate with neurological status in the injured human brain. Science 321:1221–1224

Buckner RL, Sepulcre J, Talukdar T, Krienen FM, Liu H, Hedden T, Andrews-Hanna JR, Sperling RA, Johnson KA (2009) Cortical hubs revealed by intrinsic functional connectivity: mapping, assessment of stability, and relation to Alzheimer's disease. J Neurosci 29:1860–1873

Buckner RL, Snyder AZ, Shannon BJ, LaRossa G, Sachs R, Fotenos AF, Sheline YI, Klunk WE, Mathis CA, Morris JC, Mintun MA (2005) Molecular, structural, and functional characterization of Alzheimer's disease: evidence for a relationship between default activity, amyloid, and memory. J Neurosci 25:7709–7717

Cirrito JR, Kang JE, Lee J, Stewart FR, Verges DK, Silverio LM, Bu G, Mennerick S, Holtzman DM (2008) Endocytosis is required for synaptic activity-dependent release of amyloid-beta in vivo. Neuron 58:42–51

Cirrito JR, Yamada KA, Finn MB, Sloviter RS, Bales KR, May PC, Schoepp DD, Paul SM, Mennerick S, Holtzman DM (2005a) Synaptic activity regulates interstitial fluid amyloid-beta levels in vivo. Neuron 48:913–922

Cirrito JR, May PC, O'Dell MA, Taylor JW, Parsadanian M, Cramer JW, Audia JE, Nissen JS, Bales KR, Paul SM, DeMattos RB, Holtzman DM (2003) In vivo assessment of brain interstitial fluid with microdialysis reveals plaque-associated changes in amyloid-beta metabolism and half-life. J Neurosci 23:8844–8853

Cirrito JR, Deane R, Fagan AM, Spinner ML, Parsadanian M, Finn MB, Jiang H, Prior JL, Sagare A, Bales KR, Paul SM, Zlokovic BV, Piwnica-Worms D, Holtzman DM (2005b) P-glycoprotein deficiency at the blood-brain barrier increases amyloid-beta deposition in an Alzheimer disease mouse model. J Clin Invest 115:3285–3290

Cirrito JR, Disabato BM, Restivo JL, Verges DK, Goebel WD, Sathyan A, Hayreh D, D'Angelo G, Benzinger T, Yoon H, Kim J, Morris JC, Mintun MA, Sheline YI (2011) Serotonin signaling is associated with lower amyloid-{beta} levels and plaques in transgenic mice and humans. Proc Natl Acad Sci U S A 108:14968–14973

Csernansky JG, Dong H, Fagan AM, Wang L, Xiong C, Holtzman DM, Morris JC (2006) Plasma cortisol and progression of dementia in subjects with Alzheimer-type dementia. Am J Psychiatry 163:2164–2169

Davis AA, Fritz JJ, Wess J, Lah JJ, Levey AI (2010) Deletion of M1 muscarinic acetylcholine receptors increases amyloid pathology in vitro and in vivo. J Neurosci 30:4190–4196

DeMattos RB, Cirrito JR, Parsadanian M, May PC, O'Dell MA, Taylor JW, Harmony JA, Aronow BJ, Bales KR, Paul SM, Holtzman DM (2004) ApoE and clusterin cooperatively suppress abeta levels and deposition. evidence that ApoE regulates extracellular abeta metabolism in vivo. Neuron 41:193–202

Dong H, Goico B, Martin M, Csernansky CA, Bertchume A, Csernansky JG (2004) Modulation of hippocampal cell proliferation, memory, and amyloid plaque deposition in APPsw (Tg2576) mutant mice by isolation stress. Neuroscience 127:601–609

Farris W, Schutz SG, Cirrito JR, Shankar GM, Sun X, George A, Leissring MA, Walsh DM, Qiu WQ, Holtzman DM, Selkoe DJ (2007) Loss of neprilysin function promotes amyloid plaque formation and causes cerebral amyloid angiopathy. Am J Pathol 171:241–251

Fazeli MS, Breen K, Errington ML, Bliss TV (1994) Increase in extracellular NCAM and amyloid precursor protein following induction of long-term potentiation in the dentate gyrus of anaesthetized rats. Neurosci Lett 169:77–80

Gordon-Krajcer W, Salinska E, Lazarewicz JW (2002) N-methyl-d-aspartate receptor-mediated processing of beta-amyloid precursor protein in rat hippocampal slices: in vitro–superfusion study. Folia Neuropathol 40:13–17

Gouras GK, Relkin NR, Sweeney D, Munoz DG, Mackenzie IR, Gandy S (1997) Increased apolipoprotein E epsilon 4 in epilepsy with senile plaques. Ann Neurol 41:402–404

Hartley DM, Walsh DM, Ye CP, Diehl T, Vasquez S, Vassilev PM, Teplow DB, Selkoe DJ (1999) Protofibrillar intermediates of amyloid beta-protein induce acute electrophysiological changes and progressive neurotoxicity in cortical neurons. J Neurosci 19:8876–8884

Holtzman DM, Morris JC, Goate AM (2011) Alzheimer's disease: the challenge of the second century. Sci Transl Med 3:77sr71

Hong S, Quintero-Monzon O, Ostaszewski BL, Podlisny DR, Cavanaugh WT, Yang T, Holtzman DM, Cirrito JR, Selkoe DJ (2011) Dynamic analysis of amyloid beta-protein in behaving mice reveals opposing changes in ISF versus parenchymal Abeta during age-related plaque formation. J Neurosci 31:15861–15869

Kang JE, Lim MM, Bateman RJ, Lee JJ, Smyth LP, Cirrito JR, Fujiki N, Nishino S, Holtzman DM (2009) Amyloid-beta dynamics are regulated by orexin and the sleep-wake cycle. Science 326:1005–1007

Klunk WE, Bacskai BJ, Mathis CA, Kajdasz ST, McLellan ME, Frosch MP, Debnath ML, Holt DP, Wang Y, Hyman BT (2002) Imaging Abeta plaques in living transgenic mice with multiphoton microscopy and methoxy-X04, a systemically administered Congo red derivative. J Neuropathol Exp Neurol 61:797–805

Klyubin I, Walsh DM, Cullen WK, Fadeeva JV, Anwyl R, Selkoe DJ, Rowan MJ (2004) Soluble arctic amyloid beta protein inhibits hippocampal long-term potentiation in vivo. Eur J Neurosci 19:2839–2846

Koo EH, Squazzo SL (1994) Evidence that production and release of amyloid beta-protein involves the endocytic pathway. J Biol Chem 269:17386–17389

Lah JJ, Levey AI (2000) Endogenous presenilin-1 targets to endocytic rather than biosynthetic compartments. Mol Cell Neurosci 16:111–126

Lammich S, Kojro E, Postina R, Gilbert S, Pfeiffer R, Jasionowski M, Haass C, Fahrenholz F (1999) Constitutive and regulated alpha-secretase cleavage of Alzheimer's amyloid precursor protein by a disintegrin metalloproteinase. Proc Natl Acad Sci U S A 96:3922–3927

Lesne S, Ali C, Gabriel C, Croci N, MacKenzie ET, Glabe CG, Plotkine M, Marchand-Verrecchia C, Vivien D, Buisson A (2005) NMDA receptor activation inhibits alpha-secretase and promotes neuronal amyloid-beta production. J Neurosci 25:9367–9377

Mackenzie IR, Miller LA (1994) Senile plaques in temporal lobe epilepsy. Acta Neuropathol (Berl) 87:504–510

Magnoni S, Esparza TJ, Conte V, Carbonara M, Carrabba G, Holtzman DM, Zipfel GJ, Stocchetti N, Brody DL (2012) Tau elevations in the brain extracellular space correlate with reduced amyloid-beta levels and predict adverse clinical outcomes after severe traumatic brain injury. Brain 135:1268–1280

Marcello E, Gardoni F, Mauceri D, Romorini S, Jeromin A, Epis R, Borroni B, Cattabeni F, Sala C, Padovani A, Di Luca M (2007) Synapse-associated protein-97 mediates alpha-secretase ADAM10 trafficking and promotes its activity. J neurosci 27:1682–1691

Marquez-Sterling NR, Lo AC, Sisodia SS, Koo EH (1997) Trafficking of cell-surface beta-amyloid precursor protein: evidence that a sorting intermediate participates in synaptic vesicle recycling. J Neurosci 17:140–151

Meyer-Luehmann M, Stalder M, Herzig MC, Kaeser SA, Kohler E, Pfeifer M, Boncristiano S, Mathews PM, Mercken M, Abramowski D, Staufenbiel M, Jucker M (2003) Extracellular amyloid formation and associated pathology in neural grafts. Nat Neurosci 6:370–377

Mintun MA, Larossa GN, Sheline YI, Dence CS, Lee SY, Mach RH, Klunk WE, Mathis CA, DeKosky ST, Morris JC (2006) [11C]PIB in a nondemented population: potential antecedent marker of Alzheimer disease. Neurology 67:446–452

Mizuno T, Nakata M, Naiki H, Michikawa M, Wang R, Haass C, Yanagisawa K (1999) Cholesterol-dependent generation of a seeding amyloid beta-protein in cell culture. J Biol Chem 274:15110–15114

Morris JC, Price AL (2001) Pathologic correlates of nondemented aging, mild cognitive impairment, and early-stage Alzheimer's disease. J Mol Neurosci 17:101–118

Nelson RL, Guo Z, Halagappa VM, Pearson M, Gray AJ, Matsuoka Y, Brown M, Martin B, Iyun T, Maudsley S, Clark RF, Mattson MP (2007) Prophylactic treatment with paroxetine ameliorates behavioral deficits and retards the development of amyloid and tau pathologies in 3 x TgAD mice. Exp Neurol 205:166–176

Newton AJ, Kirchhausen T, Murthy VN (2006) Inhibition of dynamin completely blocks compensatory synaptic vesicle endocytosis. Proc Natl Acad Sci U S A 103:17955–17960

Nitsch RM, Deng M, Tennis M, Schoenfeld D, Growdon JH (2000) The selective muscarinic M1 agonist AF102B decreases levels of total Abeta in cerebrospinal fluid of patients with Alzheimer's disease. Ann Neurol 48:913–918

Rogers J, Strohmeyer R, Kovelowski CJ, Li R (2002) Microglia and inflammatory mechanisms in the clearance of amyloid beta peptide. Glia 40:260–269

Selkoe DJ (2001) Alzheimer's disease: genes, proteins, and therapy. Physiol Rev 81:741–766

Seubert P, Vigo-Pelfrey C, Esch F, Lee M, Dovey H, Davis D, Sinha S, Schlossmacher M, Whaley J, Swindlehurst C et al (1992) Isolation and quantification of soluble Alzheimer's beta-peptide from biological fluids. Nature 359:325–327

Shankar GM, Li S, Mehta TH, Garcia-Munoz A, Shepardson NE, Smith I, Brett FM, Farrell MA, Rowan MJ, Lemere CA, Regan CM, Walsh DM, Sabatini BL, Selkoe DJ (2008) Amyloid-beta protein dimers isolated directly from Alzheimer's brains impair synaptic plasticity and memory. Nat Med 14:837–842

Shibata M, Yamada S, Kumar SR, Calero M, Bading J, Frangione B, Holtzman DM, Miller CA, Strickland DK, Ghiso J, Zlokovic BV (2000) Clearance of Alzheimer's amyloid-ss(1–40) peptide from brain by LDL receptor-related protein-1 at the blood-brain barrier. J Clin Invest 106:1489–1499

Sloviter RS, Dichter MA, Rachinsky TL, Dean E, Goodman JH, Sollas AL, Martin DL (1996) Basal expression and induction of glutamate decarboxylase and GABA in excitatory granule cells of the rat and monkey hippocampal dentate gyrus. J Comp Neurol 373:593–618

Vassar R et al (1999) Beta-secretase cleavage of Alzheimer's amyloid precursor protein by the transmembrane aspartic protease BACE. Science 286:735–741

Verges DK, Restivo JL, Goebel WD, Holtzman DM, Cirrito JR (2011) Opposing synaptic regulation of amyloid-beta metabolism by NMDA receptors in vivo. J Neurosci 31:11328–11337

Wilson RS, Evans DA, Bienias JL, Mendes de Leon CF, Schneider JA, Bennett DA (2003) Proneness to psychological distress is associated with risk of Alzheimer's disease. Neurology 61:1479–1485

Wilson RS, Barnes LL, Bennett DA, Li Y, Bienias JL, Mendes de Leon CF, Evans DA (2005) Proneness to psychological distress and risk of Alzheimer disease in a biracial community. Neurologys 64:380–382

Wyss-Coray T, Loike JD, Brionne TC, Lu E, Anankov R, Yan F, Silverstein SC, Husemann J (2003) Adult mouse astrocytes degrade amyloid-beta in vitro and in situ. Nat Med 9:453–457

Yin KJ, Cirrito JR, Yan P, Hu X, Xiao Q, Pan X, Bateman R, Song H, Hsu FF, Turk J, Xu J, Hsu CY, Mills JC, Holtzman DM, Lee JM (2006) Matrix metalloproteinases expressed by astrocytes mediate extracellular amyloid-beta peptide catabolism. J Neurosci 26:10939–10948

Chapter 10
Microdialysis in Pain Research

Martin Brunner

Abstract Microdialysis is a recognized experimental method in pain research in animals and humans. In the last 20 years a growing number of studies have provided new details on pain pathophysiology and the effect of interventions focused on pain relief such as drug treatment or training using microdialysis either alone or in combination with other methods. This chapter focuses on the use of clinical microdialysis in pain research in healthy subjects and patients and summarizes its application in several key indications such as the evaluation of chronic pain conditions in skeletal muscle, tendon tissue, or the knee, as well as the evaluation of local anesthetics and non-steroidal anti-inflammatory drugs. Thereby, it can be demonstrated that clinical microdialysis contributes to the existing knowledge of pain pathophysiology and has the potential to essentially contribute to decision making in the discovery and evaluation of novel or established pain treatment approaches.

10.1 Introduction

In 1931, Dr. Albert Schweitzer declared that "Pain is a more terrible lord of mankind than even death itself". This is still true in the twenty-first century, where the high prevalence and impact of pain-related health issues is reflected by the fact that pain affects more Americans than diabetes, heart disease, and cancer combined and is a leading cause of disability and a major contributor to healthcare costs

M. Brunner (✉)
Department of Clinical Pharmacology, Medical University of Vienna,
Waehringer Guertel 18-20, A-1090 Vienna, Austria
e-mail: martin.brunner@meduniwien.ac.at

M. Müller (ed.), *Microdialysis in Drug Development*, AAPS Advances in the Pharmaceutical Sciences Series, DOI: 10.1007/978-1-4614-4815-0_10,

(http://report.nih.gov/NIHfactsheets/Pdfs/PainManagement%28NINR%29.pdf). Although there is no standard definition, the International Association for the Study of Pain (IASP) describes pain as an unpleasant sensory and emotional experience associated with actual or potential tissue damage, or described in terms of such damage (http://www.iasp-pain.org/). Pain can be classified as acute or chronic according to its duration, or neuropathic or nociceptive according to the involved lesion or disease, i.e., the somatosensory nervous system or non-neural tissue and activation of nociceptors, respectively. While the cause of acute pain can usually be identified and treated with analgesics or anesthetics, chronic pain is regarded as a serious chronic health problem requiring access to management akin to other chronic diseases such as diabetes or chronic heart disease (http://www.iasp-pain.org/PainSummit/DeclarationOfMontreal.pdf).

Pain perception is highly heterogeneous and subjective. It involves interaction of peripheral and cortical structures and is mixed with environmental and experiential factors (Langley et al. 2008). This frequently results in diagnostic shortcomings and subsequently undertreatment of patients in acute and chronic situations. Furthermore, in spite of considerable research efforts, effective and safe treatments for many human pain disorders still remain elusive.

Pain research requires research on animals and humans. Pain research aims at supplementing and improving the existing knowledge on pathogenesis, mechanisms, diagnosis, and treatment of pain, with the intention to provide results that can ultimately be translated into therapeutic approaches for patient benefit. Microdialysis is one established experimental technique that has been employed in pain research in animal models of acute and chronic pain conditions for more than 20 years to study extracellular concentrations of pain-related mediators in different regions of the central nervous system (Stiller et al. 2003, 2007). Those studies have provided important findings describing mechanisms that contribute to pain modulation. In particular, such studies allowed correlations between behavioral signs of pain and CNS neurotransmitter release (Stiller et al. 2007). Animal pain models, however, might not always be representative for the clinical situation or a specific pain condition in humans. Thus, clinical microdialysis in healthy human subjects and patients has been increasingly used to investigate experimentally induced forms of pain, as well as several pain conditions and it has been used to elucidate pharmacokinetics and pharmacodynamics of analgesic medication.

Technical aspects of microdialysis in animal models of pain and studies describing the effects of noxious stimulation on neurotransmitter concentrations in spinal cord or cerebral pain modulatory centers have been reviewed in detail previously (Stiller et al. 2003, 2007). It could be shown that most preclinical research has been done in the context of acute nociceptive pain models describing the release of amino acids, monoamines, neuropeptides, and prostanoids, whereas more recently, microdialysis has been used to elucidate neurochemical mechanisms underlying chronic pain conditions (Stiller et al. 2003). The authors concluded that novel treatment approaches would need to be based on the results of preclinical as well as clinical pain studies in different pain conditions.

Value and necessity of animal models in pain research are currently controversially discussed (Langley et al. 2008; Mogil 2009; Mogil et al. 2010). Some call for the replacement of animal pain experiments with human volunteer studies as they argue that animal models are insufficiently representative of the multidimensional aspects of clinical pain (Langley et al. 2008). Others believe that pain research using animal models is still an ongoing necessity for the development of novel analgesics for the treatment of chronic pain, to understand fundamental mechanisms, identify new analgesic targets, and inform, guide, and follow-up human studies (Mogil et al. 2010). Data derived from microdialysis experiments are equally important for both, pre-clinical and clinical pain research. In particular, in clinical pain research microdialysis has become an important complementary experimental technique, which provides valuable insight into the pathophysiology of different pain syndromes, usually in combination with other approaches such as gene expression analysis, immunohistochemistry, ultrasonography, or laser Doppler scanning.

In the last years, the number of studies that have employed microdialysis to address pain-related issues has been steadily growing. Thus, this chapter will summarize the existing literature with the focus on the use of microdialysis in clinical pain research in healthy subjects and patients. For this purpose, the use of microdialysis in several key indications will be presented.

10.2 Clinical Microdialysis for the Evaluation of Painful Conditions in Tendon Tissue and the Knee

10.2.1 Clinical Microdialysis in Tendon Tissue

Chronic painful conditions of the Achilles or patellar tendons, usually referred to as tendinopathy, are relatively common and affect not only recreational exercisers or athletes but also inactive people. The exact etiology and pathogenesis of chronic tendon pain, however, is unknown. Tendinopathies are difficult to treat and an inflammatory component had been suspected. Consequently, anti-inflammatory strategies with non-steroidal anti-inflammatory drugs (NSAID) or corticosteroids are frequently employed, although tendon biopsies have shown an absence of inflammatory cell infiltration (Alfredson et al. 2000a).

Clinically, microdialysis was employed in patients with chronic Achilles tendinosis, to measure local concentrations of glutamate and prostaglandin E2 under resting conditions (Alfredson et al. 1999). Four times higher levels of the excitatory neurotransmitter glutamate, which is responsible for central and peripheral pain registration, were detected in painful tendons when compared with normal tendons. Prostaglandin E2, which serves an inflammation marker, however, was not elevated, which indicates a lack of local inflammation in chronic Achilles tendinosis (Alfredson et al. 1999). Those findings could be confirmed in patients with tennis elbow, where microdialysis was performed in the extensor carpi radialis brevis tendon (Alfredson et al. 2000b), although it was not clear, whether higher glutamate concentrations were the result of local hyperproduction or an increased axonal transport of glutamate into the area with tendon changes. This question was addressed in combination with immuno-histochemical analysis and enzyme histochemistry of Achilles and patellar tendon biopsies (Alfredson et al. 2001a, b). Both, free glutamate and glutamate NMDAR1 receptors were found in tendons from patients with chronic tendon pain and in controls, whereas glutamate NMDAR1 receptors were localized in association with nerve structures (Alfredson et al. 2002a). To study the influence of a treatment approach with eccentric calf-muscle training, glutamate concentrations were assessed in patients with painful chronic Achilles tendinosis before and after training with the hypothesis that eccentric training would lower glutamate concentrations (Alfredson and Lorentzon 2003). After training, patients were pain free, high intra-tendinous glutamate concentrations, however, remained unchanged, which was regarded as a possible result of a decreased sensitivity to glutamate on receptor level (Alfredson and Lorentzon 2003). Those studies demonstrated the feasibility of microdialysis in tendon tissue under resting conditions without complications. Furthermore, the findings indicated that glutamate might be involved in chronic tendon pain, whereas the constant finding of normal prostaglandin E2 concentrations points at a lack of chemical inflammation in the chronic state of tendinopathies (Alfredson and Lorentzon 2003).

To investigate, whether an ischemic component could be relevant for pain development in chronic Achilles tendinosis, local lactate concentrations were measured (Alfredson et al. 2002b). It could be demonstrated that lactate concentrations were significantly higher in painful tendons, indicating potential anaerobic conditions. Whether ischemia precedes the start of tendinosis, or whether changes in the tendon give rise to ischaemia, remains to be shown (Alfredson et al. 2002b).

Although patients with chronic tendinopathies do not show signs of local inflammation (Alfredson et al. 1999), NSAIDs are frequently prescribed. Recently, it has been shown by means of microdialysis that prostaglandin concentrations in plasma or tissue can be blocked by ingestion or local infusion of NSAIDs (Mikkelsen et al. 2008) and that NSAIDs have the potential to interfere with the adaptive activation of satellite cells of skeletal muscle in response to loading by cyclooxygenase-dependent mechanisms (Mackey et al. 2007). As the consequences of NSAID intake on tendon tissue physiology were unknown, local peritendinous prostaglandin E2 concentrations and patella tendon collagen synthesis were analyzed in response to training during NSAID intake in healthy subjects (Christensen et al. 2011). Given that prostaglandin E2 is a growth factor for collagen tissue, it was hypothesized that NSAID treatment would lead to a decrease of prostaglandin E2 and a subsequent decrease in the exercise induced increase in collagen synthesis (Christensen et al. 2011). It could be shown that intake of NSAID has the anticipated effect on collagen synthesis and that NSAID use may be detrimental to the adaptive response of the tissue to exercise.

This might also have profound implications for the optimal treatment of patients with tendinopathies and other tendon disorders (Christensen et al. 2011).

10.2.2 Clinical Microdialysis in the Knee

Clinical microdialysis has also been performed in the knee to monitor postoperative changes in biochemical markers or to assess the effect of drug therapy or cooling on postoperative pain in the synovial membrane.

Increased lactate and glycerol concentrations in the synovial membrane indicated local tissue damage (Högberg et al. 2006a). To investigate whether the local trauma response is related to subjective pain, local tissue metabolism was measured after arthroscopy and it could be demonstrated that pain was associated with increased glucose utilization and prostaglandin E2 production by the synovial membrane (Högberg et al. 2006b; Stålman et al. 2008). Intra-articular ketorolac effectively exerted an effect on inflammation and tissue metabolism by attenuating postoperative prostaglandin E2 concentrations and increasing glycerol concentrations whereby differences in prostaglandin E2 in the reference tissue furthermore indicated a potential systemic effect of ketorolac (Stålman et al. 2009). A temperature-sensitive decrease in prostaglandin E2 concentrations as well as a decreased metabolic rate of the synovial tissue were shown after local cryotherapy after knee arthroscopy, although no direct effect on postoperative pain could be detected (Stålman et al. 2011). The results indicate an anti-inflammatory rather than direct anti-nociceptive effect of cooling by the observed synovial prostaglandin E2 reduction.

So far, two studies were performed in patients with knee osteoarthritis, a degenerative joint condition, which is characterized by pain that is linked to functional impairment and disability. Intra-articular microdialysis in the knee proved to be feasible and it could be demonstrated that there are differences in local metabolite concentrations between different compartments within the knee joint (Helmark et al. 2010a). Furthermore, it could be demonstrated that exercise increases interleukin-10 concentrations intra-articularly and perisynovially in patients with knee osteoarthritis, a finding that suggests a positive effect of exercise on a chondroprotective anti-inflammatory cytokine response that might contribute to explaining the beneficial effect that exercise has on osteoarthritis (Helmark et al. 2010b).

10.3 Clinical Microdialysis in Skeletal Muscle, the Skin, and the Spinal Cord for the Evaluation of Chronic Pain Conditions

10.3.1 Chronic Myalgia

Work-related chronic pain in neck and shoulders, including trapezius myalgia, is an increasing problem in the general population in the industrialized world, with a prevalence of 10–20 % in the community (Punnett and Wegman 2004). The understanding of the pathophysiology of chronic myalgia, however, is still incomplete. The human trapezius muscle is regarded as an appropriate model muscle of chronic myalgia as it is involved in chronic neck and shoulder pain and because it is relatively easy to access for invasive investigations. Several studies have used microdialysis in the trapezius muscle to study nociceptive and metabolic mechanisms in chronic human pain conditions, such as work-related chronic trapezius myalgia (Rosendal et al. 2004, 2005; Larsson et al. 2008; Gerdle et al. 2008; Sjøgaard et al. 2010; Flodgren et al. 2010), chronic whiplash associated disorders involving the trapezius muscle (Gerdle et al. 2008), chronic myofascial pain (Shah and Gilliams 2008), fibromyalgia (Gerdle et al. 2010), or polymyalgia rheumatica (Kreiner and Galbo 2011).

The majority of these studies report altered concentrations of algesic substances together with metabolic alterations in myalgic muscles, thus implicating changes in analgesic signaling as a key event in pain development. For example, female patients with chronic trapezius myalgia had higher resting interstitial trapezius levels of lactate, pyruvate, glutamate, and serotonin than healthy controls with a more pronounced interstitial muscle lactate increase after repetitive low-force arm work (Rosendal et al. 2004), indicating increased metabolism associated with enhanced release of nociceptive substances, whereas others found no differences in glutamate or prostaglandin E2 concentrations in patients and healthy subjects (Flodgren et al. 2005; Ashina et al. 2005). Furthermore, myalgia was associated with increased interstitial concentrations of interleukin 6, LDH, potassium, alterations in collagen turn over (Rosendal et al. 2005; Larsson et al. 2008), or different interstitial concentrations of bradykinin (Gerdle et al. 2008; Shah and Gilliams 2008) or cytokines (Gerdle et al. 2008). In combination with Near-Infrared Spectroscopy (NIRS) muscle oxygenation as well as intramuscular tissue chemistry in reponse to repetitive work were assessed (Flodgren et al. 2010; Sjøgaard et al. 2010) Recently, changes in pain-inhibitory signaling molecules, such as N-acylethanolamines (NAE), a family of endogenous lipid mediators that have a diversity of actions including the regulation of inflammation and pain, have been shown to be significantly higher in myalgic patients than in healthy controls in a study that combined microdialysis with nano liquid chromatography tandem mass spectrometry (nLC-MS/MS) (Ghafouri et al. 2011).

10.3.2 Myofascial Pain Syndromes

Myofascial pain syndromes originating from myofascial trigger points (MTrPs) are among the most frequently encountered pain conditions in the general population with a prevalence of up to 85 % in people >65 years (Giamberardino et al. 2011). They are usually the result of an overuse or muscle stress syndrome that is characterized by and leads to MTrPs. They are, however, often misdiagnosed and the underlying pathophysiology is complex and still needs to be fully elucidated. MTrPs are defined as hyperirritable nodules located within a taut band of skeletal muscle that may be active and painful or latent, whereas active MTrPs activate muscle nociceptors that, upon sustained noxious stimulation, initiate motor and sensory changes in the peripheral and central nervous systems (Shah and Gilliams 2008). It is speculated that transformation of a tender nodule into a painfully activated MTrP is associated with the activation of muscle nociceptors by a variety of endogenous substances including neuropeptides, arachidonic acid derivatives, or inflammatory mediators (Shah et al. 2005). Several studies employed microdialysis to quantitatively measure tissue biochemistry in active human MTrPs. Microdialysis was performed in latent and active MTrPs in the trapezius muscle and was compared with healthy muscle tissue. Concentrations of protons, bradykinin, calcitonin gene-related peptide, substance P, tumor necrosis factor-alpha, interleukin-1beta, serotonin, norepinephrine, interleukin-6, and interleukin-8 were found to be significantly higher in the active group than either of the other two groups, which allowed to distinguish between clinically distinct groups based on measured analyte concentrations (Shah et al. 2005; Shah and Gilliams 2008). Subjects with active MTrPs in the trapezius further presented with elevated levels of inflammatory mediators, neuropeptides, catecholamines, and cytokines in the gastrocnemius muscle, suggesting that elevations of biochemicals associated with pain and inflammation may not be limited to localized areas of active MTrPs (Shah and Gilliams 2008). It was recommended that future microdialysis studies should compare MTrPs with various degrees of activity, which would increase the understanding of the nature of MTrPs.

Increased muscle tenderness is a consistent finding in patients with tension-type headache and has been ascribed to local inflammation and release of inflammatory mediators. To study interstitial concentrations of prostaglandin E2, adenosine 5′-triphosphate, glutamate, bradykinin in a tender point of patients with chronic tension-type headache, microdialysis was performed in the trapezius muscle in patients and healthy controls and inflammatory mediators were measured at rest and after static exercise. No difference could be seen between patients and controls providing evidence that inflammatory mediators and metabolites are comparable to the healthy state, which indicates that tender points are not sites of ongoing inflammation (Ashina et al. 2003).

10.3.3 Fibromyalgia

One important differential diagnosis of myofascial pain syndromes is fibromyalgia, a chronic, widespread muscle tenderness syndrome that is often accompanied by multiple somatic syndromes. Although fibromyalgia affects approximately 2 % of the populations with females being affected to a larger extent, the etiology of the disease is largely unknown. To test for metabolic alterations in the trapezius muscle of female patients with fibromyalgia microdialysis was performed in the symptomatic trapezius and tibialis anterior muscles (Gerdle et al. 2010). Lactate, pyruvate, and glutamate concentrations were compared between symptomatic patients and healthy controls and were correlated with pressure pain thresholds of both muscles (Gerdle et al. 2010). In patients higher interstitial lactate and pyruvate concentrations could be detected that correlated more strongly with pain pressure thresholds than in controls (Gerdle et al. 2010). Leukotriene B4 and prostaglandin E2 concentrations were furthermore measured in the masseter muscle of patients with fibromyalgia. Leukotriene B4 was elevated and prostaglandin E2 was positively correlated to muscular pain, which indicates that pain is partly of peripheral inflammatory origin in this patient population (Hedenberg-Magnusson et al. 2001). In summary, the use of microdialysis in fibromyalgia increased knowledge of how metabolic and pro- and anti-nociceptive substances are altered and how they are related to chronic pain, which might influence the design of effective new rehabilitation interventions (Gerdle et al. 2010).

10.3.4 Polymyalgia Rheumatica

Recently, microdialysis has been employed to elucidate disease mechanisms in symptomatic muscles of patients suffering from polymyalgia rheumatica, a frequently observed chronic inflammatory condition of elderly patients, who present with muscle , tenderness, and stiffness. Although muscle symptoms predominate clinically, the pathophysiology of polymyalgia is thought to be dominated by inflammation in synovial structures (Healey 2006). Interstitial concentrations of pro-inflammatory and anti-inflammatory cytokines were measured in patients before and after abolition of their clinical symptoms by 14 days of treatment with prednisolone in symptomatic vastus lateralis and trapezius muscles (Kreiner et al. 2010). After prednisolone, clinical symptoms as well as initially increased interstitial cytokine concentrations were normalized. The study introduced a novel aspect into the pathophysiology of polymyalgia rheumatica, indicating that increased levels of pro-inflammatory cytokines in symptomatic muscle contribute to the disease, and that cytokines may be released locally, as indicated by constantly higher interstitial concentrations as compared to plasma concentrations (Kreiner et al. 2010). The notion that intramuscular disease mechanisms are an important aspect in polymyalgia rheumatica is also supported by an increase in the concentrations of the potentially pain-inducing molecules glutamate and presumably prostaglandin E2 in symptomatic muscle that varied in parallel with disease activity, whereas ischemia did not contribute to muscle pain (Kreiner and Galbo 2011)

10.3.5 Complex Regional Pain Syndrome

Complex regional pain syndromes (CRPS) are disorders that develop after a trauma affecting a limb or after peripheral nerve lesions. Characteristic symptoms are pain, hyperalgesia, impairment of motor function, edema, increased skin temperature, trophic changes, or reddening (Leis et al. 2004). The combination of such symptoms suggests inflammation. Classical inflammation, however, has not been proven and thus, neurogenic inflammation is assumed, in which neuropeptides are released and lead to vasodilatation and plasma protein extravasasion (Leis et al. 2004). Increased neuropeptide release could be verified by intradermal microdialysis in CRPS patients, which suggested facilitated neurogenic inflammation in CRPS (Weber et al. 2001). Furthermore, there is evidence that impairment of neuropeptide degradation in CRPS patients is not restricted to the affected tissue, because exogenous substance P application induced significantly increased protein extravasation on both affected and unaffected limbs, irrespective of CRPS activity (Leis et al. 2003, 2004). To investigate whether facilitated neurogenic inflammation may be a possible risk factor for CRPS, electrically evoked neurogenic inflammatory response was examined in patients with clinically resolved CRPS on the contralateral, previously unaffected limb (Leis et al. 2004). No protein extravasation could be detected, which indicates that an initiating trauma is necessary to induce neuropeptide synthesis (Leis et al. 2004). Intracutaneous delivery of peptidase inhibitors by microdialysis was employed to explore the effect of neutral endopeptidase inhibition on neurogenic inflammation as insufficient degradation of neuropeptides might contribute to CRPS (Krämer et al. 2005). It could be demonstrated that neutral endopeptidase affects neurogenic vasodilation and calcitonin gene-related peptide concentrations in the human skin, an important finding for the mechanistic understanding of CRPS, which has been associated with increased activity of gene-related peptide concentrations (Krämer et al. 2005).

10.3.6 Clinical Microdialysis in the Spinal Cord

An in vivo microdialysis approach was developed allowing to measure extracellular changes of several neurotransmitter and nonneurotransmitter amino acids shown to be involved in pain control in the spinal cord dorsal horn, an important structure responsible for pain control processing (Mertens et al. 2000, 2001;

Parrot et al. 2004). Microdialysis using custom-made probes was performed during a specific microsurgical procedure in the dorsal root entry zone, which aims at treating selected cases of chronic pain in pharmacoresistant patients. It could be demonstrated that direct access to the dorsal horn is feasible and safe in humans and provides essential information on functional biochemistry of chronic neuropathic pain in the spinal cord with more anatomical specificity than previously available by cerebrospinal fluid sampling (Mertens et al. 2000, 2001; Parrot et al. 2004).

10.4 Clinical Microdialysis for the Evaluation of Local Anesthetics and Non-Steroidal Anti-inflammatory Drugs (NSAIDs)

10.4.1 Local Anesthetics

The effect of needle insertion trauma and anesthesia on skin perfusion, erythema, and skin thickness was studied by intracutaneous microdialysis in healthy subjects in the dorsal forearm skin (Groth and Serup 1998). Subjects received local anesthesia with lidocaine in both forearms and a microdialysis probe was inserted in one of the arms. Insertion trauma and the effect of anesthesia on skin blood flow and erythema were studied by laser Doppler perfusion imaging, trauma-induced edema and effects on skin thickness were studied by ultrasound imaging. Microdialysis probes were also inserted in four subjects without prior anesthesia and the effects on skin blood flow and erythema were investigated. It could be demonstrated that microdialysis probe insertion induced significant increases in skin blood flow, erythema and skin thickness, whereas local anesthesia prior to the insertion reduced traumatic effects (Groth and Serup 1998). Those results provided evidence that approximately 2 h are required after probe insertion to allow the vascular reaction to needle trauma to return to baseline conditions (Groth and Serup 1998).

Additives to local anesthetics for peripheral and central neuraxial blockades have been used for many years to prolong and enhance the anesthetic effect. The addition of epinephrine is thought to decrease local blood flow, delay clearance from the injection site and consequently to affect the anesthetic effect. In a study in healthy subjects who received lidocaine with and without the addition of epinephrine, it could be shown that epinephrine prolonged the sensory block by decreasing local blood flow and slowing clearance (Bernards and Kopacz 1999). In a similar study, addition of clonidine to lidocaine also prolonged the duration of the peripheral nerve block by affecting blood flow and clearance, which suggested that the observed changes were at least in part pharmacokinetically mediated (Kopacz and Bernards 2001). To investigate how a mild surgical trauma such as arthroscopy followed by intra-articular injection of bupivacain with adrenaline alters synovial and adipose tissue metabolism as well as local blood flow post-operatively, glucose, lactate, and glycerol were monitored with microdialysis (Högberg et al. 2006a). Although pronounced effects on metabolism and blood flow in the synovial membrane of the knee could be detected, they were not mediated by the addition of adrenaline to the local anesthetic (Högberg et al. 2006a).

As it is known that the sympathetic nervous activity, which can be attenuated by epidural anesthesia, influences lipolysis and the release of glycerol, the impact of postoperative thoracic epidural analgesia on extracellular glycerol concentrations and glucose metabolism in subcutaneous adipose tissue was assessed using microdialysis in patients after major abdominal or thoraco-abdominal surgery (Ederoth et al. 2002). Differences in local lipolysis could be shown between epidural analgesia and intravenous analgesia, which might be explained by a regional metabolic influence of thoracic epidural analgesia, possibly via the sympathetic nervous system (Ederoth et al. 2002). To study whether interrupting neuronal signaling using a peripheral nerve block modifies local nociceptive and inflammatory processes in inflammatory skin of healthy subjects, a femoral nerve block or a sham block was established before inducing an experimental UVB burn (Carvalho et al. 2010). Single point regional anesthesia minimally affected wound hyperalgesia and inflammation indicating that other anesthetic techniques might be better suited to achieve analgesia (Carvalho et al. 2010). Recently, pharmacokinetics of the local anesthetic ropivacaine were assessed after transversus abdominis plane (TAP) block in healthy volunteers as it was unknown to which extent systemic absorption and redistribution to the abdominal wall contribute to the anesthetic effect (Latzke et al. 2012). It could be confirmed that the TAP block led to high local, although also highly variable, ropivacaine concentrations in the abdominal wall, whereas systemic concentrations were comparable between subjects (Latzke et al. 2012).

Local anesthetics have also been locally applied by iontophoresis with the aim to enhance local drug delivery to the desired treatment area. What was unknown so far, however, was, whether the anesthetic reached the tissue beneath the application site in sufficiently large quantities to exert a therapeutic effect. Quantifiable lidocaine concentrations could only be detected in superficially located subcutaneous tissue when lidocaine was applied alone (Coglianese et al. 2011). When lidocaine was combined with epinephrine, however, iontophoretic drug delivery could be increased to deeper tissue layers, thereby confirming the vasoconstrictor effect of epinephrine on local drug pharmacokinetics as described previously (Draper et al. 2011).

In addition to blocking nociceptive input from surgical sites, long acting local anesthetics might directly modulate inflammation (Gordon et al. 2008). To address this question, the proinflammatory effects of bupivacaine on local prostaglandin E2 production and cyclooxygenase gene expression were assessed and it could be shown that bupivacaine stimulates cyclooxygenase-2 gene expression after tissue injury, which is associated with higher prostaglandin E2 production and pain after the local anesthetic effect dissipates in healthy subjects (Gordon et al. 2008).

10.4.2 Non-Steroidal Anti-Inflammatory Drugs

A critical issue in the treatment with topical drug formulations designed to act locally in the tissue beneath the application site is the uncertainty whether tissue concentrations high enough to account for the reported analgesic effects are really attained. To address the question of target site penetration of analgesic substances, microdialysis has frequently been used to assess tissue pharmacokinetics of several topical NSAIDs in the dermis, subcutaneous adipose tissue and skeletal muscle tissue. For diclofenac, ketoprofen and ibuprofen a high interindividual variability in tissue penetration could be observed after single (Müller et al. 1997; Tegeder et al. 1999, 2001) and multiple dose administration (Dehghanyar et al. 2004; Müller et al. 1998) in healthy subjects. Furthermore, it could be demonstrated that one factor governing skin penetration are the specific galenic properties of the formulation (Brunner et al. 2005, 2011; Müller et al. 1998).

To relate tissue concentrations to anti-inflammatory or anti-nociceptive actions and to elucidate the pathophysiology of NSAID treatment at the site of action, the effect of NSAIDs on cyclooxygenase inhibition was assessed in different clinical pain models. Most studies were performed in the human dental pain model, as pain following removal of impacted third molar teeth has proven to be a useful clinical model for evaluating oral analgesics (Cooper and Desjardins 2010). In this model micodialysis probes are inserted into extraction sockets to collect exudates for measuring biochemical mediators of pain or drug levels at the site of injury (Cooper and Desjardins 2010). It could be demonstrated that oral NSAIDs such as flurbiprofen, ibuprofen or ketorolac altered bradykinin (Swift et al. 1993), prostaglandin E2 (O'Brien et al. 1996; Roszkowski et al. 1997), or thromboxane B2 (Gordon et al. 2002) concentrations in inflamed tissue at the surgical site. Thereby, alterations of prostanoid concentrations were related to target site concentrations of NSAIDs and led to measurable pain reduction. Results from the dental pain model provided insight into peripheral mechanisms of inflammatory mediator release and mode of action of NSAIDs. A recent study examined messenger RNA (mRNA) expression of cyclooxygenase isoforms after treatment of rofecoxib and naproxen and found that both drugs were equally effective against postoperative pain while acting with different time patterns to modulate prostaglandin E2 release at the surgical site, which is likely the result of the pharmacologic modulation of cyclooxygenase-2 constitutively active at distinct locations from the surgical area (Fornai et al. 2006). The question whether the antihyperalgesic affect of diclofenac may include both peripheral and central sites of actions was assessed in a human experimental pain model, in which inflammation was mechanically induced (Burian et al. 2003). At comparable tissue concentrations oral diclofenac was more effective than topical diclofenac, suggesting that not only peripheral but also central mechanisms are involved (Burian et al. 2003).

Further information on the pathophysiology of cyclooxygenase inhibition was derived from the measurement of cytokines in the interstitial fluid collected from non-inflamed and experimentally inflamed skin during ibuprofen treatment, which suggested that ibuprofen exerts anti-inflammatory effects by downregulating tissue cytokine levels (Angst et al. 2008). The authors summarized that the combined use of microdialysis and multiplex immunoassay technology for measuring cytokine arrays in tissue provides valuable insight into the complex interactions of biochemical markers mediating pain and inflammation and that this approach can be used for exploring analgesic and anti-inflammatory actions of known and novel drugs in the future (Angst et al. 2008).

10.5 Summary

The pathophysiology of human pain is highly versatile. Consequently, a variety of different techniques is required to better define the mechanisms contributing to this complex phenomenon. In the last 20 years, microdialysis has become an established technique in pain research in animals and in humans. In animals, direct in vivo access to pain-related CNS regions allows the online measurement of neurotransmitter concentrations in different brain areas and the spinal cord and thus provided new insight into the mechanisms how neurotransmitter release is responsible for transmitting and modulating pain signals.

In humans, clinical microdialysis has been used to describe previously unknown details on alterations of metabolic, inflammatory and pro- or anti-nociceptive mediators in several chronic pain conditions. By characterizing new details on disease pathophysiology, for example showing a lack of an inflammatory component in chronic painful tendinopathies, current treatment approaches, such as the administration of NSAIDs in such conditions, might be critically re-evaluated. Furthermore, by relating information on tissue biochemistry to clinical information on pain modification, existing therapies can be optimized or novel effective interventions can be designed and tested. In the design and discovery of new analgesic substances, microdialysis derived data on drug pharmacokinetics and pharmacodynamics can be used to facilitate the decision whether a novel analgesic compound warrants further development. As microdialysis can be performed in healthy subjects and also patients with the target disease, drug evaluation can be performed considering clinically relevant pathophysiological mechanisms, which is usually not feasible in experimental animal studies. Furthermore, the combination of microdialysis with other approaches such as gene expression analysis, immunohistochemistry, ultrasonography, laser Doppler scanning, NIRS, or neuroimaging provides a comprehensive assessment of the pathophysiological status of a certain pain condition.

Limitations of the technique in clinical pain research are derived from the fact that for the investigation of certain pain conditions, such as joint osteoarthritis, microdialysis is an invasive surgical procedure under strict sterile conditions, which contrasts to the usually semi-invasive nature of the technique. The use in combination with surgery also limits the sampling duration and poses challenges with regard to in vivo recovery assessment. In many instances, it is also not possible to perform clinical microdialysis using commercially available instrumentation. This could be due to technical issues, as for example new probes had to be designed for the use in the human spinal cord, or due to analytical issues, when highly sensitive analytical methods have to be adapted for analyte detection. Furthermore, sampling of certain biomarkers requires the use of microdialysis probes with a high molecular mass cutoff that might not be readily available for clinical use.

In conclusion, microdialysis is a recognized experimental instrument in pain research that complements the existing knowledge of pain pathophysiology and has the potential to essentially contribute to decision making in the discovery and evaluation of novel or established pain treatment approaches.

References

Alfredson H, Thorsen K, Lorentzon R (1999) In situ microdialysis in tendon tissue: high levels of glutamate, but not prostaglandin E2 in chronic Achilles tendon pain. Knee Surg Sports Traumatol Arthrosc 7:378–381

Alfredson H, Lorentzon R (2000) Chronic achilles tendinosis: recommendations for treatment and prevention. Sports Med 29:135–146

Alfredson H, Ljung BO, Thorsen K, Lorentzon R (2000) In vivo investigation of ECRB tendons with microdialysis technique–no signs of inflammation but high amounts of glutamate in tennis elbow. Acta Orthop Scand 71:475–479

Alfredson H, Forsgren S, Thorsen K, Fahlström M, Johansson H, Lorentzon R (2001a) Glutamate NMDAR1 receptors localised to nerves in human Achilles tendons. Implications for treatment? Knee Surg Sports Traumatol Arthrosc 9:123–226

Alfredson H, Forsgren S, Thorsen K, Lorentzon R (2001b) In vivo microdialysis and immunohistochemical analyses of tendon tissue demonstrated high amounts of free glutamate and glutamate NMDAR1 receptors, but no signs of inflammation, in Jumper's knee. J Orthop Res 19:881–886

Alfredson H, Lorentzon R (2002) Chronic tendon pain: no signs of chemical inflammation but high concentrations of the neurotransmitter glutamate. Implications for treatment? Curr Drug Targets 3:43–54

Alfredson H, Bjur D, Thorsen K, Lorentzon R, Sandström P (2002) High intratendinous lactate levels in painful chronic Achilles tendinosis. An investigation using microdialysis technique. J Orthop Res 20:934–938

Alfredson H, Lorentzon R (2003) Intratendinous glutamate levels and eccentric training in chronic Achilles tendinosis: a prospective study using microdialysis technique. Knee Surg Sports Traumatol Arthrosc 11:196–199

Angst MS, Clark JD, Carvalho B, Tingle M, Schmelz M, Yeomans DC (2008) Cytokine profile in human skin in response to experimental inflammation, noxious stimulation, and administration of a COX-inhibitor: a microdialysis study. Pain 139:15–27

Ashina M, Stallknecht B, Bendtsen L, Pedersen JF, Schifter S, Galbo H, Olesen J (2003) Tender points are not sites of ongoing inflammation -in vivo evidence in patients with chronic tension-type headache. Cephalalgia 23:109–116

Ashina M, Jorgensen M, Stallknecht B, Mork H, Bendtsen L, Pedersen JF, Olesen J, Jensen R (2005) No release of interstitial glutamate in experimental human model of muscle pain. Eur J Pain 9:337–343

Bernards CM, Kopacz DJ (1999) Effect of epinephrine on lidocaine clearance in vivo: a microdialysis study in humans. Anesthesiology 91:962–968

Brunner M, Dehghanyar P, Seigfried B, Martin W, Menke G, Müller M (2005) Favourable dermal penetration of diclofenac after administration to the skin using a novel spray gel formulation. Br J Clin Pharmacol 60:573–577

Brunner M, Davies D, Martin W, Leuratti C, Lackner E, Müller M (2011) A new topical formulation enhances relative diclofenac bioavailability in healthy male subjects. Br J Clin Pharmacol 71:852–859

Burian M, Tegeder I, Seegel M, Geisslinger G (2003) Peripheral and central antihyperalgesic effects of diclofenac in a model of human inflammatory pain. Clin Pharmacol Ther 74:113–120

Carvalho B, Aleshi P, Horstman DJ, Angst MS (2010) Effect of a preemptive femoral nerve block on cytokine release and hyperalgesia in experimentally inflamed skin of human volunteers. Reg Anesth Pain Med 35:514–519

Christensen B, Dandanell S, Kjaer M, Langberg H (2011) Effect of anti-inflammatory medication on the running-induced rise in patella tendon collagen synthesis in humans. J Appl Physiol 110:137–141

Coglianese M, Draper DO, Shurtz J, Mark G (2011) Microdialysis and delivery of iontophoresis-driven lidocaine into the human gastrocnemius muscle. J Athl Train 46:270–276

Cooper SA, Desjardins PJ (2010) The value of the dental impaction pain model in drug development. Methods Mol Biol 617:175–190

Dehghanyar P, Mayer BX, Namiranian K, Mascher H, Müller M, Brunner M (2004) Topical skin penetration of diclofenac after single- and multiple-dose application. Int J Clin Pharmacol Ther 42:353–359

Draper DO, Coglianese M, Castel C (2011) Absorption of iontophoresis-driven 2 % lidocaine with epinephrine in the tissues at 5 mm below the surface of the skin. J Athl Train 46:277–281

Ederoth P, Flisberg P, Ungerstedt U, Nordström CH, Lundberg J (2002) Local metabolic changes in subcutaneous adipose tissue during intravenous and epidural analgesia. Acta Anaesthesiol Scand 46:585–591

Flodgren GM, Crenshaw AG, Alfredson H, Fahlström M, Hellström FB, Bronemo L, Djupsjöbacka M (2005) Glutamate and prostaglandin E2 in the trapezius muscle of female subjects with chronic muscle pain and controls determined by microdialysis. Eur J Pain 9:511–515

Flodgren GM, Crenshaw AG, Hellström F, Fahlström M (2010) Combining microdialysis and near-infrared spectroscopy for studying effects of low-load repetitive work on the intramuscular chemistry in trapezius myalgia. J Biomed Biotechnol 513803

Fornai M, Colucci R, Graziani F, Cei S, Antonioli L, Tonelli M, Vassalle C, Blandizzi C, Gabriele M, Del Tacca M (2006) Cyclooxygenase-2 induction after oral surgery does not entirely account for analgesia after selective blockade of cyclooxygenase 2 in the preoperative period. Anesthesiology 104(1):152–157

Gerdle B, Hilgenfeldt U, Larsson B, Kristiansen J, Søgaard K, Rosendal L (2008) Bradykinin and kallidin levels in the trapezius muscle in patients with work-related trapezius myalgia, in patients with whiplash associated pain, and in healthy controls.A microdialysis study of women. Pain 139:578–587

Gerdle B, Söderberg K, Salvador Puigvert L, Rosendal L, Larsson B (2010) Increased interstitial concentrations of pyruvate and lactate in the trapezius muscle of patients with fibromyalgia: a microdialysis study. J Rehabil Med 42:679–687

Ghafouri N, Ghafouri B, Larsson B, Turkina MV, Karlsson L, Fowler CJ, Gerdle B (2011) High levels of N-palmitoylethanolamide and N-stearoylethanolamide in microdialysate samples from myalgic trapezius muscle in women. PLoS One 6:e27257

Giamberardino MA, Affaitati G, Fabrizio A, Costantini R (2011) Myofascial pain syndromes and their evaluation. Best Pract Res Clin Rheumatol 25:185–198

Gordon SM, Brahim JS, Rowan J, Kent A, Dionne RA (2002) Peripheral prostanoid levels and nonsteroidal anti-inflammatory drug analgesia: replicate clinical trials in a tissue injury model. Clin Pharmacol Ther 72:175–183

Gordon SM, Chuang BP, Wang XM, Hamza MA, Rowan JS, Brahim JS, Dionne RA (2008) The differential effects of bupivacaine and lidocaine on prostaglandin E2 release, cyclooxygenase gene expression and pain in a clinical pain model. Anesth Analg 106:321–327

Groth L, Serup J (1998) Cutaneous microdialysis in man: effects of needle insertion trauma and anaesthesia on skin perfusion, erythema and skin thickness. Acta Derm Venereol 78:5–9

Healey LA (2006) Polymyalgia rheumatica is the result of synovitis. J Clin Rheumatol 12:165–166

Hedenberg-Magnusson B, Ernberg M, Alstergren P, Kopp S (2001) Pain mediation by prostaglandin E2 and leukotriene B4 in the human masseter muscle. Acta Odontol Scand 59:348–355

Helmark IC, Mikkelsen UR, Krogsgaard MR, Belhage B, Petersen MC, Langberg H, Kjaer M (2010a) Early osteoarthritis and microdialysis: a novel in vivo approach for measurements of biochemical markers in the perisynovium and intraarticularly. Knee Surg Sports Traumatol Arthrosc 18:1617–1623

Helmark IC, Mikkelsen UR, Børglum J, Rothe A, Petersen MC, Andersen O, Langberg H, Kjaer M (2010b) Exercise increases interleukin-10 levels both intraarticularly and peri-synovially in patients with knee osteoarthritis: a randomized controlled trial. Arthritis Res Ther 12:R126

Högberg E, Wredmark T, Dungner E, Arner P, Felländer-Tsai L (2006a) Changes in metabolism and blood flow following catecholamine stimulation in the synovial membrane measured with microdialysis. Knee Surg Sports Traumatol Arthrosc 14:133–140

Högberg E, Stålman A, Wredmark T, Tsai JA, Arner P, Felländer-Tsai L (2006b) Opioid requirement after arthroscopy is associated with decreasing glucose levels and increasing PGE2 levels in the synovial membrane. Acta Orthop 77:657–661

Kopacz DJ, Bernards CM (2001) Effect of clonidine on lidocaine clearance in vivo: a microdialysis study in humans. Anesthesiology 95:1371–1376

Krämer HH, Schmidt K, Leis S, Schmelz M, Sommer C, Birklein F (2005) Inhibition of neutral endopeptidase (NEP) facilitates neurogenic inflammation. Exp Neurol 195:179–184

Kreiner F, Langberg H, Galbo H (2010) Increased muscle interstitial levels of inflammatory cytokines in polymyalgia rheumatica. Arthritis Rheum 62:3768–3775

Kreiner F, Galbo H (2011) Elevated muscle interstitial levels of pain-inducing substances in symptomatic muscles in patients with polymyalgia rheumatica. Pain 152:1127–1132

Langley CK, Aziz Q, Bountra C, Gordon N, Hawkins P, Jones A, Langley G, Nurmikko T, Tracey I (2008) Volunteer studies in pain research–opportunities and challenges to replace animal experiments: the report and recommendations of a focus on alternatives workshop. Neuroimage 42(2):467–473

Larsson B, Rosendal L, Kristiansen J, Sjøgaard G, Søgaard K, Ghafouri B, Abdiu A, Kjaer M, Gerdle B (2008) Responses of algesic and metabolic substances to 8 h of repetitive manual work in myalgic human trapezius muscle. Pain 140:479–490

Latzke D, Marhofer P, Kettner SC, Koppatz K, Turnheim K, Lackner E, Sauermann R, Müller M, Zeitlinger M (2012) Pharmacokinetics of the local anesthetic ropivacaine after transversus abdominis plane block in healthy volunteers. Eur J Clin Pharmacol 68:419–425

Leis S, Weber M, Isselmann A, Schmelz M, Birklein F (2003) Substance-P-induced protein extravasation is bilaterally increased in complex regional pain syndrome. Exp Neurol 183:197–204

Leis S, Weber M, Schmelz M, Birklein F (2004) Facilitated neurogenic inflammation in unaffected limbs of patients with complex regional pain syndrome. Neurosci Lett 359:163–166

Mackey AL, Kjaer M, Dandanell S, Mikkelsen KH, Holm L, Døssing S, Kadi F, Koskinen SO, Jensen CH, Schrøder HD, Langberg H (2007) The influence of anti-inflammatory medication on exercise-induced myogenic precursor cell responses in humans. J Appl Physiol 103:425–431

Mertens P, Ghaemmaghami C, Bert L, Perret-Liaudet A, Sindou M, Renaud B (2000) Amino acids in spinal dorsal horn of patients during surgery for neuropathic pain or spasticity. NeuroReport 11:1795–1798

Mertens P, Ghaemmaghami C, Bert L, Perret-Liaudet A, Guenot M, Naous H, Laganier L, Later R, Sindou M, Renaud B (2001) Microdialysis study of amino acid neurotransmitters in the spinal dorsal horn of patients undergoing microsurgical dorsal root entry zone lesioning. Tech note J Neurosurg 94(1 Suppl):165–173

Mikkelsen UR, Helmark IC, Kjaer M, Langberg H (2008) Prostaglandin synthesis can be inhibited locally by infusion of NSAIDS through microdialysis catheters in human skeletal muscle. J Appl Physiol 104:534–537

Mogil JS (2009) Animal models of pain: progress and challenges. Nat Rev Neurosci 10:283–294

Mogil JS, Davis KD, Derbyshire SW (2010) The necessity of animal models in pain research. Pain 151:12–17

Müller M, Mascher H, Kikuta C, Schäfer S, Brunner M, Dorner G, Eichler HG (1997) Diclofenac concentrations in defined tissue layers after topical administration. Clin Pharmacol Ther 62:293–299

Müller M, Rastelli C, Ferri P, Jansen B, Breiteneder H, Eichler HG (1998) Transdermal penetration of diclofenac after multiple epicutaneous administration. J Rheumatol 25:1833–1836

O'Brien TP, Roszkowski MT, Wolff LF, Hinrichs JE, Hargreaves KM (1996) Effect of a non-steroidal anti-inflammatory drug on tissue levels of immunoreactive prostaglandin E2, immunoreactive leukotriene, and pain after periodontal surgery. J Periodontol 67:1307–1316

Parrot S, Sauvinet V, Xavier JM, Chavagnac D, Mouly-Badina L, Garcia-Larrea L, Mertens P (2004) Capillary electrophoresis combined with microdialysis in the human spinal cord: a new tool for monitoring rapid peroperative changes in amino acid neurotransmitters within the dorsal horn. Electrophoresis 25:1511–1517

Punnett L, Wegman DH (2004) Work-related musculoskeletal disorders: the epidemiologic evidence and the debate. J Electromyogr Kinesiol 14:13–23

Rosendal L, Larsson B, Kristiansen J, Peolsson M, Søgaard K, Kjaer M, Sørensen J, Gerdle B (2004) Increase in muscle nociceptive substances and anaerobic metabolism in patients with trapezius myalgia: microdialysis in rest and during exercise. Pain 112:324–334

Rosendal L, Kristiansen J, Gerdle B, Søgaard K, Peolsson M, Kjaer M, Sörensen J, Larsson B (2005) Increased levels of interstitial potassium but normal levels of muscle IL-6 and LDH in patients with trapezius myalgia. Pain 119:201–209

Roszkowski MT, Swift JQ, Hargreaves KM (1997) Effect of NSAID administration on tissue levels of immunoreactive prostaglandin E2, leukotriene B4, and (S)-flurbiprofen following extraction of impacted third molars. Pain 73:339–345

Shah JP, Phillips TM, Danoff JV, Gerber LH (2005) An in vivo microanalytical technique for measuring the local biochemical milieu of human skeletal muscle. J Appl Physiol 99:1977–1984

Shah JP, Gilliams EA (2008) Uncovering the biochemical milieu of myofascial trigger points using in vivo microdialysis: an application of muscle pain concepts to myofascial pain syndrome. J Bodyw Mov Ther 12:371–384

Sjøgaard G, Rosendal L, Kristiansen J, Blangsted AK, Skotte J, Larsson B, Gerdle B, Saltin B, Søgaard K (2010) Muscle oxygenation and glycolysis in females with trapezius myalgia during stress and repetitive work using microdialysis and NIRS. Eur J Appl Physiol 108:657–669

Stålman A, Tsai JA, Wredmark T, Dungner E, Arner P, Felländer-Tsai L (2008) Local inflammatory and metabolic response in the knee synovium after arthroscopy or arthroscopic anterior cruciate ligament reconstruction. Arthroscopy 24:579–584

Stålman A, Tsai JA, Segerdahl M, Dungner E, Arner P, Felländer-Tsai L (2009) Ketorolac but not morphine exerts inflammatory and metabolic effects in synovial membrane after knee arthroscopy: a double-blind randomized prospective study using the microdialysis technique. Reg Anesth Pain Med 34:557–564

Stålman A, Berglund L, Dungnerc E, Arner P, Felländer-Tsai L (2011) Temperature-sensitive release of prostaglandin E and diminished energy requirements in synovial tissue with postoperative cryotherapy: a prospective randomized study after knee arthroscopy. J Bone Joint Surg Am 93:1961–1968

Stiller CO, Taylor BK, Linderoth B, Gustafsson H, Warsame Afrah A, Brodin E (2003) Microdialysis in pain research. Adv Drug Deliv Rev 55:1065–1079

Stiller CO, Brodin E, Bradley K (2007) In: Westerink B and Cremers T (eds) Microdialysis in pain research Elsevier, London UK

Swift JQ, Garry MG, Roszkowski MT, Hargreaves KM (1993) Effect of flurbiprofen on tissue levels of immunoreactive bradykinin and acute postoperative pain. J Oral Maxillofac Surg 51:112–116

Tegeder I, Muth-Selbach U, Lötsch J, Rüsing G, Oelkers R, Brune K, Meller S, Kelm GR, Sörgel F, Geisslinger G (1999) Application of microdialysis for the determination of muscle and subcutaneous tissue concentrations after oral and topical ibuprofen administration. Clin Pharmacol Ther 65:357–368

Tegeder I, Lötsch J, Kinzig-Schippers M, Sörgel F, Kelm GR, Meller ST, Geisslinger G (2001) Comparison of tissue concentrations after intramuscular and topical administration of ketoprofen. Pharm Res 18:980–986

Weber M, Birklein F, Neundörfer B, Schmelz M (2001) Facilitated neurogenic inflammation in complex regional pain syndrome. Pain 91:251–257

Chapter 11
Microdialysis in Ocular Drug Development

Mitesh R. Patel, Nanda K. Mandava and Ashim K. Mitra

Abstract Development of novel treatments for ocular diseases demands a complete understanding of ocular drug pharmacokinetics. Microdialysis is a continuous sampling technique that has been frequently applied for assessing ocular pharmacokinetic parameters. This sampling technique has gained considerable attention in ocular pharmacokinetics since it considerably reduces the requirement of large number of animals needed to obtain a complete pharmacokinetic profile. Remarkable progress has been made in ocular anterior and posterior segment microdialysis in recent years. These developments have simplified sampling of ocular fluids such as vitreous and aqueous humor. In this book chapter, the principle and application of microdialysis in measuring drug disposition in various segments of the eye will be discussed. This chapter will also highlight on recent progress on the development of various animal models and probe designs in ocular microdialysis.

11.1 Introduction

Development of a highly efficient drug delivery system for the treatment of ocular diseases requires a thorough understanding of ocular pharmacokinetic profile of therapeutic agents. Prompt therapeutic intervention is highly essential to prevent vision loss. Topical drug administration is one of the most commonly used methods for the treatment of anterior segment indications. However, this route of

---

M. R. Patel · N. K. Mandava · A. K. Mitra (✉)
Division of Pharmaceutical Sciences, School of Pharmacy,
University of Missouri-Kansas City, 2464 Charlotte Street, Kansas City, MO 64108, USA
e-mail: mitraa@umkc.edu

M. Müller (ed.), *Microdialysis in Drug Development*, AAPS Advances in the Pharmaceutical Sciences Series, DOI: 10.1007/978-1-4614-4815-0_11,

administration faces major challenges for drug absorption resulting in an ocular bioavailability of less than 5 % (Urtti 2006). Factors attributing to this low bioavailability include rapid blinking (6–15 times/min), tear turnover rate (0.5–2.2 μL/min), high systemic clearance, and low permeability across corneal epithelium (Boddu et al. 2010; Thrimawithana et al. 2011). On the other hand, ocular drug absorption from systemic circulation is limited by blood aqueous and blood retinal barrier (BRB) (Hornof et al. 2005; Hughes et al. 2005). Thus, the complex anatomy and physiology of the eye poses one of the most formidable challenges in developing efficient drug delivery systems.

Drug delivery strategies intended to treat ocular diseases require a continuous observation of therapeutic drug concentrations in different ocular tissues. Microdialysis is one such experimental technique that has acquired considerable attention due to its unique ability of obtaining drug pharmacokinetic profiles in various ocular tissues while minimizing the number of animals. Remarkable progress made in microdialysis in recent years has simplified drug sampling in inaccessible ocular tissues such as vitreous and aqueous humor (Dias and Mitra 2003). Moreover, this sampling technique is highly preferable over*in vivo* perfusion method as the inserted probe causes negligible damage to the surrounding tissues (Chefer et al. 2009; Shippenberg and Thompson 2001).

11.2 Obstacles to Gathering Consistent Ocular Pharmacokinetic Data

One of the major limitations in assessing ocular pharmacokinetic profile is the difficulty in generating a continuous sampling procedure of the ocular fluids. Ocular tissues, such as retina, choroid, and Bruch's membrane are the primary targets for most of the posterior segment diseases. In such cases, it is nearly impossible to assess drug pharmacokinetic profiles without sacrificing animals and quantifying drug concentrations in these ocular tissues. Thus, for obtaining drug pharmacokinetic profile, it is necessary to determine ocular drug concentrations by using at least 6–20 rabbit eyes at each time point (Boddu et al. 2010). Besides, an adequate number of time points must be selected for clearly depicting the absorption, distribution, and elimination processes (Anand et al. 2006; Katragadda et al. 2008). A total of 120–150 rabbits must be consumed to produce a single dose pharmacokinetic profile. In such situation, ocular microdialysis can serve as a potential sampling procedure which can limit the use of significant number of animals.

Microdialysis was first employed by Kalant in 1957 for quantifying steroid concentration in peripheral blood plasma (Kalant 1958). Subsequently, tremendous modifications were made to make microdialysis a powerful tool for measuring drug concentrations in different tissues. In 1974, microdialysis was employed to measure dopamine concentrations in rat brain (Ungerstedt and

Pycock 1974). In mid 1980s, microdialysis was applied for the first time by Gunnarson et al. (1987) for quantitative analysis of endogenous amino acids in vitreous humor in albino rats. Since then, microdialysis has been applied for obtaining ocular pharmacokinetic profiles of several therapeutic agents and endogenous substances.

Due to similar globe anatomy and aqueous and vitreous volumes, rabbits are the most commonly selected animal model in ocular microdialysis (Boddu et al. 2010). In addition to rabbits, ocular microdialysis studies in cats, dogs, birds, and rats in ocular microdialysis have also been reported (Adachi et al. 1995; Ben-Nun et al. 1988; Rittenhouse et al. 1998; Torto et al. 1998). Till date, microdialysis has been utilized for measuring drug concentrations in adrenal, heart, liver, adipose tissues, muscle, blood, eye, ovary, and uterus (Rittenhouse and Pollack 2000; Ungerstedt 1991). These studies clearly indicate that microdialysis has become one of the most powerful tools for assessing drug pharmacokinetic profiles.

11.3 Conventional Method for Generating Ocular Pharmacokinetic Profiles

Difficulties encountered in assessing ocular pharmacokinetic profiles present a major barrier to improving drug efficacy. Previously, pharmacokinetic parameters were calculated from data obtained from several subjects to generate a single point estimate (Drusano et al. 1995). However, results obtained with such approaches can be misleading owning to high inter- and intra-subject variability (Beal and Sheiner 1982). Additionally, such investigation requires at least 100 animals to provide complete pharmacokinetic profile (Boddu et al. 2010; Sheiner et al. 1977).

In order to reduce the number of animals, novel continuous microsampling technique was established in rabbits (Miller et al. 1992; Perkins et al. 1995). For obtaining aqueous humor samples, a 30G needle connected to a 20 μL capillary tube was placed in the anterior chamber of the eye. Approximately, 7 μL of aqueous humor samples were acquired at predetermined time points. For vitreous sampling, a 28G needle was placed carefully in the vitreous cavity (4 mm from the limbus) and 20 μL samples were withdrawn at time points similar to aqueous humor sampling. Ability of sequential sampling of aqueous and vitreous humor significantly reduced the number of animals needed for determining a complete ocular pharmacokinetic profile of a therapeutic agent. However, continuous loss of aqueous and vitreous humor might disturb the fluid dynamics and can therefore significantly influence drug pharmacokinetic parameters (Macha and Mitra 2001a). Thus, adequate modifications are needed to develop an efficient microdialysis technique to study ocular pharmacokinetics of therapeutic agents.

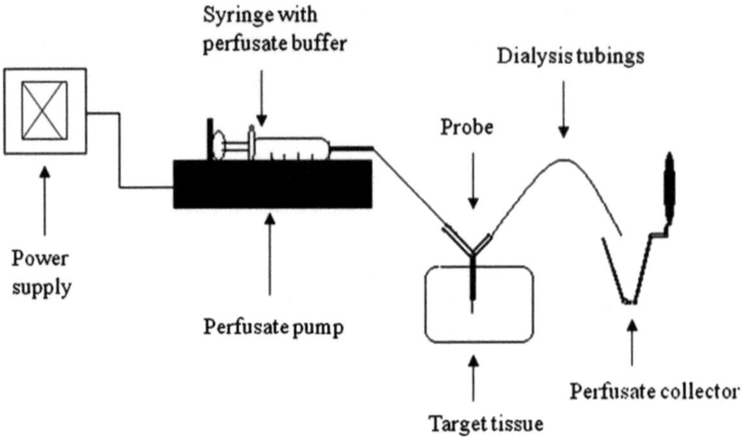

Fig. 11.1 Basic instrumentation of ocular microdialysis. (Figure referred from Boddu et al. (2010))

11.4 Instrumentation and Working

Figure 11.1 represents a general design of microdialysis instrumentation used for studying ocular pharmacokinetic profiles. It is composed of a perfusion pump, syringe, probe, dialysis tubing, and perfusate collector (Boddu et al. 2010). The syringe is filled with perfusate buffer which is circulated at a steady flow rate with the help of an infusion pump. The isotonicity of the perfusate buffer must be maintained similar to the physiological fluid in contact with the probe membrane. A minor difference in the composition can significantly alter the ionic strength of the ocular fluid surrounding the membrane. The probe is placed in the target tissue or fluid in which the accumulation of solute needs to be calculated (de Lange et al. 1999; Gunaratna et al. 2004; Macha and Mitra 2001a). The syringe, probe, and perfusate collector are connected in series with polyethylene tubings.

The basic mechanism behind the working of microdialysis is similar to dialysis where, the therapeutic drug molecules diffuse along the concentration gradient. The microdialysis probe is designed in such a way that it nearly simulates blood capillaries (Ciechanowska et al. 2008). The probe contains a semipermeable membrane through which perfusate buffer is continuously circulated at a steady flow rate. Due to continuous flow of perfusate buffer through the probe, solute molecules fail to reach true equilibrium across the probe membrane. Consequently, the concentration of solute in perfusate and targeted fluid (vitreous or aqueous humor) differs significantly at any given time point. The solute transports from the target tissue or fluid of interest into the perfusate buffer which is later collected for analysis.

Although movement of the drug is concentration dependent, the molecular cutoff of the dialysis membrane can also hinder drug diffusion. Solutes with molecular mass lesser than 20 kDa can easily diffuse through the probe membrane.

Therefore, the concentration of solutes in the targeted tissue or fluid of interest is determined from probe recovery or extraction efficiency.

11.4.1 Probe Design

The probe used in ocular microdialysis contains a tubular semipermeable membrane of 200–500 μm in diameter. This hollow fiber membrane is connected to an impermeable inlet and outlet through which the perfusate buffer enters and leaves the probe. Solutes with low molecular mass have a greater tendency to move bidirectionally across the probe membrane.

Till now, four varieties of probe have been used in microdialysis. These include linear, concentric, loop, and side-by-side probes. These probes greatly differ in shape and size. Selection of a probe is highly based on the target tissue or fluid where the solute concentration has to be measured. Linear polyacrylonitrile probes (MD-2000, 0.32 × 10 mm) with 0.22 mm tubing have been previously selected for anterior segment microdialysis. These probes are flexible and have a cylindrical shaped semipermeable membrane which is attached to small bore tubings on either side (Boddu et al. 2010). Polycarbonate concentric probes (CMA/20 with 0.5 × 10 mm) are commonly employed for performing microdialysis in the posterior segment of the eye. These probes are constructed in such a manner that the inlet and the outlet ends are parallel to each other. The perfusate buffer enters through the inlet, passes through the dialysis membrane, and is eventually collected from the outlet at predetermined time intervals (Chaurasia 1999).

Polyethersulfone, cuprophan, polyacronitriles, polycarbonate ethers, regenerated cellulose, and polyamide polymers have been employed for manufacturing semipermeable probe membrane (Hsiao et al. 1990; Wang and Stenken 2006). The molecular weight cutoff for the permeants is imparted by the design of the probe polymer (5–100 kDa). Extraction fraction, permeability coefficients, temperature stability, and enzyme interaction of these membranes have been previously investigated by Torto et al. (1998) with amylopectin and oligosaccharides. Membranes prepared with polyamide and polyethersulfone polymers exhibited lower extraction fraction and non-specific protein interaction relative to polysulfone membranes. A 250-fold increase in the saccharide concentration did not produce any significant effect on the extraction fraction of these membranes. Furthermore, the texture of membranes remained unaffected when exposed to 90 °C suggesting its usability at high temperatures.

In order to prevent contamination of samples with proteins, membranes with lower molecular weight cutoff (5–30 kDa) can be applied for microdialysis. This might be due to hydrostatic pressure which pushes the perfusate buffer containing the solute of interest against the semipermeable membrane. As a result, large molecular weight molecules such as proteins and peptides fail to adequately permeate the membrane. Such low molecular weight cutoff probes also provides higher recovery of solutes relative to 100 kDa membranes. Previously, polyethylene membrane with molecular weight cutoff of 3,000 kDa has been selected for measuring brain levels of interleukins and nerve growth factor (Winter et al. 2002). Due to high efficacy, macromolecules, such as nucleic acids, protein, and peptides have been indicated in the treatment of ocular diseases (Biswas et al. 2010; Fattal and Bochot 2006; Hosseini et al. 2012). Therefore, large size membrane cutoff polymers could serve as a potential tool for measuring ocular concentration of these macromolecules.

11.4.2 Probe Recovery or Extraction Efficiency

Probe recovery is an important parameter in microdialysis which reflects the relationship of the membrane permeability with fraction recovery of the solute of interest. During microdialysis, the perfusate buffer carries only a small portion of the actual solute concentration in the target tissue or fluid through the dialysis membrane. Hence, in order to estimate the actual concentration in the target tissue or fluid, it is very crucial to consider the probe recovery of the solute. Such probe recovery or relative recovery is represented as the ratio of solute concentration in dialysate (C_{out}) to the physiological fluid surrounding the probe (C_{in}). In vitro probe recovery of solute can be determined by (11.1).

$$\text{Recovery}_{\text{in vitro}} = \frac{C_{\text{out}}}{C_{\text{in}}} \tag{11.1}$$

Once in vitro probe recovery is calculated, the actual concentration of solute in the target tissue or fluid can be estimated as shown in (11.2).

$$C_{\text{in}} = \frac{C_{\text{out}}}{\text{Recovery}_{\text{in vitro}}} \tag{11.2}$$

C_{in} represents the concentration of solute in the target of interest and C_{out} is the concentration of solute in dialysate collected at preset time intervals.

The absolute recovery reflects the amount of the solute over predetermined time periods. The absolute recovery of the solute can be calculated according to (11.3) (Wages et al. 1986).

$$\text{Absolute recovery} = C * R * F \tag{11.3}$$

C represents the concentration of solute, F denotes perfusion flow rate, and R is the relative recovery. Generally, the perfusate buffer is pumped through the probe at a steady flow rate of 2 μL/min in ocular microdialysis. The percent recovery of solute might be altered by several factors including perfusion flow rate, temperature, length and diameter of the membrane, molecular mass, weight, and charge of the solute. Therefore, determination of in vitro probe recovery (prior to microdialysis) is highly essential for accurate analysis of*in vivo* data. For linear

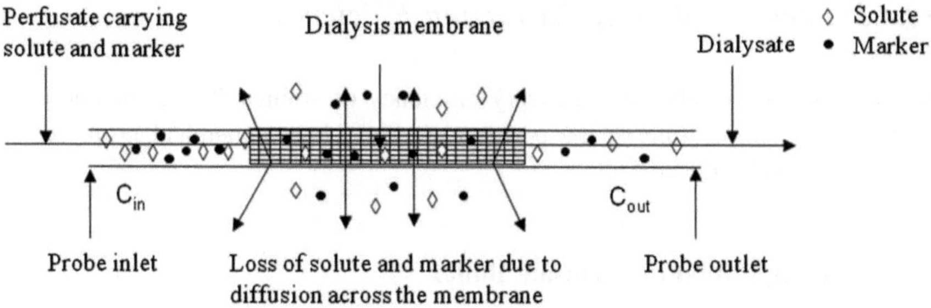

Fig. 11.2 Retrodialysis schematics. Diffusion of ganciclovir (*solute*) and mannitol (*marker*) across semipermeable probe membrane. Ganciclovir and mannitol are schematically represented as *diamond* and *filled circle* in the figure. *Arrows* represent the direction of transport of ganciclovir and mannitol. (Figure referred from Dias and Mitra (2003))

probes, percent recovery ranges from 15 to 18 % while concentric probes have recovery efficiency of 20–25 % (Boddu et al. 2010).

Probe recovery can also be determined by reverse microdialysis, a process commonly known as retrodialysis. This technique is used to examine the integrity or recovery efficiency of the probe following ocular implantation. A retrodialysis marker (internal standard) with diffusion properties close to solute of interest is circulated through the probe at a constant flow rate (Katragadda et al. 2008). The dialysate is collected and the amount of marker and solute lost during perfusion is calculated by (11.4).

$$\text{Recovery} = \frac{(C_{\text{in}} - C_{\text{out}})}{C_{\text{in}}} \tag{11.4}$$

C_{in} is the known concentration of marker entering and C_{out} is the concentration leaving the probe. Recovery of these two compounds can be compared by taking ratio of individual recovery as shown in (11.5).

$$\text{Recovery ratio} = \frac{\text{Recovery}_{\text{marker}}}{\text{Recovery}_{\text{solute of interest}}} \tag{11.5}$$

The principle behind retrodialysis is based on the fact that the amount of solute lost from the perfusion buffer is the amount accumulated in the target of interest. Mannitol has been suggested to be an excellent marker for studying ocular pharmacokinetics of ganciclovir by microdialysis in conscious rabbit model (Dias and Mitra 2003). In this investigation, mannitol and ganciclovir have been circulated through the probe at a constant flow rate of 2 μL/min (Fig. 11.2). The probe is placed in phosphate buffer saline solution isotonic with physiological fluid and samples are collected every 20 min. Following probe insertion in the target fluid of interest (aqueous or vitreous humor), the relative in vivo recovery of the marker is also evaluated. Several reports, published in the literature, also indicate the use of reverse microdialysis in determining ocular disposition of therapeutic agents (Anand et al. 2004; Duvvuri et al. 2007; Katragadda et al. 2008; Wei et al. 2006).

11.4.3 Factors Affecting Extraction Efficiency

Several factors can alter the recovery efficiency of solutes. These include composition and temperature of the perfusate buffer, physicochemical properties of solute, and tissue-related factors.

11.4.3.1 Composition of Perfusate Buffer

In microdialysis, perfusion buffer is constantly circulated through the probe due to which it is in continuous contact with the surrounding physiological fluid. Retina and iris-ciliary body have high blood capillary network and are mainly responsible for preserving the intraocular fluid homeostasis. Hence, perfusate buffer with iso-osmosity similar to plasma is usually preferred. The recovery efficiency of a given solute can vary dramatically with changes in iso-osmosity of perfusion buffer. Lehmann (1989) studied the effect of buffer composition (osmolarity) on the extracellular levels of several amino acids in rat hippocampus. Perfusion buffer containing lower concentration of sodium chloride generated higher levels of glutamate, alanine, aspartate, and gamma-aminobutyrate in dialysate from hippocampus. Surprisingly, the same hypoosmolar buffer significantly reduced methionine, leucine, and glutamate levels in the dialysate. This result clearly suggests that osmolarity of perfusion fluid might play a considerable role in altering the recovery efficiency of solutes. In ocular microdialysis, IPBS buffer maintained at pH 7.4 is generally considered as ideal perfusion fluid.

11.4.3.2 Temperature of Perfusate Buffer

Temperature of perfusate buffer has a direct effect on recovery efficiency as it significantly alters the diffusion properties of solutes. Even a slight increase in the temperature (1 °C) can significantly elevate solute diffusion by 1–2 % (Benveniste and Huttemeier 1990). The effect of temperature on 3,4-Dihydroxyphenylacetic acid recovery was studied by Wages et al. (1986). This article describes that the recovery of 3,4-Dihydroxyphenylacetic acid is significantly higher at 37 °C relative to 23 °C (Fig. 11.3).

Hence, it is very crucial to estimate in vitro probe recovery of solutes at temperature similar to the tissue or fluid of interest. For ocular microdialysis, the temperature of perfusate buffer is maintained at 34 °C (Boddu et al. 2010).

11.4.3.3 Flow Rate of Perfusate Buffer

Recovery is inversely proportional to the flow rate of perfusion buffer. A 100 % relative recovery can be obtained at perfusion rates lesser than 0.1 μL/min

Fig. 11.3 Percentage recovery of 3,4-Dihydroxyphenylacetic acid as a function of temperature at various perfusate flow rate. (Figure reproduced with permission from Wages et al. (1986))

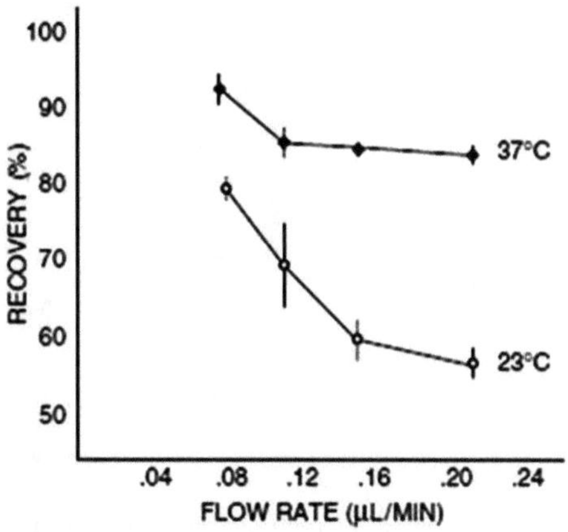

(Van Wylen et al. 1986). However, lower flow rates are not usually preferred as the process becomes very time-consuming. A dramatic reduction in the absolute recovery of solutes is observed with increase in perfusion flow rate up to 2 μL/min (Boddu et al. 2010).

The effect of perfusion flow rate on the recovery of zidovudine (analyte) and 3′-azido-2′,3′-dideoxyuridine (calibrator) was studied by Wang et al. (1993) by retrodialysis. A significant fall in the recovery of zidovudine and internal standard was observed simultaneously with rise in perfusion flow rate from 0.5 to 5 μL/min (Fig. 11.4). Several reports have been published in the literature demonstrating the inverse relationship of relative recovery with perfusion flow rate (Fu et al. 2011; Hutchinson et al. 2000; Zhang et al. 2005). Therefore, proper selection of perfusion flow rate is highly critical in optimization of microdialysis parameters. A perfusion flow rate of 2 μL/min is commonly selected in most ocular microdialysis experiments (Vaishya 2011).

11.4.3.4 Tissue-Related Factors

The diffusion of solutes from the target tissue can be restricted by a majority of factors. These include reduced fluid volume, high tortuosity, and protein adsorption on the surface of probe membrane (Amberg and Lindefors 1989; Nicholson and Rice 1986; Rice et al. 1985).

Eye globe is generally divided into two compartments i.e., anterior and posterior segments. Anterior segment consists of cornea, iris, ciliary body, and lens. Aqueous humor is present in the anterior and posterior chamber which is located between cornea and lens. Nearly 250 μL of aqueous humor is present in the anterior chamber and 55 mL in the posterior chamber (Boddu et al. 2010). It is produced and secreted from ciliary epithelial cells at a rate of about 2–3 μL/min (Tang-Liu et al. 1984). Due to its continuous formation and rapid elimination, anterior segment of the eye is highly dynamic in nature. The flow dynamics of

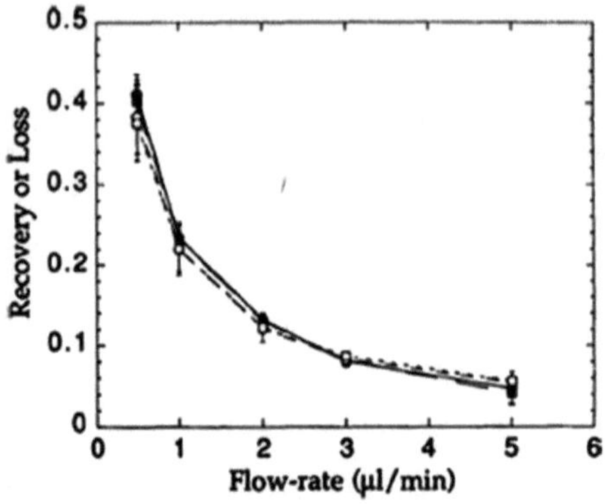

Fig. 11.4 Effect of perfusate flow rate on in vitro recovery of zidovudine and 3′-azido-2′,3′-dideoxyuridine during retrodialysis. *Filled squares* and *circles* represent loss of zidovudine and 3′-azido-2′,3′-dideoxyuridine. *Empty squares* and *circles* show recovery of zidovudine (analyte) and 3′-azido-2′,3′-dideoxyuridine. (Figure reproduced with permission from Wang et al. (1993))

aqueous humor plays a vital role in modulating ocular disposition of therapeutic drugs. Aqueous humor drains from the eye by flowing into the canal of Schlemm through trabecular meshwork. In addition to this route, aqueous humor also drains through uveoscleral passage which is a pressure independent process.

Vitreous humor is present in the posterior segment of the eye. It represents the major sampling site for posterior segment microdialysis. It is composed of ~99 % water and contains chondroitin sulfates, hyaluronates, collagens, and glycosaminoglycans. The volume of vitreous humor is ~4 mL. The posterior segment of the eye is composed of retina, choroid, vitreous humor, and optic nerve. Retina receives nutrients from choroidal and retinal blood capillaries (Vaishya 2011). The transport of nutrients such as amino acids and neurotransmitters from blood to retina is primarily regulated by specific influx transporters. Different routes for drug elimination to and within vitreous humor are explained by Fig. 11.5. However, the movement of these molecules to and within vitreous humor is highly dependent on simple diffusion (Vaishya 2011). In comparison to aqueous humor, vitreous humor can be considered as an unstirred static fluid (Hughes et al. 1996).

11.4.3.5 Physicochemical Properties of Analytes

Molecular weight of analytes is one of the major factors that can drastically alter probe recovery. Previously, Kendrick investigated the effect of molecular weight ranging from 400 to 4,500 Da on the in vitro probe recovery of 40 neuropeptides (Kendrick 1990). This study revealed that log percent in vitro probe recovery is inversely proportional to the molecular weight of analytes. In another study, the same researcher observed that the relative recovery for amino acids ranged from 33 to 40 % compared to monoamines (22–30 %) and neuropeptides (1.5–24 %)

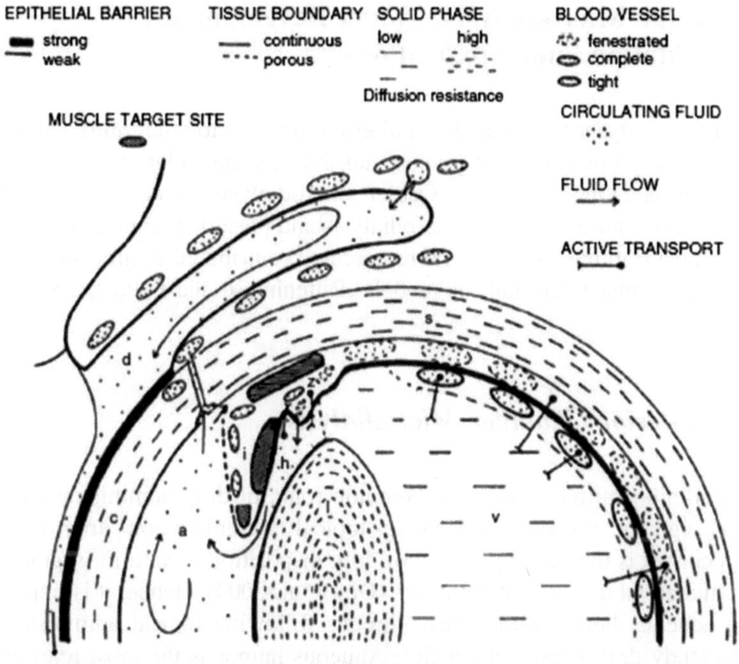

Fig. 11.5 Diagrammatic representation of various routes/pathways for elimination of therapeutic drugs from vitreous chamber. (Figure reproduced with permission from Gunnarson et al. (1987))

(Kendrick 1989). This result clearly suggests that probe recovery can reduce dramatically with increasing molecular weight of analytes.

Partition coefficient (log *P*) of analytes can also alter probe recovery. Recovery of lipophilic analytes can be lower due to their adsorption on the probe membrane. Estradiols, neurotensin, and neuropeptides *Y* have produced lower probe recovery due to their high lipophilicity (Boddu et al. 2010). Such hydrophobic analytes may require prolonged equilibration time for achieving steady level of probe recovery (Thompson et al. 1995).

The passage of selected antibiotics, corticosteroids, cytostatics, and three different molecular weight molecules: formic acid (MW 70 Da), glucose (MW 189 Da), and insulin (MW 5,200 Da) through different microdialysis membranes was reported by Waga and Ehinger (1995). This study demonstrates that molecules with high lipophilicity had greater tendency to bind to polycarbonate but not to polyamide membranes. Beside insulin, all molecules exhibited probe recovery of about 10–20 %. Furthermore, variations in dimensions of polyamide or polycarbonate membrane generated negligible effect on the transport characteristics of all experimented molecules.

11.5 Ocular Microdialysis: Model Development, Validation, and Applications

Ocular microdialysis technique is gaining momentum to study intraocular disposition of endogenous and exogenous substrates, intraocular drug delivery and ocular pharmacokinetic studies. Ocular microdialysis models can be broadly classified into anterior chamber microdialysis and posterior segment microdialysis based on the sampling sites. Several researchers contributed significantly to design and validate ocular microdialysis models (Rittenhouse and Pollack 2000)

11.5.1 Anterior Chamber Microdialysis

Topical administration is the most common route of drug administration to treat anterior segment diseases. Following topical administration drug absorption through cornea is the major pathway of drug disposition to anterior segment other than conjunctival and scleral pathways (Dey et al. 2004). Hence, it is important to sample aqueous humor rather than vitreous following topical administration in order to study drug pharmacokinetics. Aqueous humor is the most relevant sampling site to study corneal absorption and distribution of topically administered drugs. Earlier techniques such as paracentesis have been used for aqueous humor sampling at each time point. Several adverse events are associated with paracentesis sampling, such as disruption of blood-aqueous barrier, fibrin formation in aqueous humor, and increase in aqueous humor protein influx. All these events severely affect pharmacokinetics and pharmacodynamics of drugs (Liu et al. 2007; Rittenhouse and Pollack 2000).

In order to overcome these issues, pharmacokinetic study approach should involve single subject sampling as terminal procedure. However, considerably large number of animals may be required for obtaining sufficient sample pool to estimate pharmacokinetics of the drug. Utilizing microdialysis technique for aqueous humor sampling address all the issues associated with earlier techniques. Moreover, aqueous humor concentration verses time profile can be obtained from single animal. Utilization of microdialysis to study anterior chamber drug absorption and regional drug distribution has been well explored in the past decade.

A typical pharmacokinetic study design includes sampling of circulation fluid in the central compartment and measuring drug concentrations at various time points. Several researchers applied similar approach to study ocular pharmacokinetics following topical administration (Lee and Robinson 1979; Miller et al. 1980).

Several compartmental models have been studied since several ocular tissues and fluids are involved in ocular distribution of drugs. Considering all possible rate limiting barriers and tissue compartments into single pharmacokinetic model results in very complex equations and utilization of such models in clinical application becomes difficult (Boddu et al. 2010). For example, Makoid and Robinson (1979) described the ocular disposition of pilocarpine using a four compartment (precorneal area, the cornea, the aqueous humor, lens and vitreous) classical model represented by a quaternary exponential equation yielding eight parameters. This compartmental model fails in accurately assigning the absorption, distribution, or elimination parameters (Dey et al. 2004). Moreover, penetration of drugs across cornea is rate limiting step in disposition of topically applied drugs and absorption through cornea is slower than elimination from eye. This suggests high possibility of assigning erroneous slopes (Boddu et al. 2010).

In order to study corneal absorption, researchers used to study transport of drugs across cornea intact and cornea without epithelium. Antonie et al. (1984) studied pharmacokinetics of topically administered phenylephrine hydrochloride. In order to study role of corneal epithelium on drug disposition, permeability studies were performed in rabbits with intact corneas and corneas with epithelium removed. The results clearly demonstrated that maximum aqueous humor concentrations were achieved in less time and 13-fold higher in eyes with epithelium removed. Moreover, low concentrations of phenylephrine metabolites in corneas without epithelium compared intact corneas suggesting that corneal epithelium is not only a barrier for drug absorption but also contributes to the metabolic degradation. Altogether penetration of drugs across cornea is rate limiting step in disposition of topically applied drugs and absorption through cornea is slower than elimination from eye. This suggests high possibility of assigning erroneous slopes (Boddu et al. 2010). Estimating absorption rate constants from previously developed pharmacokinetic compartmental models may be difficult. Moreover, failure to recognize parallel first-order non-absorptive processes such as drainage from the eye may result in an over estimation of transcorneal absorption rate constant (Sieg and Robinson 1977). The model depicting the precorneal and intraocular drug movement following topical dosing is shown in Fig. 11.6.

Eller et al. (1985) developed a topical infusion model to provide a constant rate into aqueous humor in anesthetized rabbits to determine absorption rate constant of ethoxzolamide analogues. In this topical infusion model, effect of precorneal elimination pathways were obviated and constant drug concentrations were maintained through use of a plastic cylindrical well with base shaped like contact lens. This procedure permitted comparison of corneal absorption rate, aqueous humor elimination rate, steady state volume of distribution, and total ocular clearance. However, this procedure has some limitations such as large number of animals required and probability of inter-subject variability is very high which makes precise pharmacokinetic analysis difficult (Boddu et al. 2010; Katragadda et al. 2008).

During past two decades, drug disposition in anterior segment has been explored by anterior chamber microdialysis (Hughes and Mitra 1993). A microdialysis probe containing a dialysis membrane is implanted into the tissue of interest. Probes are continuously perfused with sampling solution and small molecules can diffuse into the probe from the extracellular fluid. Drug concentrations can be measured in

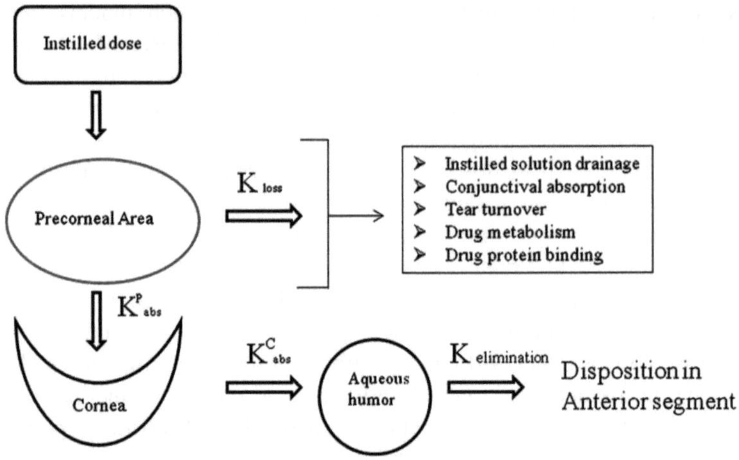

Fig. 11.6 Figure depicting precorneal and intraocular drug movement following topical administration. (Figure referred from Boddu et al. (2010))

collected dialysates (Waga et al. 1991). Several researchers contributed significantly in developing anterior chamber microdialysis. Fukuda et al. (1995) and Sato et al. (1996) demonstrated the utility of anterior chamber microdialysis to examine drug disposition in anterior segment. Sato et al. (1996) examined aqueous humor concentrations of fluoroquinolones (norfloxacin and lomefloxacin) following oral administration and topical instillation using microdialysis sampling technique in anesthetized rabbits. From this study, authors concluded that fluoroquinolones penetrated better into aqueous humor and eliminated faster following topical instillation. No drug concentrations were found in aqueous humor following oral administration. Later pharmacokinetics of topical β-adrenergic antagonists, cartelol, and timolol, was evaluated by Ohtori et al. (1998) in pigmented rabbits. Linear microdialysis probe with 5 mm length and 0.2 mm diameter was implanted in aqueous humor and drugs were administered topically shortly after recovery from anesthesia. In this study, authors could obtain ocular pharmacokinetic parameters with very few number of animals compared to conventional sampling methods.

Rittenhouse et al. (1998) used aqueous humor microdialysis to examine ocular absorption and disposition of propranolol and studied interspecies variability of propranolol pharmacokinetics in rabbits and dogs. Intraocular tissue availability of propranolol in rabbits was approximately 10-fold higher than in dogs. These studies clearly demonstrated the utility of aqueous humor microdialysis sampling in studying ocular pharmacokinetics of drugs with very few number of animals than conventional sampling methods. Moreover, the probe has a specific molecular weight cutoff it excludes cellular protein from diffusing inside and complicating the analysis process (Boddu et al. 2010).

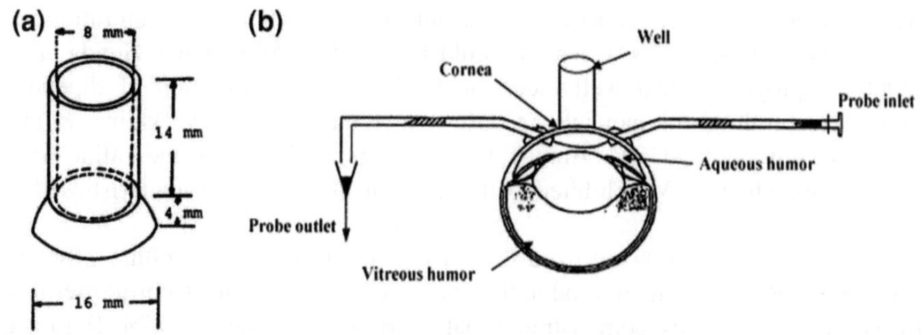

Fig. 11.7 Custom designed topical well (**a**). Reproduced with permission from Dey et al. (2004). Representation of topical well and aqueous humor microdialysis probe (**b**). (Reproduced with permission from Katragadda et al. (2008))

11.5.1.1 Anesthetized Topical Well Model for Anterior Chamber Microdialysis

Mitra and colleagues successfully combined topical well infusion model developed by Eller et al. (1985) and aqueous humor microdialysis sampling to study corneal permeability characteristics and ocular disposition of drugs with less interindividual variability (Dey et al. 2004). The dimensions of custom designed topical well are given in Fig. 11.7a. The experimental details include anesthesia in rabbits by intramuscular administration of ketamine hydrochloride and xylazine (35–50 mg/kg and 3–5 mg/kg respectively). One to two drops of 1 % tropicamide solution was applied topically to dilate pupil. A 25G needle is inserted across the anterior chamber just above corneal–scleral limbus. Outlet of the linear probe is inserted into bevel end edge of the needle and needle is retracted leaving probe in anterior chamber. Then the probe is positioned such that dialysis membrane stays in middle of anterior chamber.

Figure 11.7b represents probe implanted in anterior chamber. The probe was perfused with buffer at a flow rate of 2–3 μL/min using CMA/100 microinjection pump. Animals were allowed for 2 h in order to stabilize the probe. This stabilization period of 2 h allows restoration of intraocular pressure (IOP) and replenishing aqueous humor lost during probe implantation. After stabilizing for 2 h, eyelids were retracted with Colibri retractors. A topical well is placed in over with help of an adhesive such that a well is formed over cornea and drug solution can remain in contact with cornea. Then drug solution was added to well and drugs were allowed to diffuse through cornea for 2 h followed by aspiration of drug solution and washing eye with distilled water to remove any drug solution remained. Sampling is continued for 4–6 h after removal of topical well (Dey et al. 2004; Gunda et al. 2006; Tirucherai et al. 2002).

Katragadda et al. (2008) utilized this model to study corneal absorption of a series of acyclovir (ACV) amino acid prodrugs (L-alanine-ACV, L-serine-ACV, L-isoleucine-ACV, γ-glutamate-ACV, L-valine-ACV). Pharmacokinetic parameters were obtained for all the prodrugs and area under aqueous humor concentration and time curve of prodrugs is at least twofold higher than ACV. Later Gunda et al. (2006) employed topical well model to study corneal absorption of dipeptide monoester prodrugs of ganciclovir (GCV), glycine-valine-GCV, valine–valine-GCV, tyrosine-valine-GCV. Among the prodrugs studied, tyrosine-valine-GCV and valine–valine-GCV exhibited better corneal absorption characteristics than GCV.

Hariharan et al. (2009) utilized this model to study effect of efflux proteins, P-gp and MRP2, and their modulation on ocular disposition of corticosteroids. Results from this study demonstrated that steroids were able to inhibit P-gp and MRP2 mediated efflux of erythromycin (Hariharan et al. 2009). Thus, anesthetized topical well model has been successfully employed to study not only drug disposition but also to study role of efflux proteins and nutrient transporters on ocular disposition following topical administration.

11.5.1.2 Conscious Animal Model for Anterior Chamber Microdialysis

In anesthetized animal models, using anterior chamber microdialysis, dialysate is generally collected for 8–10 h (Katragadda et al. 2008). This technique is good for drugs with shorter elimination half-lives. A conscious animal model has been developed to study pharmacokinetics of drugs with longer half-lives. This model is also employed for studying release of drugs from sustained drug delivery systems in aqueous humor. Moreover, adequate recovery period following probe implantation allows minimizing effect several events following surgical insult, such as protein infiltration, blood–aqueous barrier disruption on pharmacokinetics of drugs. Results from aforementioned microdialysis experiments resulted in high variability attributed to high protein influx into aqueous humor following surgical trauma.

In order to overcome all these issues, conscious animal microdialysis has been developed (Duvvuri et al. 2005; Lopez-Cortes et al. 2001; Ohtori et al. 1998; Wang et al. 1993). Rittenhouse et al. (1999) studied the effect of anesthesia and aqueous humor protein concentrations on ocular disposition of highly protein bound drug propranolol. Pharmacokinetics and pharmacodynamics of propranolol were compared in anesthetized and conscious animals. Results from this study clearly demonstrated that anesthesia and protein binding significantly alter pharmacokinetics and IOP lowering effect of propranolol. A 5-day recovery period following surgical insult restored normal endogenous protein levels in aqueous humor.

Later, Rittenhouse et al. (2000) employed this technique to study rate of ascorbate secretion into posterior chamber from ciliary body pigmented epithelium and clearance from aqueous humor. In this study, microdialysis probes were implanted in both anterior and posterior chambers. This study concluded that no saturable uptake of ascorbate from blood to aqueous humor was observed at physiologic ascorbate blood concentrations. Based on these results, pharmacodynamics of propranolol was studied choosing aqueous humor ascorbate as a marker for aqueous humor turnover. This pharmacodynamic study was facilitated by conscious animal microdialysis. Nonlinear least square regression analysis of aqueous humor concentrations of ascorbate and propranolol at various time points provided estimates of aqueous humor outflow and intraocular pressure lowering effects of propranolol. All these studies demonstrate utility of conscious animal microdialysis is viable tool to delineate pharmacokinetics and pharmacodynamics of topically applied drugs.

11.5.2 Posterior Segment Microdialysis

Treatment of posterior segment diseases, such as macular edema, retinoblastoma, age-related macular degeneration, require therapeutic concentrations of drugs in vitreous humor, and other posterior segment tissues. Various physiological barriers are included in disposition of drugs into vitreous following systemic or periocular administration (Palestine et al. 1991). Among all the routes, intravitreal injection is proven to be the best to achieve higher vitreal concentrations. Vitreous chamber is the sampling site for posterior segment microdialysis. However, drugs are rapidly cleared from vitreous based on polarity and molecular weight (Pavan et al. 1994; Stempels et al. 1993). Microdialysis sampling technique has been utilized for vitreous sampling which serves as robust sampling technique to study vitreal kinetics of compounds (Hughes et al. 1996; Waga and Ehinger 1997).

Vitreous microdialysis was first reported by Gunnarson et al. (1987). Microdialysis probe was implanted in pre-retinal vitreous space to measure concentrations of endogenous amino acids in albino rabbits. The probe used in this study consisted of a 3 mm loop of cuprofan B4 membrane matrix with 3,000 KDa cutoff. In this study appearance of radiolabeled water [^{3}HOH] and [^{14}C] mannitol was completely excluded. Among the amino acids determined, concentration of glutamate is similar to that of plasma and cerebrospinal fluid (CSF). Vitreal concentration of taurine is fourfold higher than in CSF (Gunnarson et al. 1987).

Later on Ben-Nun et al. (1988) performed vitreous microdialysis in cats. This study was conducted to study pharmacokinetics of gentamicin after subconjunctival and intravitreal injections. No concentrations of gentamicin were found in vitreous after subconjunctival injection up to 8 h. Following intravitreal injection, vitreal concentrations of gentamicin ranged from 30 to 80 mg/L in 4–8 h and 26–22 mg/L in 9–16 h. Thus, Ben-Nun et al. (1988) demonstrated the utility of vitreous microdialysis for studying vitreal pharmacokinetics of drugs. Louzada-Junior et al. (1992) employed vitreous microdialysis to study effect of ischemia on the release of excitatory amino acids into vitreous. Microdialysis probe was implanted close to retina and release amino acids during ischemia were studied.

Hughes et al. (1996) determined pharmacokinetics of ACV and GCV after intravitreal injection using vitreal microdialysis in anesthetized New Zealand white rabbits. In this study, a concentric probe (CMA-12, 4 mm, CMA/AB

Microdialysis, Stockholm, Sweden) was inserted into vitreous. Vitreous elimination of GCV and ACV were 2.62 and 2.98 h, respectively. Effect of ocular pigmentation on pharmacokinetics was also determined in this study. Rate of elimination of GCV and ACV in pigmented rabbits ($t_{1/2}$ is 5.59 and 8.63 h for GCV and ACV) is slower than elimination from Dutch Belted rabbits ($t_{1/2}$ is 2.62 and 2.96 h for GCV and ACV). Knowledge of vitreal elimination kinetics is important for designing dosing regimen and microdialysis is best technique to study elimination kinetics of drugs.

11.5.2.1 Anesthetized Dual Probe Microdialysis

Drugs are eliminated from vitreous by transport across BRB or through aqueous humor following diffusion across retrozonular space (Boddu et al. 2010). Determination of either aqueous humor or vitreous humor concentration of drugs may not yield conclusive results. Hence, Macha and Mitra (2001a) developed and validated a dual probe anesthetized model which enables determination of complete ocular pharmacokinetics of a therapeutic drug. A linear probe (MD-2000, 0.32 × 10 mm polyacrylonitrile membrane and 0.22 mm tubing) was implanted in aqueous humor of anesthetized New Zealand albino rabbits using 25G needle. A concentric vitreous probe (CMA/20 with 0.5 × 10 mm polycarbonate membrane and 14 mm shaft) was implanted in the vitreous of the same rabbit. Initially, a 22G needle was inserted into the eye approximately 3 mm below corneal–scleral limbus through pars plana. The needle was removed and probe was inserted immediately and adjusted such that dialysis membrane resides in mid vitreous. Location of probe was confirmed by microscopic examination. Probes were then perfused with isotonic phosphate buffer at flow rate of 2 μL/min for 2 h before dosing. A diagrammatic representation of probes implanted was given in Fig. 11.8.

In this study, IOP and protein concentrations in vitreous and aqueous humor were measured to ascertain any tissue damage following probe implantation. IOP levels were increased slightly after probe implantation returned to baseline after 2 h recovery period. No significant increase in protein levels was found in aqueous and vitreous humor. In this study, aqueous and vitreous elimination kinetics of fluorescein was studied following systemic and intravitreal administration. High fluorescein levels were observed in aqueous humor than vitreous after systemic administration suggesting diffusion of fluorescein to aqueous humor from iris/ciliary blood capillaries. Permeability index of anterior chamber is 9.48 % where as vitreous chamber is 1.99 % suggesting free movement of fluorescein between blood and aqueous humor. Thus, dual probe microdialysis clearly demonstrated its utility for simultaneous measurement of drug concentrations in aqueous and vitreous chambers.

Atluri and Mitra (2003) utilized this technique to study effect of lipophilicity on vitreal pharmacokinetics. Short-chain aliphatic alcohols (^{14}C-methanol, ^{14}C-1-propanol, ^{14}C-1-pentanol, ^{14}C-l-heptanol) with log P ranging from −0.77 to +2.7 were administered intravitreally. Theoretically, elimination from vitreous and

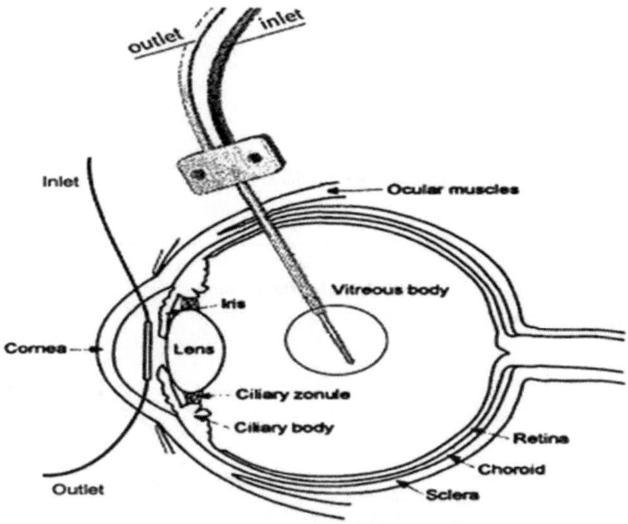

Fig. 11.8 Schematic representation of dual probe model. Microdialysis probes implanted in vitreous humor of the eye. (Figure reproduced with permission from Macha and Mitra (2001))

aqueous should be faster with increase in lipophilicity due to increase in tissue permeability. But vitreous elimination half-life increased with increment in chain length and lipophilicity from methanol to heptanol (Fig. 11.9).

Macha et al. (2004) studied ocular disposition of acyl diester prodrugs of GCV following intravitreal injection. Ocular disposition of prodrugs (acetate, propionate, butyrate) was dependant on enzymatic degradation by esterases and peptidases. These acyl diester prodrugs were degraded to respective monoester prodrugs and subsequently to GCV. Mean residence time of regenerated GCV from GCV diacetate, dipropionate, and dibutyrate (356 ± 16, 341 ± 11 and 324 ± 19 min) was approximately twice longer than that of GCV (185 ± 28 min). Despite of increase in lipophilicity, vitreal elimination half-lives of these prodrugs decreased from GCV diacetate → GCV dipropionate → GCV dibutyrate.

Later, Macha and Mitra (2001b) studied ocular pharmacokinetics of cephalosporins in rabbit vitreous using this technique. Duvvuri et al. (2003) utilized dual probe microdialysis to delineate involvement of multi-drug resistance protein, P-glycoprotein (MDR1/P-gp) on ocular disposition of drugs. Quinidine, a model substrate for P-gp, was administered by systemic and intravitreal injection. The AUC in vitreous humor of quinidine (19.21 ± 3.73 min. μg/mL) is significantly lower than quinidine in presence of verapamil (39.27 ± 6.47 min. μg/mL). Permeability of xenobiotics across blood retinal and blood aqueous barrier is largely affected by presence of efflux protein and nutrient transporters. Dias et al. (2002) synthesized peptidomimetic prodrugs of ACV (valine-ACV (V-ACV), valine–valine-ACV(VV-ACV)), and utilized dual probe microdialysis to determine ocular

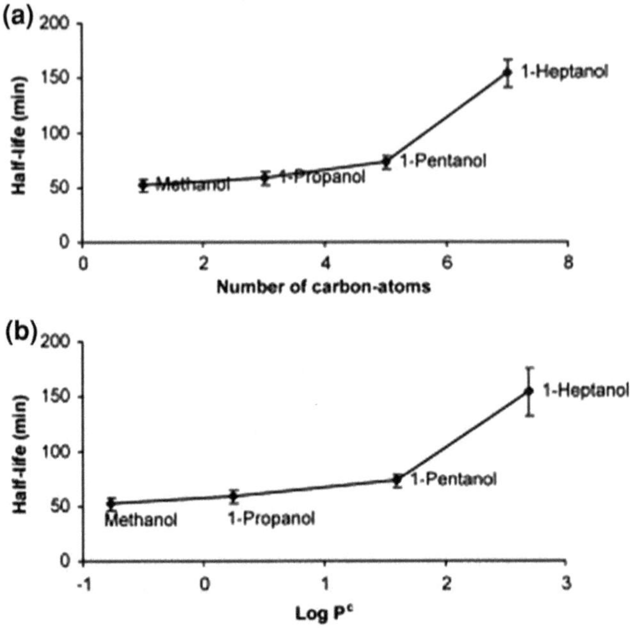

Fig. 11.9 Vitreal half-life as function of number of carbon atoms in alcohols (**a**). Vitreal half-life of alcohols as function of log Pc (**b**). (Reproduced with permission from Atluri and Mitra (2003))

pharmacokinetics of these prodrugs. Anterior segment area under concentration time curve values of V-ACV and VV-ACV are about three and sevenfold higher than ACV respectively.

11.5.2.2 Conscious Animal Model

Waga et al. (1991) developed conscious animal model using rabbits. Polycarbonate and polyamide dialysis membranes were studied for net dialysis following probe implantation with various substrates. Probes remained intact for 2–3 weeks. This model was used to study disposition of antibiotics, corticosteroids, and cytostatics by Waga and Ehinger (1995). Later, Dias and Mitra (2003) developed relatively simple surgical procedure and improved the conscious model. Rabbits were anesthetized and topical tropicamide was applied to dilate pupil. The eye was proptosed, 25G needle was inserted diametrically across posterior segment approximately 8 mm below sclera-limbus junction. A linear probe (MD-2000: Bioanalytical Systems) was inserted into needle and pulled through eye by withdrawing needle. Probe was positioned such that dialysis membrane lies completely in vitreous. Then probe was fixed to conjunctiva with a 6-0 surgical chrome gut suture. Area between rabbit's ears was shaved and an incision of 4 mm was made. A subcutaneous tunnel is made between incisioned area and upper eyelid with help

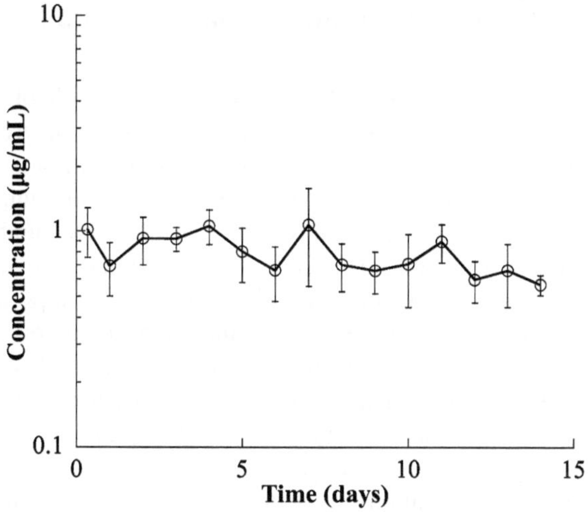

Fig. 11.10 Vitreal concentration of ganciclovir (GCV) versus time profile following intravitreal administration of the mixture formulation (196 μg of GCV). (Reproduced with permission from Duvvuri et al (2007))

of forceps. A small incision was made on upper eyelid close to eyeball. Both ends of probe were pulled through the tunnel and incision was closed. Then probe was secured to skin with 3-0 suture. Probe was perfused with buffer throughout the procedure. 0.1 % dexamethasone phosphate was applied at incision site to prevent any inflammation. Animals were allowed to recover before any experimentation.

Using this model, Dias and Mitra studied effect of anesthesia, probe implantation, and recovery period on ocular disposition of GCV following systemic administration (Dias and Mitra 2003). Animals were divided into three groups. Group I animals were dosed without any recovery period. Group II and III animals were allowed 5 days for recovery. Group II animals were conscious during experiment where as group III animals were anesthetized. Results clearly demonstrated that anesthesia caused an improved bioavailability. AUCs of group I, II, and III obtained were 650.71 ± 264.70, 95.16 ± 78.58, and 564.34 ± 227.67. AUCs of group I and III are approximately six to seven times higher than group II. This difference in AUC is attributed to anesthesia and high protein levels in vitreous.

Further, Anand et al. (2004) validated this model by studying integrity of BRB using fluorescein disposition and variation in intravitreal protein levels. This study demonstrated that intravitreal protein levels return to baseline within 2 days following surgery. Duvvuri et al. (2007) employed conscious vitreous microdialysis model to examine drug release following inravitreal administration of GCV loaded poly(d,L-lactide-co-glycolide (PLGA) microparticles suspended in PLGA-PEG-PLGA thermosensitive gel. Following intravitreal administration, gel forms a depot and releases microparticles and drug into vitreous humor. The experiment was performed 14 days and levels of GCV approximately 0.79 ± 0.17 μg/mL were maintained. Only about 17 % of total GCV was released in 14 days (Fig. 11.10). Thus, conscious animal vitreous microdialysis can be employed to study drug release from polymeric implants into vitreous and vitreal kinetics of released drugs.

11.6 Conclusion and Future Directions

The assessment of ocular pharmacokinetic profile of therapeutic agents is important in determining the dosage regimen required in the treatment of a particular disease. The use of microdialysis has provided a continuous sampling procedure of inaccessible tissues such as aqueous and vitreous humor. Moreover, this sampling technique has considerably reduced use of large number of animals. In recent years, ocular microdialysis has underwent remarkable transformations and provided valuable insights for the development of novel approaches in enhancing ocular drug absorption. Several articles highlighting the importance of microdialysis in determining ocular drug disposition have been published in the literature.

Ocular microdialysis has so far been performed in animals following topical, intravitreal, and systemic drug administration of small molecular drugs and endogenous substances. In some cases, ocular drug delivery might generate lower drug concentrations in ocular tissue or fluid of interest. For instance, subconjunctival administration generates significantly lower vitreal drug concentrations. In such situations, an insufficient amount of drug (below limit of quantification) may be generated in the dialysate. Therefore, it is highly preferable that microdialysis samples are analyzed by a highly robust and sensitive analytical method for generating accurate results. Quantification of samples using mass spectrometry would be ideal as it can analyze lower drug concentrations and is highly reliable and reproducible. The application of such sophisticated analytical techniques has further encouraged researchers to use microdialysis in determining novel therapeutic agents intended for the treatment of ocular diseases. Furthermore, refinement in analytical techniques and probe membrane technology with high molecular weight cutoff might be useful in studying ocular disposition of macromolecules such as antibodies and nucleic acids by microdialysis.

Acknowledgments Research performed in Dr. Ashim K. Mitra's laboratory was supported by National Institute of Health (NIH) Grants R01EY09171 and R01 EY 10659.

References

Adachi A, Hasegawa M, Ebihara S (1995) Measurement of circadian rhythms of ocular melatonin in the pigeon by in vivo microdialysis. NeuroReport 7:286–288

Amberg G, Lindefors N (1989) Intracerebral microdialysis: II. Mathematical studies of diffusion kinetics. J Pharmacol Methods 22:157–183

Anand BS, Atluri H, Mitra AK (2004) Validation of an ocular microdialysis technique in rabbits with permanently implanted vitreous probes: systemic and intravitreal pharmacokinetics of fluorescein. Int J Pharm 281:79–88

Anand BS, Katragadda S, Gunda S, Mitra AK (2006) In vivo ocular pharmacokinetics of acyclovir dipeptide ester prodrugs by microdialysis in rabbits. Mol Pharm 3:431–440

Antoine ME, Edelhauser HF, O'Brien WJ (1984) Pharmacokinetics of topical ocular phenylephrine HCl. Invest Ophthalmol Vis Sci 25:48–54

Atluri H, Mitra AK (2003) Disposition of short-chain aliphatic alcohols in rabbit vitreous by ocular microdialysis. Exp Eye Res 76:315–320

Beal SL, Sheiner LB (1982) Estimating population kinetics. Crit Rev Biomed Eng 8:195–222

Ben-Nun J, Cooper RL, Cringle SJ, Constable IJ (1988) Ocular dialysis. A new technique for in vivo intraocular pharmacokinetic measurements. Arch Ophthalmol 106:254–259

Benveniste H, Huttemeier PC (1990) Microdialysis–theory and application. Prog Neurobiol 35:195–215

Biswas NR, Das GK, Dubey AK (2010) Monoclonal antibodies in ophthalmology. Nepal Med Coll J 12:264–271

Boddu SH, Gunda S, Earla R, Mitra AK (2010) Ocular microdialysis: a continuous sampling technique to study pharmacokinetics and pharmacodynamics in the eye. Bioanalysis 2:487–507

Chaurasia CS (1999) In vivo microdialysis sampling: theory and applications. Biomed Chromatogr 13:317–332

Chefer VI, Thompson AC, Zapata A, Shippenberg TS (2009) Overview of brain microdialysis. Curr Protoc Neurosci. (Chapter 7, Unit7 1)

Ciechanowska A, Sabalinska S, Gutowska M, Wójcicki JM (2008) Modern application of membrane technique in therapeutic and diagnostic medical systems. Biocybern Biomed Eng 28:21–33

de Lange EC, de Boer BA, Breimer DD (1999) Microdialysis for pharmacokinetic analysis of drug transport to the brain. Adv Drug Deliv Rev 36:211–227

Dey S, Gunda S, Mitra AK (2004) Pharmacokinetics of erythromycin in rabbit corneas after single-dose infusion: role of P-glycoprotein as a barrier to in vivo ocular drug absorption. J Pharmacol Exp Ther 311:246–255

Dias C, Nashed Y, Atluri H, Mitra A (2002) Ocular penetration of acyclovir and its peptide prodrugs valacyclovir and val-valacyclovir following systemic administration in rabbits: an evaluation using ocular microdialysis and LC-MS. Curr Eye Res 25:243–252

Dias CS, Mitra AK (2003) Posterior segment ocular pharmacokinetics using microdialysis in a conscious rabbit model. Invest Ophthalmol Vis Sci 44:300–305

Drusano GL, Liu W, Perkins R, Madu A, Madu C, Mayers M, Miller MH (1995) Determination of robust ocular pharmacokinetic parameters in serum and vitreous humor of albino rabbits following systemic administration of ciprofloxacin from sparse data sets by using IT2S, a population pharmacokinetic modeling program. Antimicrob Agents Chemother 39: 1683–1687

Duvvuri S, Gandhi MD, Mitra AK (2003) Effect of P-glycoprotein on the ocular disposition of a model substrate, quinidine. Curr Eye Res 27:345–353

Duvvuri S, Janoria KG, Pal D, Mitra AK (2007) Controlled delivery of ganciclovir to the retina with drug-loaded Poly(d, L-lactide-co-glycolide) (PLGA) microspheres dispersed in PLGA-PEG-PLGA Gel: a novel intravitreal delivery system for the treatment of cytomegalovirus retinitis. J Ocul Pharmacol Ther 23:264–274

Duvvuri S, Rittenhouse KD, Mitra AK (2005) Microdialysis assessment of drug delivery systems for vitreoretinal targets. Adv Drug Deliv Rev 57:2080–2091

Eller MG, Schoenwald RD, Dixson JA, Segarra T, Barfknecht CF (1985) Topical carbonic anhydrase inhibitors IV: relationship between excised corneal permeability and pharmacokinetic factors. J Pharm Sci 74:525–529

Fattal E, Bochot A (2006) Ocular delivery of nucleic acids: antisense oligonucleotides, aptamers and siRNA. Adv Drug Deliv Rev 58:1203–1223

Fu X, Ling J, Wu X (2011) Study on in vitro microdialysis recovery of Shuanghuanglian. Zhongguo Zhong Yao Za Zhi 36:1951–1954

Fukuda S, Mikitani M, Ueda T, Inatomi M, Koide R, Kurata N, Uchida E, Yasuhara H, Uchida N, Kanda Y et al (1995) Application of microdialysis for pharmacokinetic study in rabbit anterior chamber. Nihon Ganka Gakkai Zasshi 99:400–405

Gunaratna PC, Kissinger PT, Kissinger CB, Gitzen JF (2004) An automated blood sampler for simultaneous sampling of systemic blood and brain microdialysates for drug absorption, distribution, metabolism, and elimination studies. J Pharmacol Toxicol Methods 49:57–64

Gunda S, Hariharan S, Mitra AK (2006) Corneal absorption and anterior chamber pharmacokinetics of dipeptide monoester prodrugs of ganciclovir (GCV): in vivo comparative evaluation of these prodrugs with Val-GCV and GCV in rabbits. J Ocul Pharmacol Ther 22:465–476

Gunnarson G, Jakobsson AK, Hamberger A, Sjostrand J (1987) Free amino acids in the preretinal vitreous space. Effect of high potassium and nipecotic acid. Exp Eye Res 44:235–244

Hariharan S, Gunda S, Mishra GP, Pal D, Mitra AK (2009) Enhanced corneal absorption of erythromycin by modulating P-glycoprotein and MRP mediated efflux with corticosteroids. Pharm Res 26:1270–1282

Hornof M, Toropainen E, Urtti A (2005) Cell culture models of the ocular barriers. Eur J Pharm Biopharm 60:207–225

Hosseini H, Nowroozzadeh MH, Salouti R, Nejabat M (2012) Anti-VEGF therapy with bevacizumab for anterior segment eye disease. Cornea 31:322–334

Hsiao JK, Ball BA, Morrison PF, Mefford IN, Bungay PM (1990) Effects of different semipermeable membranes on in vitro and in vivo performance of microdialysis probes. J Neurochem 54:1449–1452

Hughes PM, Krishnamoorthy R, Mitra AK (1996) Vitreous disposition of two acycloguanosine antivirals in the albino and pigmented rabbit models: a novel ocular microdialysis technique. J Ocul Pharmacol Ther 12:209–224

Hughes PM, Mitra AK (1993) Effect of acylation on the ocular disposition of acyclovir II: Corneal permeability and anti-HSV 1 activity of 2′-esters in rabbit epithelial keratitis. J Ocul Pharmacol 9:299–309

Hughes PM, Olejnik O, Chang-Lin JE, Wilson CG (2005) Topical and systemic drug delivery to the posterior segments. Adv Drug Deliv Rev 57:2010–2032

Hutchinson PJ, O'Connell MT, Al-Rawi PG, Maskell LB, Kett-White R, Gupta AK, Richards HK, Hutchinson DB, Kirkpatrick PJ, Pickard JD (2000) Clinical cerebral microdialysis: a methodological study. J Neurosurg 93:37–43

Kalant H (1958) A microdialysis procedure for extraction and isolation of corticosteroids from peripheral blood plasma. Biochem J 69:99–103

Katragadda S, Gunda S, Hariharan S, Mitra AK (2008) Ocular pharmacokinetics of acyclovir amino acid ester prodrugs in the anterior chamber: evaluation of their utility in treating ocular HSV infections. Int J Pharm 359:15–24

Kendrick KM (1989) Use of microdialysis in neuroendocrinology. Methods Enzymol 168: 182–205

Kendrick KM (1990) Microdialysis measurement of in vivo neuropeptide release. J Neurosci Methods 34:35–46

Lee VH, Robinson JR (1979) Mechanistic and quantitative evaluation of precorneal pilocarpine disposition in albino rabbits. J Pharm Sci 68:673–684

Lehmann A (1989) Effects of microdialysis-perfusion with anisoosmotic media on extracellular amino acids in the rat hippocampus and skeletal muscle. J Neurochem 53:525–535

Liu Z, Yang XG, Li X, Pan W, Li J (2007) Study on the ocular pharmacokinetics of ion-activated in situ gelling ophthalmic delivery system for gatifloxacin by microdialysis. Drug Dev Ind Pharm 33:1327–1331

Lopez-Cortes LF, Pastor-Ramos MT, Ruiz-Valderas R, Cordero E, Uceda-Montanes A, Claro-Cala CM, Lucero-Munoz MJ (2001) Intravitreal pharmacokinetics and retinal concentrations of ganciclovir and foscarnet after intravitreal administration in rabbits. Invest Ophthalmol Vis Sci 42:1024–1028

Louzada-Junior P, Dias JJ, Santos WF, Lachat JJ, Bradford HF, Coutinho-Netto J (1992) Glutamate release in experimental ischaemia of the retina: an approach using microdialysis. J Neurochem 59:358–363

Macha S, Duvvuri S, Mitra AK (2004) Ocular disposition of novel lipophilic diester prodrugs of ganciclovir following intravitreal administration using microdialysis. Curr Eye Res 28:77–84

Macha S, Mitra AK (2001a) Ocular pharmacokinetics in rabbits using a novel dual probe microdialysis technique. Exp Eye Res 72:289–299

Macha S, Mitra AK (2001b) Ocular pharmacokinetics of cephalosporins using microdialysis. J Ocul Pharmacol Ther 17:485–498

Makoid MC, Robinson JR (1979) Pharmacokinetics of topically applied pilocarpine in the albino rabbit eye. J Pharm Sci 68:435–443

Miller MH, Madu A, Samathanam G, Rush D, Madu CN, Mathisson K, Mayers M (1992) Fleroxacin pharmacokinetics in aqueous and vitreous humors determined by using complete concentration-time data from individual rabbits. Antimicrob Agents Chemother 36:32–38

Miller SC, Gokhale RD, Patton TF, Himmelstein KJ (1980) Pilocarpine ocular distribution volume. J Pharm Sci 69:615–616

Nicholson C, Rice ME (1986) The migration of substances in the neuronal microenvironment. Ann N Y Acad Sci 481:55–71

Ohtori R, Sato H, Fukuda S, Ueda T, Koide R, Kanda Y, Kiuchi Y, Oguchi K (1998) Pharmacokinetics of topical beta-adrenergic antagonists in rabbit aqueous humor evaluated with the microdialysis method. Exp Eye Res 66:487–494

Palestine AG, Polis MA, De Smet MD, Baird BF, Falloon J, Kovacs JA, Davey RT, Zurlo JJ, Zunich KM, Davis M et al (1991) A randomized, controlled trial of foscarnet in the treatment of cytomegalovirus retinitis in patients with AIDS. Ann Intern Med 115:665–673

Pavan PR, Oteiza EE, Hughes BA, Avni A (1994) Exogenous endophthalmitis initially treated without systemic antibiotics. Ophthalmology 101:1289–1296 (discussion 1296–1287)

Perkins RJ, Liu W, Drusano G, Madu A, Mayers M, Madu C, Miller MH (1995) Pharmacokinetics of ofloxacin in serum and vitreous humor of albino and pigmented rabbits. Antimicrob Agents Chemother 39:1493–1498

Rice ME, Gerhardt GA, Hierl PM, Nagy G, Adams RN (1985) Diffusion coefficients of neurotransmitters and their metabolites in brain extracellular fluid space. Neuroscience 15:891–902

Rittenhouse KD, Peiffer RL Jr, Pollack GM (1998) Evaluation of microdialysis sampling of aqueous humor for in vivo models of ocular absorption and disposition. J Pharm Biomed Anal 16:951–959

Rittenhouse KD, Peiffer RL Jr, Pollack GM (1999) Microdialysis evaluation of the ocular pharmacokinetics of propranolol in the conscious rabbit. Pharm Res 16:736–742

Rittenhouse KD, Peiffer RL Jr, Pollack GM (2000) Assessment of ascorbate ocular disposition in the conscious rabbit: an approach using the microdialysis technique. Curr Eye Res 20: 351–360

Rittenhouse KD, Pollack GM (2000) Microdialysis and drug delivery to the eye. Adv Drug Deliv Rev 45:229–241

Sato H, Fukuda S, Inatomi M, Koide R, Uchida N, Kanda Y, Kiuchi Y, Oguchi K (1996) Pharmacokinetics of norfloxacin and lomefloxacin in aqueous humour analysed by microdialysis. Nihon Ganka Gakkai Zasshi 100:513–519

Sheiner LB, Rosenberg B, Marathe VV (1977) Estimation of population characteristics of pharmacokinetic parameters from routine clinical data. J Pharmacokinet Biopharm 5:445–479

Shippenberg TS, Thompson AC (2001) Overview of microdialysis. Curr Protoc Neurosci Chapter 7, Unit 7 1

Sieg JW, Robinson JR (1977) Vehicle effects on ocular drug bioavailability II: evaluation of pilocarpine. J Pharm Sci 66:1222–1228

Stempels N, Tassignon MJ, Sarre S (1993) A removable ocular microdialysis system for measuring vitreous biogenic amines. Graefes Arch Clin Exp Ophthalmol 231:651–655

Tang-Liu DD, Liu SS, Weinkam RJ (1984) Ocular and systemic bioavailability of ophthalmic flurbiprofen. J Pharmacokinet Biopharm 12:611–626

Thompson AC, Justice JB Jr, McDonald JK (1995) Quantitative microdialysis of neuropeptide Y. J Neurosci Methods 60:189–198

Thrimawithana TR, Young S, Bunt CR, Green C, Alany RG (2011) Drug delivery to the posterior segment of the eye. Drug Discov Today 16:270–277

Tirucherai GS, Dias C, Mitra AK (2002) Corneal permeation of ganciclovir: mechanism of ganciclovir permeation enhancement by acyl ester prodrug design. J Ocul Pharmacol Ther 18:535–548

Torto N, Bang J, Richardson S, Nilsson GS, Gorton L, Laurell T, Marko-Varga G (1998) Optimal membrane choice for microdialysis sampling of oligosaccharides. J Chromatogr A 806: 265–278

Ungerstedt U (1991) Microdialysis–principles and applications for studies in animals and man. J Intern Med 230:365–373

Ungerstedt U, Pycock C (1974) Functional correlates of dopamine neurotransmission. Bull Schweiz Akad Med Wiss 30:44–55

Urtti A (2006) Challenges and obstacles of ocular pharmacokinetics and drug delivery. Adv Drug Deliv Rev 58:1131–1135

Vaishya R, Ananthula HK, and Mitra A (2011) Microdialysis for vitreal pharmacokinetics. In: Kompella UB, Edelhauser HF (ed) Drug product development for the back of the eye AAPS Advances in pharmaceutical sciences series, vol 2. Springer, New York, pp 21–45

Van Wylen DG, Park TS, Rubio R, Berne RM (1986) Increases in cerebral interstitial fluid adenosine concentration during hypoxia, local potassium infusion, and ischemia. J Cereb Blood Flow Metab 6:522–528

Waga J, Ehinger B (1995) Passage of drugs through different intraocular microdialysis membranes. Graefes Arch Clin Exp Ophthalmol 233:31–37

Waga J, Ehinger B (1997) Intravitreal concentrations of some drugs administered with microdialysis. Acta Ophthalmol Scand 75:36–40

Waga J, Ohta A, Ehinger B (1991) Intraocular microdialysis with permanently implanted probes in rabbit. Acta Ophthalmol (Copenh) 69:618–624

Wages SA, Church WH, Justice JB Jr (1986) Sampling considerations for on-line microbore liquid chromatography of brain dialysate. Anal Chem 58:1649–1656

Wang X, Stenken JA (2006) Microdialysis sampling membrane performance during in vitro macromolecule collection. Anal Chem 78:6026–6034

Wang Y, Wong SL, Sawchuk RJ (1993) Microdialysis calibration using retrodialysis and zero-net flux: application to a study of the distribution of zidovudine to rabbit cerebrospinal fluid and thalamus. Pharm Res 10:1411–1419

Wei G, Ding PT, Zheng JM, Lu WY (2006) Pharmacokinetics of timolol in aqueous humor sampled by microdialysis after topical administration of thermosetting gels. Biomed Chromatogr 20:67–71

Winter CD, Iannotti F, Pringle AK, Trikkas C, Clough GF, Church MK (2002) A microdialysis method for the recovery of IL-1beta, IL-6 and nerve growth factor from human brain in vivo. J Neurosci Methods 119:45–50

Zhang Y, Zhou L, Wang X, Wang L (2005) Study of the relative recovery of sinomenine microdialysis system and it's influential factor. Zhong Yao Cai 28:1064–1067

Chapter 12
Microdialysis in Metabolic Research

Per-Anders Jansson

Abstract Clinical metabolic research needs tools to get a footprint of human physiology and pathophysiology in vivo. The novel idea of monitoring interstitial fluid by microdialysis in the brain of laboratory animals was a complement to ordinary blood sampling. Semipermeable hollow fibers were implemented in the brain-tissue mimicking characteristics of artificial blood vessels communicating with freely diffusing molecules in situ. Subcutaneous microdialysis for measurements of interstitial concentrations of any low molecular compound was introduced in 1987. Each microdialysis catheter required calibration in situ by the "no net flux" method to get the "true" interstitial glucose, lactate, adenosine, and glycerol concentrations. Addition of ^{133}Xe-clearance assessments of adipose tissue blood flow (ATBF) allowed adoption of Fick's principle to get an estimate of substrate release. By the time, technical improvements of probes and recovery techniques have made it possible to monitor interstitial insulin and cytokines/chemokines in both sc adipose tissue and muscle. In fact, almost every organ in the human body may be accessible for monitoring of small and large molecules in the interstitial fluid. Thus, the future for the microdialysis technique in metabolic research looks bright.

P.-A. Jansson (✉)
The Wallenberg Laboratory, Department of Molecular and Clinical Medicine, Institute of Medicine, Sahlgrenska Academy at University of Gothenburg, Sahlgrenska University Hospital, Bruna Stråket 16, Level 2/3, S-413 45 Gothenburg, Sweden
e-mail: per-anders.jansson@medic.gu.se

M. Müller (ed.), *Microdialysis in Drug Development*, AAPS Advances in the Pharmaceutical Sciences Series, DOI: 10.1007/978-1-4614-4815-0_12,

12.1 Historical Background

Adipose tissue was our rationale to start up metabolic studies. The opinion that adipose tissue was a passive player in metabolism was a strong voice back in time (Toldt 1870). Our goal was to test hypotheses proving that adipose tissue was a metabolically active organ. Available methods in 1985 were based on investigations on adipocyte function in isolated cell systems (Rodbell 1964). The currently used needle biopsy technique to obtain human adipose tissue for these studies was originally described more than 50 years ago (Hirsch et al. 1960) and during the 1960s the collagenase method for liberation of fat cells from a tissue specimen and means to study fat cell size and number in vitro were reported (Hirsch et al. 1960; Hirsch and Gallian 1968; Knittle and Hirsch 1970). In fact, at that time information on adipose tissue function in vivo in humans could only be obtained from circulating FFA levels (Dole 1956) and via animal studies addressing lipolysis through A–V balance studies of glycerol or FFA turnover (Carlson and Orö 1963; Havel 1965). Another example suggested was to collect lymphatic fluid which has a similar composition as the interstitial tissue water, but unfortunately lymph cannot be used to specifically reflect the adipose tissue (Rasio et al. 1967). Moreover, access to the interstitial fluid was possible via the blister suction techniques for measurements of substances in the ultrafiltrate, but there were several reported drawbacks also with this method (Kiistala 1968). Furthermore, enzyme electrodes might be an alternative to monitor glucose in the subcutaneous interstitial fluid (Updike and Hicks 1967; Shichiri et al. 1984). However, proper calibration of the electrodes was a problem because proteins bind to the electrode membrane and, in addition, regional steady state conditions were difficult to maintain due to varying oxygen tension and temperature in the tissue. Finally, the subcutaneous fluid might be analyzed by using a cotton thread introduced in the tissue, the so called wick technique (Scholander et al. 1968), but this procedure was also judged to be inconvenient in human studies. To summarize, the briefly mentioned methods to study adipose tissue metabolism displayed insufficiencies and could only partly reflect metabolic events at the cellular level in humans and, therefore, other techniques were required, such as microdialysis.

12.2 The Advent of Microdialysis

Animal studies within experimental brain-research made the first advancements to sample from the extracellular environment. The "push–pull" technique described how liquid is injected directly into the brain tissue through one cannula and withdrawn from the tissue at the same speed by an adjacent cannula (Gaddum 1961), but this procedure might harm the tissue and was difficult to operate. Another suggestion was to implant "dialysis sacs" containing dextran into subcutaneous tissue and the brain of dogs (Bito et al. 1966). The sacs were surgically removed after >2 months for monitoring of the substrate concentration of interest. Apparently, this was not a method for kinetic studies and again the tissue might be harmed. The first paper describing the tissue microdialysis technique was published by (Delgado et al. 1972). His group applied the technique in brain-research on Rhesus Monkeys, aiming for unique information of neurochemistry after administration of drugs via this device to specific brain structures. Shortly after this important innovation, Ungerstedt and Pycock reported on successful experiments with hollow fibers implanted into the brain of freely moving rats for studies of dopamine neurons (Ungerstedt and Pycock 1974), and during the next two decades Ungerstedt developed the microdialysis technique and became opinion leader in the field (Ungerstedt 1991). Moreover, Ungerstedt established a company distributing microdialysis articles for in vivo studies in mice and men. A true methodological achievement was the "needle probe" with a few mm long hollow fiber membrane at its tip (Ungerstedt 1984) but important input on brain microdialysis was also added from other groups (Hamberger et al. 1983). We concluded that microdialysis had a potential to obtain information at a cellular level, might be easy to implement and, in addition to that, provided a protein-free sample for analysis and, accordingly, we set out to elucidate its usefulness for research in subcutaneous adipose tissue.

12.3 Subcutaneous Microdialysis

In the mid-1980s, we started pilot experiments on ourselves to find out the optimal study design. We looked for a suitable membrane based on the following criteria: (1) Diffusible pores should be significantly larger (×10) than the molecule of interest; (2) Substrates measured and the membrane material should not interact physically or chemically; (3) Diffusible membrane area should be kept constant and well defined to eliminate risk of obstruction. Our decision fell on a membrane introduced to us by Hamberger and Nyström, Department of Histology, University of Gothenburg. This was a hollow fiber from a Cobe HF 130 dialyzer (Cobe Lab Inc, Lakewood, Colorado, USA) having a membrane (Cuprophane B4 AH, molecular cutoff 3 kDa) made of hemicellulose, with an outer diameter of 0.3 mm showing high diffusion properties and low capacity for ultrafiltration. The microdialysis catheters were manufactured in our laboratory using inert polyethylene tubings and cyanoacrylate to get a linear design profiled for insertion by a cannula adopting an "in-and-out fashion" into subcutaneous tissue. After tests of multiple prototypes our decision fell on a dialysis membrane of 30 mm (to optimize recovery) and the catheters were sent to ethylene oxide sterilisation. We kept the intraluminal pressure low by adopting a 50–70 cm long tubing connecting the catheter with the syringe-pump. The perfusion rate of the pump was kept slow to minimize the outward pressure and thereby improve diffusion of molecules across the membrane inward to the perfusion liquid. A compromise was chosen at 2.5 μl/min for the flow rate. This enabled enough volume for analyzing metabolites and monitoring of time kinetics. The perfusion liquid was based on isotonic saline mimicking the characteristics of the interstitial fluid. Historically, the extraction fraction, i.e., relative recovery, of a molecule X from the interstitium to the perfusion liquid was based on calibration experiments of the membrane in vitro (Zetterström et al. 1982). However, by the time the research field became aware of that mass transport of a molecule in the extracellular water space is lower than that in aqueous solutions in vitro, and as a consequence, the higher in vitro recovery would underestimate the interstitial metabolite concentration (Lerma et al. 1986; Rice et al. 1985). To overcome this problem, different methods have been suggested to calibrate microdialysis catheters in vivo. The first attempts were based on extrapolations to zero flow and theoretical considerations on mass transport in vivo (Jacobson et al. 1985; Amberg and Lindefors 1989; Benveniste and Hüttemeier 1990). However, these methods were not easy to adapt for our purposes in adipose tissue. Therefore, we developed the no "net flux" method for in vivo calibrations of glucose, or any other small molecular compound, in human subcutaneous tissue. A hypothesis was tested that perfusion of the catheter in situ with varying concentrations of glucose enables detection of the concentration at which glucose is similar in the perfusion liquid (before dialysis in situ) and the dialysate (fluid collected ex vivo after dialysis in situ). This procedure was presented in the first paper applying the microdialysis technique for studies of metabolism in human subjects (Lönnroth et al. 1987). During the next two decades less time-consuming recovery methods were presented for estimations of interstitial concentrations of small and large molecules (Scheller and Kolb 1991; Bolinder et al. 1992; Jansson et al. 1994; Lönnroth and Strindberg 1995; Strindberg and Lönnroth 2000).

12.4 Microdialysis of Small Molecular Compounds in Human Adipose Tissue

12.4.1 Glucose

Conflicting results had been presented regarding the interstitial glucose concentration in adipose tissue (Claremont et al. 1986). Our first aim was to determine the "true" glucose concentration in the subcutaneous interstitial fluid by calibrating each catheter with the "no net flux" method. We followed glucose kinetics and our results indicated that interstitial glucose was similar to venous plasma in the postabsorptive phase as well as after an oral glucose load (Lönnroth et al. 1987). This was the starting point for pursuing the idea that microdialysis might be used in the development of a glucose sensor for long-term measurements in diabetes patients (Bolinder et al. 1989, 1993; Mader et al. 2012). We only observed a slight time delay of glucose appearance in the extracellular water compared to plasma after glucose ingestion. However, rapid increase of a glucose infusion revealed $\approx$10 min delay of glucose in the extracellular water (Jansson et al. 1988).

Moreover, it was evident that the dialysis procedure could create a zone in situ next to the membrane where glucose may be depleted. Accordingly, we recommended addition of glucose to the perfusion liquid for reliable measurements of glucose related events in situ (Lönnroth et al. 1987).

12.4.2 Adenosine

Adenosine has a very short half-life in plasma and measurements of the "true" interstitial concentration have been a challenge (Klabunde 1983). We measured tissue adenosine in situ with microdialysis to get an idea about its potentially regulatory role. Data revealed that interstitial adenosine levels were 60 % of the circulating level in young healthy subjects (Lönnroth et al. 1989). Laboratory experiments suggested that adenosine at that level was considered significant for a decrease of basal and stimulated lipolysis but not for an increase of glucose transport (Kather et al. 1985; Joost and Steinfelder 1982). The observations indicated that under normal conditions interstitial adenosine was not determined by the adipose cells and adenosine might not serve as an autoregulator of lipolysis. Furthermore, we noticed stable sampling conditions during the in vivo calibration of adenosine. Because adenosine is formed from the breakdown of ATP this was an indirect evidence that our microdialysis technique was unlikely to cause damage of the cells after the initial cannula-induced trauma. Thus, the recommendation from us, and the Karolinska-group as well, was to wait at least 30 min from catheter insertion until start of dialysate collection (Bolinder et al. 1989; Lönnroth et al. 1989).

12.4.3 Lactate

Adipocytes are effective generators of lactate from glucose in vitro (Kashiwagi et al. 1983). Whether subcutaneous adipose tissue may be a significant site for glucose uptake and conversion to lactate in vivo has been less clear. In healthy participants, we observed increased lactate levels in the interstitial fluid compared to plasma both in the fasting and the postprandial state suggesting that adipose tissue is a significant source of lactate (Jansson et al. 1990), and similar data were also reported from (Hagström et al. 1990; Frayn et al. 1989). As a corollary to our results, lactate generation was blunted when glucose was depleted from the dialysis area (Jansson et al. 1990). Moreover, obese subjects showed higher fasting blood lactate in line with the view that adipose tissue contributes to whole body lactate turnover (Doar and Cramp 1970). Indeed, adipose tissue lactate production might be of importance for the impaired glucose homeostasis in obese patients (Lovejoy et al. 1990). First, glucose uptake in adipose tissue may be more significant than previously thought (Mårin et al. 1987) and, second, lactate may be a precursor of hepatic glucose production and possibly dyslipidemia (Consoli et al. 1990). Unfortunately, extraordinary measures are required to apply the microdialysis technique for monitoring of deeper subcutaneous depots in humans (Enevoldsen et al. 2001). Thus, the efficiency of lactate production from visceral fat depots in vivo in humans was unknown, but on the other hand available data at that time indicated that total body fat was important for lactate turnover in obese patients (Doar and Cramp 1970; Lovejoy et al. 1990).

12.4.4 Glycerol

Glycerol is an attractive molecule to monitor since it is an excellent marker of lipolysis. Our first study to assess subcutaneous lipolysis in situ in obese subjects showed increased glycerol levels in the abdominal and femoral sites compared to lean healthy subjects (Jansson et al. 1990). Interstitial glycerol was highest in the abdominal subcutaneous depot in obese subjects which fitted perfectly well with the strong association between waist circumference and the metabolic syndrome (Krotkiewski et al. 1983). Moreover, we noticed how critical it was that some hollow fibers used were checked for contamination with glycerol prior to sampling. In recent protocols, we have perfused the microdialysis probes with saline in vitro to get rid of glycerol before using them. If this might be neglected, the estimated interstitial glycerol levels are overestimated (Maggs et al. 1995). Unfortunately, no microdialysis techniques of today can monitor interstitial FFA because of protein binding, however, tracer techniques might be used to shed light on glycerol and FFA turnover in vivo in humans (Bortz et al. 1972; Jensen 1991).

A breakthrough for studies of adipose tissue metabolism in man was the concept to cannulate epigastric veins to get blood draining abdominal subcutaneous tissue (Frayn et al. 1989). This was an approach to come closer to metabolic events in adipose tissue and A–V balance studies of fatty acids and triglycerides could be undertaken in the postprandial period. On the negative side was the fact that the information obtained was indirect due to admixture of blood from various regions, different kinetics and degree of protein binding for substances in the blood and the interstitial space. Head-to-head comparisons between microdialysis and vein cannulation were reported (Simonsen et al. 1994; Summers et al. 1998), and a fair compromise and conclusion is that the methods complement each other.

In the next two decades, several groups in the metabolic field started up in different constellations to conduct microdialysis studies on lipolysis during exercise (Arner et al. 1990; Borsheim et al. 2000; Stallknecht et al. 2000; Hickner 2000), in obesity research (Lafontan 2008; Hickner et al. 2011; Koppo et al. 2012), during insulin deprivation as well as hypoglycemia (Hagstrom-Toft et al. 1991; Maggs et al. 1997) and when modulating the sympathetic tone (Dodt et al. 2000; Wellhöner et al. 2007). Adrenergic regulation of subcutaneous lipolysis was studied in detail by the Karolinska-group and the protocols often described how pharmacological agents known to interfere with lipolysis were added to the perfusion medium (Arner et al. 1990). However, among the compounds studied several of them were vasoactive. We hypothesized that interstitial glycerol levels may not only depend of appearance but also depend of clearance from the subcutaneous tissue. Accordingly, we set up the ^{133}Xe-clearance technique for measurements of adipose tissue blood flow (ATBF) after training in Bülows laboratory at the Bispebjerg Hospital in Copenhagen.

12.5 Subcutaneous Adipose Tissue Blood Flow and Substrate Release

We purchased a monitoring system with NaI scintillation detectors for Xenon wash out and followed the procedures as originally described by Larsen (Larsen et al. 1966). The tissue to blood partition coefficient for Xenon was assumed to be 10.0 ml/g both in lean and obese participants (Jansson and Lönnroth 1995). Our first observations were the decreased ATBF in obese as compared to lean subjects and the blunted increase of ATBF in obese subjects after oral glucose ingestion (Jansson et al. 1992). This was later confirmed and, indeed, the blunted response was proposed to be a facet of the metabolic syndrome (Jansson et al. 1998; Karpe et al. 2002). The observed inverse relationship between ATBF and fat cell size has also been shown in a recent study (Sandqvist et al. 2011). Data confirmed the old report indicating that the number of blood vessels per cell in a fat depot is constant (DiGirolamo and Esposito 1975).

The Xenon washout procedure is a quantitative method to assess blood flow. This means that we could apply Fick's principle for estimation of apparent subcutaneous glycerol and lactate release in lean and obese individuals (Jansson et al. 1992, 1994). The larger fat cells released more glycerol and lactate per cell but on the other hand substrate release per 100 g tissue was similar in lean and obese subjects (Jansson et al. 1992, 1994; Sandqvist et al. 2001). From this follows that rate of appearance of glycerol and lactate is directly proportional to body fat mass and emphasizes the metabolic impact of the peripheral fat depots (Blackard et al. 1993). Somewhat unexpectedly we could not show any increase in subcutaneous glycerol and lactate release in type 2 diabetic patients compared with non-diabetic controls (Jansson et al. 1995). The molecular mechanisms behind these observations are far from clear but an interesting autocrine lactate loop was recently described in detail (Ahmed et al. 2010).

From now on, a majority of research groups worldwide included means to assess ATBF or ATBF change in the study protocols, and these methods were the microdialysis ethanol technique (Hickner et al. 1991; Maggs et al. 1997; Rooyackers et al. 2004; Karpe et al. 2002), the Doppler technique (Wellhöner et al. 2006) or PET (Virtanen et al. 2001). The Xenon wash out is for us a mandatory method to include when an intervention may affect both vascular and cellular functions in adipose tissue e.g., after administration of agents that amplify NO signaling (Ardilouze et al. 2004;

Murdolo et al. 2012). It is worth noticing that Xenon wash out was superior to the ethanol method for monitoring of ATBF after an oral glucose load (Karpe et al. 2002).

12.6 Membranes with Large Pores for Measurements of Insulin, Growth Factors, and Cytokines in Subcutaneous Tissue and Skeletal Muscle

We developed a method for assessment of interstitial insulin in abdominal adipose tissue using a polypropylene membrane (Jansson et al. 1993). The design chosen was a linear probe having a membrane length of ca 45 mm and a pore size of ca 0.2 μm. Inulin which is a carbohydrate molecule of the same size as insulin was used as reference substance throughout the experiments. Data from obese healthy volunteers suggested presence of an endothelial barrier for insulin whereas this was not shown for inulin (Jansson et al. 1993). Moreover, there was a delay of insulin appearance in the interstitial fluid compared to plasma of ca 20 min but insulin showed similar kinetics in both compartments (Jansson et al. 1993). Insulin recovery in situ was only a few percent in these experiments and that prompted us to infuse insulin to a circulating level of 300–600 mU/l. Data from our group also support the presence of an endothelial barrier in skeletal muscle when using a custom-made concentric probe (MW cutoff 100 kD) (Sjöstrand et al. 1999, 2000). However, study results regarding the importance of the endothelial barrier for insulin differ between laboratories (Mokshagundam et al. 1996; Herkner et al. 2003). Recently, we were also able to estimate interstitial insulin in the periphery during a meal (Sandqvist et al. 2006) and during a glucose tolerance test (Sandqvist et al. 2011) due to the more sensitive insulin kits and use of the Schmelz-probe (below). In fact, during physiological hyperinsulinemia the transfer of insulin might be receptor regulated and not mediated via passive diffusion (Sandqvist et al. 2011; Herkner et al. 2003). The role of vascular insulin resistance for this observation remains to be elucidated in humans (Yuen et al. 2009; Kubota et al. 2011). Interestingly, microdialysis was also successfully implemented for characterization of insulin-like growth factors and vascular endothelial growth factor in human muscle (Berg et al. 2006; Hansen et al. 2010).

Through collaboration with Martin Schmelz, University of Heidelberg, we started up subcutaneous measurements using a custom-made linear needle probe (Plasmaflo OP–02, Asahi Medical Co. Ltd, Oita, Japan, MW cutoff 3 MDa). We perfused the device slowly (1.0 μl/min) with physiologic saline containing 1 % albumin. In our first study using this probe, healthy subjects displayed unexpectedly high IL-6 concentrations in the interstitial fluid (Sopasakis et al. 2004). The membranes were calibrated in situ according to the endogenous reference technique and that resulted in a tissue level of IL-6 that was 100-fold higher than that in plasma (Sopasakis et al. 2004). A few years later, we used this probe for subcutaneous measurements of MCP-1 in a comparative study between lean and obese subjects (Murdolo et al. 2007). Again substantially increased interstitial levels compared to plasma were demonstrated [10-fold increase of MCP-1). We were unable to show any differences in the interstitial MCP-1 levels between the groups and no regulation of MCP-1 by insulin was observed (Murdolo et al. 2007). To gain knowledge about the impact of tissue trauma on the sampled adipokines, we studied the time profile of a battery of cytokines/Chemokines in situ after catheter insertion (Murdolo et al. 2008). We took advantage of the Luminex technique and the results were striking. 90 min after tissue disruption there was a prompt increase of IL-6, MCP-1 and IL-8 while IL-1-alpha, RANTES and leptin remained stable or decreased (Murdolo et al. 2008). Thus, it is probable that IL-6 and MCP-1 data from our first studies were affected by the procedure-induced acute tissue response, at least to some extent. Based on these observations it is strongly recommended to conduct a pilot study to map the time profile for the chemokine in question prior to a fullscale metabolic study.

Furthermore, concerns have been raised how to interprete results when there may be an interrelationship between the tissue response and the analyte of interest (Stenken et al. 2010). However, we have arguments against this notion from a study of adiponectin and its isoforms in situ in lean and obese subjects (Murdolo et al. 2009). Interstitial HMW adiponectin was suppressed in obese subjects and also during a glucose clamp in lean and obese subjects (Murdolo et al. 2009). Second, monitoring of subcutaneous interstitial insulin levels allow statements about different kinetics of insulin transfer to adipose cells in lean and obese females (Sandqvist et al. 2011). In summary, in our hands microdialysis is a minimally invasive technique but caution is warranted when the method is used to assess tissue inflammation.

Similar experiences from adipose tissue were shared by the microdialysis groups in Graz and Copenhagen (Pachler et al. 2007; Clausen et al. 2009). Moreover, Höjbjerre et al. reported on an inflamed adipose tissue in sedentary type 2 diabetes relatives (Höjbjerre et al. 2011). Furthermore, skeletal muscle and tendon microdialysis measurements of IL-6 were conducted following exercise (Langberg et al. 2002; Andersen et al. 2011) and in studies of trapezius myalgia (Rosendal et al. 2005). Importantly, microdialysis revealed increased levels of pain-inducing substances in the symptomatic muscles in patients with polymyalgia rheumatica (Kreiner and Galbo 2011), but this inflammatory pattern was not observed in tender points in patients with tension-type headache (Ashina et al. 2003).

12.7 Microdialysis Studies on Skeletal Muscle Metabolism

Skeletal muscle is the major tissue responsible for glucose uptake after a glucose load, and, indeed, in the 1990s muscle microdialysis became validated for measurements of hormones and metabolic substrates (Sjöstrand et al. 1999;

Grönlund et al. 1991; Müller et al. 1995, 1996; Hagström-Toft et al. 1997). Usually concentric microdialysis probes were inserted in the vastus lateralis and brachioradialis muscles. Feasibility of the microdialysis procedure for metabolic purposes was first shown by Muller et al. (1996) who monitored glucose and lactate in the femoral muscle in healthy participants and lipolysis regulation in muscle was later addressed by Qvisth et al. (2004) in two elegant microdialysis studies (Qvisth et al. 2005). However, Holmäng, Niklasson, Sjöstrand, Gudbjörnsdottir and Lönnroth mainly wrote the metabolic road map in muscle using this technique. Based on microdialysis work, insulin resistance in muscle was induced by glucosamine, epinephrine and manipulation of steroid hormones and improved by exercise and induction of ischemia (Niklasson et al. 1998, 2000; Holmäng et al. 1999, 2001). The experimental design allowed separation as well as integration of vascular and metabolic effects and these studies deepened our understanding and set the agenda for future research on the impact of insufficient insulin and glucose delivery to the muscle cell in insulin resistance. Parallel investigations on the role of nutritive blood flow for exchange of nutrients in rat skeletal muscle included microdialysis and supported the significance of substrate delivery to keep insulin sensitivity in muscle intact (Rattigan et al. 1997; Newman et al. 2001, 2002). However, the picture is far from clear, Weinhandl and co-workers reported results suggesting less impact of insulin-induced capillary recruitment for delivery of macromolecules in healthy human subjects (Weinhandl et al. 2007).

Our group assessed the available capillaries in the muscle of interest via a commonly used formula within capillary physiology. The variable PS was used as a surrogate for capillary recruitment and PS glucose equals the permeability surface area product for glucose in a tissue bed. The formula employed for the calculations: $PS = \ln(V - A/I - A)\ x - Q$, includes variables obtained via muscle microdialysis, cannulation of deep forearm vein and radial artery, and muscle plethysmography for plasma flow (Gudbjornsdottir et al. 2003). We used this study design and assumed that PS mirrors the functionality of muscle capillaries to evaluate two types of interventions to improve transfer of insulin and glucose from the plasma to the interstitium in insulin-resistant human muscle. First, metacholine was infused in a forearm artery, while the contralateral arm served as control, in insulin-resistant obese and healthy subjects (Murdolo et al. 2008). Amplification of the NO-signal resulted in improvements in the obese but not in the lean subjects lending support to the idea that a vascular defect is part of muscle insulin resistance in obesity. Further, glucose uptake and PSglucose improved in the fasting state in type 2 diabetic patients when distal NO-signaling was augmented by tadalafil, a phosphodiesterase-5 inhibitor (Jansson et al. 2010). Work is in progress to find out the role of this pharmacological principle to increase muscle glucose uptake in the postprandial state.

12.8 Future Perspectives

The progress of microdialysis in clinical experimental research in humans is impressive. The first paper on the technique in humans was profiled for monitoring of substrates in subcutaneous interstitial fluid and 25 years later the number of hits for "microdialysis and human" exceeds 3000 in a PubMed search. In fact, applications of microdialysis can be found for almost every organ in humans (Müller 2002) as exemplified during surgery where interstitial sampling was accomplished in myocardium (Mantovani et al. 2010), lung (Zeitlinger et al. 2005) and the gastrointestinal tract (Suzuki et al. 2003; Solligård et al. 2007).

I would like to summarize some personal reflections on clinical and experimental microdialysis for the future. First, the potential benefit to have a probe implemented in a tissue fluid or in a peripheral vein for continuous sampling of a protein-free dialysate of glucose, lactate, or pyruvate without frequent blood testing (Rooyackers et al. 2010). Second, monitoring of the interstitial fluid gets access to metabolic events closer to the cells compared to plasma and therefore may effectively predict ischemia or rejection (Smit et al. 2010; Haugaa et al. 2012). Third, when considering microdialysis as a research tool even more complex protocols may be seen in the near future covering knowledge gaps in integrative human physiology (Eriksson et al. 1999; Vestergaard et al. 2008; Höjbjerre et al. 2010; Goossens et al. 2011; Mortensen et al. 2011). Fourth, large pore membranes may be used to explore the proteome obtained in peripheral human tissues for subsequent hypothesis testing (Gill et al. 2011). Fifth, efforts to increase recovery of interstitial insulin by treatment of the catheter material may improve the reliability of microdialysis measurements of sticky molecules (Dahlin et al. 2012).

In the light of numerous ongoing microdialysis projects and new interesting applications of the technique it is anticipated that the literature on microdialysis will grow fast. However, the more important trend strikes me, namely that microdialysis has been moving out from the lab bench to a bedside position and nowadays may support the clinician for optimal patient care in neurosurgery (Goodman and Robertson 2009; Clausen et al. 2012), liver transplantation (Haugaa et al. 2012), plastic surgery (Smit et al. 2010; Samuelsson et al. 2006) and even might be used for monitoring of metabolic control in diabetic patients (Mader et al. 2012).

Acknowledgments In the mid-1980s, I was fortunate to start my research career in Ulf Smiths laboratory when Peter Lönnroth decided to test the feasibility of microdialysis, a well-known procedure in neurobiology, for monitoring of adipose tissue metabolism in humans. I am grateful to both of you for the fact that we, more or less, have managed to walk the long and winding road together. Also thank you to all research friends at the University of Gothenburg and all over the world for nice microdialysis collaborations. Finally, I would like to thank my research group, in particular Lena Strindberg whose expertise in running clinical microdialysis experiments is unique.

References

Ahmed K, Tunaru S, Tang C, Müller M, Gille A, Sassmann A, Hanson J, Offermanns S (2010) An autocrine lactate loop mediates insulin-dependent inhibition of lipolysis through GPR81. Cell Metab 11:311–319

Amberg G, Lindefors N (1989) Intracerebral microdialysis: II. Mathematical studies of diffusion kinetics. J Pharm Meth 22:157–183

Andersen MB, Pingel J, Kjaer M, Langberg H (2011) Interleukin-6: a growth factor stimulating collagen synthesis in human tendon. J Appl Physiol 110:1549–1554

Ardilouze JL, Fielding BA, Currie JM, Frayn KN, Karpe F (2004) Nitric oxide and beta-adrenergic stimulation are major regulators of preprandial and postprandial subcutaneous adipose tissue blood flow in humans. Circulation 109:47–52

Arner P, Kriegholm E, Engfeldt P, Bolinder J (1990a) Adrenergic reulation of lipolysis in situ at rest and during exercise. J Clin Invest 85:893–898

Arner P, Kriegholm E, Engfeldt P (1990b) In situ studies of catecholamine-induced lipolysis in human adipose tissue using microdialysis. J Pharmacol Exp Ther 254:284–288

Ashina M, Stallknecht B, Bendtsen L, Pedersen JF, Schifter S, Galbo H, Olesen J (2003) Tender points are not sites of ongoing inflammation—in vivo evidence in patients with chronic tension-type headache. Cephalalgia 23:109–116

Benveniste H, Hüttemeier PC (1990) Microdialysis—theory and application. Prog Neurobiol 35:195–215

Berg U, Gustafsson T, Sundberg CJ, Carlsson-Skwirut C, Hall K, Jakeman P, Bang P (2006) Local changes in the insulin-like growth factor systemin human skeletal muscle assessed by microdialysis and arterio-venous difference technique. Growth Horm IGF Res 16:217–223

Bito L, Davson H, Levin EM, Murray M, Snider N (1966) The concentration of free amino acids and other electrolytes in cerebrospinal fluid in vivo dialysate of brain and blood plasma of the dog. J Neurochem 13:1057–1067

Blackard WG, Clore JN, Glickman PS, Nestler JE, Kellum JM (1993) Insulin sensitivity of splanchnic and peripheral adipose tissue in vivo in morbidly obese man. Metabolism 42:1195–1200

Bolinder J, Hagström E, Ungerstedt U, Arner P (1989) Microdialysis of subcutaneous adipose tissue in vivo for continuous glucose monitoring in man. Scand J Clin Lab Invest 49:465–474

Bolinder J, Ungerstedt U, Arner P (1992) Microdialysis measurement of the absolute glucose concentration in subcutaneous adipose tissue allowing glucose monitoring in diabetic patients. Diabetologia 35:1177–1180

Bolinder J, Ungerstedt U, Arner P (1993) Long-term continuous glucose monitoring with microdialysis in ambulatory insulin-dependent diabetic patients. Lancet 342:1080–1085

Borsheim E, Lönnroth P, Knardahl S, Jansson P-A (2000) No difference in the lipolytic response to beta-adrenoceptor stimulation in situ but a delayed increase in adipose tissue blood flow in moderately obese compared with lean men in the postexercise period. Metabolism 49: 579–587

Bortz WM, Paul P, Haff AC, Holmes WL (1972) Glycerol turnover and oxidation in man. J Clin Invest 51:1537–1546

Carlson LA, Orö L (1963) Studies on the relationship between the concentration of plasma free fatty acids and glycerol in vivo. Metabolism 12:132–142

Claremont DJ, Sambrook IE, Penton C, Pickup JC (1986) Subcutaneous implantation of a ferrocene-mediated glucose sensor in pigs. Diabetologia 29:817–821

Clausen TS, Kaastrup P, Stallknecht B (2009) Proinflammattory tissue response and recovery of adipokines during four days of subcutaneouss large-pore microdialysis. J Pharmacol Toxicol Methods 60:281–287

Clausen F, Marklund N, Lewén A, Enblad P, Basu S, Hillered L (2012) Interstitial F(2)-Isoprostane 8-Iso-PGF(2alpha) as a biomarker of oxidative stress after severe human traumatic brain injury. J Neurotrauma 29:766–775

Consoli A, Nurjhan N, Reilly JJ, Bier DM, Gerich J (1990) Mechanism of increased gluconeogenesis in noninsulin-dependent diabetes mellitus. Role of alterations in systemic, hepatic, and muscle lactate and alanine metabolism. J Clin Invest 86:2038–2045

Dahlin AP, Hjort K, Hillered L, Sjödin MO, Bergquist J, Wetterhall M (2012) Multiplexed quantification of proteins adsorbed to surface-modified and non-modified microdialyis membranes. Anal Bioanal Chem 402:2057–2067

Delgado JMR, DeFeudis FV, Roth RH, Ryugo DK, Mitruka BM (1972) Dialytrode for long term intracerebral perfusion in awake monkeys. Arch Int Pharmacodyn 198:9–21

DiGirolamo M, Esposito J (1975) Adipose tissue blood flow and cellularity in the growing rabbit. Am J Physiol 229:107–112

Doar JWH, Cramp DG (1970) The effect of obesity and maturity-onset diabetes mellitus on L(+) lactic acid metabolism. Clin Sci 76:509–516

Dodt C, Lönnroth P, Fehm HL, Elam M (2000) The subcutaneous lipolytic response to regional neural stimulation is reduced in obese women. Diabetes 49:1875–1879

Dole CP (1956) A relation between non-esterified fatty acids in plasma and the metabolism of glucose. J Clin Invest 35:150–154

Enevoldsen LH, Simonsen L, Stallknecht B, Galbo H, Bülow J (2001) In vivo human lipolytic activity in preperitoneal and subdivisions of subcutaneous abdominal adipose tissue. Am J Physiol 28:E1110–E1114

Eriksson JW, Smith U, Waagstein F, Wysocki M, Jansson PA (1999) Glucose turnover and adipose tissue lipolysis are insulin-resistant in healthy relatives of type 2 diabetes patients: is cellular insulin resistance a secondary phenomenon? Diabetes 48:1572–1578

Frayn KN, Coppack SW, Humphreys SM, Whyte PL (1989) Metabolic characteristics of human adipose tissue in vivo. Clin Sci 76:509–516

Gaddum JH (1961) Push–pull cannulae. J Physiol 155:1–2

Gill C, Parkinson E, Church MK, Skipp P et al (2011) A qualitative and quantitative proteomic study of human microdialysate and the cutaneous response to injury. AAPS J 13:309–317

Goodman JC, Robertson CS (2009) Microdialysis: is it ready for prime time? Curr Opin Crit Care 15:110–117

Goossens GH, Bizzarri A, Venteclef N, Essers Y et al (2011) Increased adipose tissue oxygen tension in obese compared with lean men is accompanied by insulin-resistance, impaired adipose tissue capillarization a, and inflammation. Circulation 124:67–76

Grönlund B, Astrup A, Bie P, Christensen NJ (1991) Noradrenaline release in skeletal muscle and in adipose tissue studied by microdialysis. Clin Sci 80:595–598

Gudbjornsdottir S, Sjöstrand M, Strindberg L, Wahren J, Lönnroth P (2003) Direct measurements of the permeability surface area for insulin and glucose in human skeletal muscle. J Clin Endocrinol Metab 88:4559–4564

Hagström E, Arner P, Ungerstedt U, Bolinder J (1990) Subcutaneous adipose tissue: a source of lactate production after glucose ingestion in humans. Am J Physiol 258:E888–E893

Hagstrom-Toft E, Arner P, Näslund B, Ungerstedt U, Bolinder J (1991) Effect of insulin deprivation and replacement on in vivo subcutaneous adipose tissue substrate metabolism in humans. Diabetes 40:666–672

Hagström-Toft E, Enoksson S, Moberg E, Bolinder J, Arner P (1997) Absolute concentrations of glycerol and lactate in human skeletal muscle, adipose tissue, and blood. Am J Physiol 273:E584–E592

Hamberger A, Berthold CH, Karlsson B, Lehmann A, Nyström B (1983) Extracellular GABA, glutamate and glutamine in vivo—perfusion-dialysis of the rabbit hippocampus. In: Glutamine, glutamate and GABA in the central nervous system. Alan R Liss Inc., New York, pp 119–139

Hansen AH, Nielsen JJ, Saltin B, Hellsten Y (2010) Exercise training normalizes skeletal muscle vascular endothelial growth factor in patients with essential hypertension. J Hypertens 28:1176–1185

Haugaa H, Thorgersen EB, Pharo A, Boberg KM et al (2012) Early bedside detection of ischemia and rejection in liver transplants by microdialysis. Liver Transpl [Epub ahead of print]

Havel R (1965) Some influences of the sympathetic nervous system and insulin on mobilization of fat from adipose tissue: studies of the turnover rates of free fatty acids and glycerol. Ann N Y Acad Sci 13:91–101

Herkner H, Klein N, Joukhadar C, Lackner E et al (2003) Transcapillary insulin transfer in human skeletal muscle. Eur J Clin Invest 33:141–146

Hickner R (2000) Application of the microdialysis technique in sports medicine. Exerc Sport Sci Rev 28:117–122

Hickner RC, Rosdahl H, Borg I, Ungerstedt U, Jorfeldt L, Henriksson J (1991) Ethanol may be used with the microdialysis technique to monitor blood flow changes in skeletal muscle: dialysate glucose concentration is blood-flow-dependent. Acta Physiol Scand 143:355–356

Hickner RC, Kemeny G, Clark PD, Gavin VB, McIver KL, Evans CA, Carper MJ, Garry JP (2011) In vivo nitric oxide suppression of lipolysis in subcutaneous abdominal adipose tissue is greater in obese than in lean women. Obesity 20:1174–1178

Hirsch J, Gallian E (1968) Methods for the determination of adipose cell size in man and animals. L Lipid Res 9:110–119

Hirsch J, Farquhar JW, Ahrens EH, Peterson ML, Stoffel W (1960) Studies of adipose tissue in man. A microtechnic for sampling and analysis. Am J Clin Nutr 8:499–511

Höjbjerre L, Sonne MP, Alibegovic AC, Dela F, Vaag A, Meldgaard JB, Christensen KB, Stallknecht B (2010) Impact of physical inactivity on subcutaneous adipose tissue metabolism in healthy young male offspring of patients with type 2 diabetes. Diabetes 59:2790–2798

Höjbjerre L, Sonne MP, Alibegovic AC, Dela F, Bruun JM, Stallknecht B (2011) Impact of physical inactivity on adipose tissue low-grade inflammation in first-degree relatives of type two diabetic patients. Diabetes Care 34:2265–2272

Holmäng A, Nilsson C, Niklasson M, Larsson BM, Lönnroth P (1999) Induction of insulin resistance by glucosamine reduces blood flow but not interstitial levels of either glucose or insulin. Diabetes 48:106–111

Holmäng A, Niklasson M, Rippe B, Lönnroth P (2001) Insulin insensitivity and delayed transcapillary of insulin in oophrectomized rats treated with testosterone. Acta Physiol Scand 171:427–438

Jacobson I, Sandberg M, Hamberger A (1985) Mass transfer in brain dialysis devices—a new method for the estimation of extracellular amino acid concentrations. J Neurosci Meth 15:263–268

Jansson P-A, Lönnroth P (1995) Comparison of two methods to assess the tissue/blood partition coefficient for Xenon in subcutaneous adipose tissue in man. Clin Physiol 15:47–55

Jansson PA, Fowelin J, Smith U, Lönnroth P (1988) Characterization by microdialysis of intercellular glucose levels in subcutaneous tissue in humans. Am J Physiol 255:E218–E220

Jansson PA, Smith U, Lönnroth P (1990a) Evidence for lactate production by human adipose tissue in vivo. Diabetologia 33:253–256

Jansson PA, Smith U, Lönnroth P (1990b) Interstitial glycerol concentration measured by microdialysis in two subcutaneous regions in man. Am J Physiol 258:E918–E922

Jansson PA, Larsson A, Smith U, Lönnroth P (1992) Glycerol production in subcutaneous adipose tissue in lean and obese humans. J Clin Invest 89:1610–1617

Jansson PA, Fowelin JP, von Schenck HP, Smith UP, Lönnroth P (1993) Measurement by microdialysis of the insulin concentration in subcutaneous interstitial fluid. Importance of the endothelial barrier for insulin. Diabetes 42:1469–1473

Jansson PA, Veneman T, Nurjhan N, Gerich J (1994a) An improved method to calculate adipose tissue interstitial substrate recovery for microdialysis studies. Life Sci 54:1621–1624

Jansson PA, Larsson A, Smith U, Lönnroth P (1994b) Lactate release from the subcutaneous tissue in lean and obese man. J Clin Invest 93:240–246

Jansson PA, Smith U, Lönnroth P (1995) Microdialysis assessment of adipose tissue metabolism in post-absorptive obese NIDDM subjects. Eur J Clin Invest 25:584–589

Jansson PA, Larsson A, Lönnroth P (1998) Relationship between blood pressure, metabolic variables and blood flow in obese subjects with or without non-insulin-dependent diabetes mellitus. Eur J Clin Invest 28:813–818

Jansson PA, Murdolo G, Sjögren L, Nyström B, Sjöstrand M, Strindberg L, Jansson PA (2010) Tadalafil increases muscle capillary recruitment and forearm glucose uptake in women with type 2 diabetes. Diabetologia 53:2205–2208

Jensen MD (1991) Regulation of forearm lipoysis in different types of obesity. In vivo evidence for adipocyte heterogenity. J Clin Invest 87:187–193

Joost HG, Steinfelder HJ (1982) Modulation of insulin sensitivity by adenosine. Mol Pharmacol 22:614–618

Karpe F, Fielding BA, Ilic V, Macdonald IA, Summers LK, Frayn KN (2002a) Impaired postprandial adipose tissue blood flow response is related to aspects of insulin sensitivity. Diabetes 51:2467–2473

Karpe F, Fielding BA, Ilic V, Humphreys SM, Frayn KN (2002b) Monitoring adipose tissue blood flow in man: a comparison between the (133)Xenon washout method and microdialysis. Int J Obes Relat Metab Disord 26:1–5

Kashiwagi A, Verso MA, Andrews J, Vasques G, Reaven G, Foley JE (1983) In vitro insulin resistance of human adipocytes isolated from subjects with noninsulin-dependent diabetes mellitus. J Clin Invest 72:146–1254

Kather H, Bieger W, Aktories MK, Jakobs KH (1985) Human fat cell lipolysis is primarily regulated by inhibitory modulators acting through distinct mechanisms. J Clin Invest 76:1559–1565

Kiistala U (1968) Suction blister device for separation of viable epidermis from dermis. J Invest Dermatol 50:129–137

Klabunde RE (1983) Dipyridamole inhibition of adenosine metabolism in human blood. Eur J Pharmacol 93:21–26

Knittle JL, Hirsch J (1970) Cellularity of obese and non-obese human adipose tissue. Fed Proc 29:1516–1521

Koppo K, Siklova-Vitkova M, Klimcakova E, Polak J et al (2012) Catecholamine and insulin control of lipolysis in subcutaneous adipose tissue during long-term diet-induced weight loss in obese women. Am J Physiol 302:E226–E232

Kreiner F, Galbo H (2011) Elevated muscle interstitial levels of pain-inducing substances in symptomatic muscles in patients with polymyalgia rheumatica. Pain 152:1127–1132

Krotkiewski M, Björntorp P, Sjöström L, Smith U (1983) Impact of obesity on metabolism in men and women. Importance of regional adipose tissue distribution. J Clin Invest 72: 1150–1162

Kubota T, Kubota N, Kumagai H, Yamaguchi S et al (2011) Impaired insulin signaling in endothelial cells reduces insulin-induced glucose uptake by skeletal muscle. Cell Metab 13:294–307

Lafontan M (2008) Advances in adipose tissue metabolism. Int J Obes 32(Suppl 7):S39–S51

Langberg H, Olesen JL, Gemmer C, Kjaer M (2002) Substantial elevation of interleukin-6 concentration in peritendinous tissue, in contrast to muscle, following prolonged exercise in humans. J Physiol 542:985–990

Larsen OA, Lassen NA, Quaade F (1966) Blood flow through human adipose tissue determined with radioactive Xenon. Acta Physiol Scand 66:337–345

Lerma J, Herranz AS, Herreras O, Abraira V, Martin del Rio R (1986) In vivo determination of extracellular concentration of amino acids in the rat hippocampus. A method based on brain dialysis and computerized analysis. Brain Res 384:145–155

Lönnroth P, Strindberg L (1995) Validation of the "internal reference technique" for calibrating microdialysis catheters in situ. Acta Physiol Scand 153:375–380

Lönnroth P, Jansson PA, Smith U (1987) A microdialysis method allowing characterization of intercellular water space in humans. Am J Physiol 253:E228–E231

Lönnroth P, Jansson PA, Fredholm BB, Smith U (1989) Measurement by microdialysis of the intercellular adenosine concentration in the subcutaneous tissue in man. Am J Physiol 256:E250–E255

Lovejoy J, Mellen B, DiGirolamo M (1990) Lactate generation following glucose ingestion: relation to obesity, carbohydrate tolerance and insulin sensitivity. Int J Obesity 14:843–855

Mader JK, Feichtner F, Bock G, Köhler G, Schaller R, Plank J, Pieber TR, Ellmerer M (2012) Microdialysis—a versatile technology to perform metabolic monitoring in diabetes and critically ill patients. Diabetes Res Clin Pract [Epup ahead of print]

Maggs DG, Jacob R, Rife F, Lange R, Leone P, During MJ, Tamborlane WV, Sherwin RS (1995) Interstitial concentrations of glycerol, glucose, and amino acids in human quadriceps muscle and adipose tissue. Evidence for significant lipolysis in skeletal muscle. J Clin Invest 96: 370–377

Maggs DG, Jacob R, Rife F, Caprio S, Tamborlane WV, Sherwin RS (1997a) Counterregulation in peripheral tissues: effect of systemic hypoglycemia on levels of substrates and catecholamines in human skeletal muscle and adipose tissue. Diabetes 46:70–76

Maggs DG, Borg WP, Sherwin RS (1997b) Microdialysis techniques in the study of of brain and skeletal muscle. Diabetologia 40(Suppl 2):S75–S82

Mantovani V, Kennergren C, Bugge M, Sala A, Lönnroth P, Berglin E (2010) Myocardial metabolism assessed by microdialysis: a prospective randomized study in on- and off-pump coronary bypass surgery. Int J Cardiol 143:302–308

Mårin P, Rebuffé-Scrive M, Smith U, Björntorp P (1987) Glucose uptake in human adipose tissue. Metabolism 36:1154–1160

Mokshagundam SP, Peiris AN, Stagner JI, Gingerich RL, Samols E (1996) Interstitial insulin during euglycemic–hyperinsulinemic clamp in obese and lean individuals. Metabolism 45:951–956

Mortensen SP, Thaning P, Nyberg M, Saltin B, Hellsten Y (2011) Local release of ATP into the arterial inflow and venous drainage of human skeletal muscle: insight from ATP determination with the intravascular microdialysis technique. J Physiol 589:1847–1857

Müller M (2002) Science, medicine, and the future. BMJ 324:588–591

Müller M, Schmid R, Nieszpaur-Los M, Fassolt A, Lönnroth P, Fasching P, Eichler HG (1995) Key metabolite kinetics in human skeletal muscle during ischaemia and reperfusion: measurement by microdialysis. Eur J Clin Invest 25:601–607

Müller M, Holmäng A, Andersson OK, Eichler HG, Lönnroth P (1996) Measurement of of interstitial muscle glucose and lactate concentrations during an oral glucose tolerance test. Am J Physiol 271:E1003–E1007

Murdolo G, Hammarstedt A, Sandqvist M, Schmelz M, Herder C, Smith U, Jansson PA (2007) Monocyte chemoattractant protein-1 in subcutaneous abdominal adipose tissue: characterization of interstitial concentration and regulation of gene expression by insulin. J Clin Endocrinol Metab 92:2688–2695

Murdolo G, Herder C, Wang Z, Rose B, Schmelz M, Jansson PA (2008a) In situ profiling of adipokines in subcutaneous microdialysates from lean and obese individuals. Am J Physiol 295:E1095–E1105

Murdolo G, Sjöstrand M, Strindberg L, Gudbjörnsdottir S, Lind L, Lönnroth P, Jansson PA (2008b) Effects of intrabrachial metacholine infusion on muscle capillary recruitment and forearm glucose uptake during physiological hyperinsulinemia in obese, insulin-resistant individuals. J Clin Endocrinol Metab 93:2764–2773

Murdolo G, Hammarstedt A, Schmelz M, Jansson PA, Smith U (2009) Acute hyperinsulinemia differentially regulates interstitial and circulating adiponectin oligomeric pattern in lean and insulin-resistant obese individuals. J Clin Endocrinol Metab 94:4508–4516

Murdolo G, Sjostrand M, Strindberg L, Lönnroth P, Jansson PA (2012) The selective phosphodiesterase-5 inhibitor tadalafil induces microvascular and metabolic effects in type 2 diabetic postmenopausal females [Epup ahead of print]

Newman JM, Di Maria CA, Rattigan S, Clark MG (2001) Nutritive blood flow affects microdialysis O/I ratio for [(14)C]ethanol and (3)H(2)O in perfused rat hindlimb. Am J Physiol 281:H2731–H2737

Newman JM, Rattigan S, Clark MG (2002) Nutritive blood flow improves interstitial glucose and lactate exchange in perfused rat hindlimb. Am J Physiol 283:H186–H192

Niklasson M, Holmäng A, Lönnroth P (1998) Induction of rat muscle insulin resistance by epinephrine is accompanied by increased interstitial glucose and lactate concentrations. Diabetologia 41:1467–1473

Niklasson M, Daneryd P, Lönnroth P, Holmäng A (2000a) Effects of exercise on insulin distribution and action in testosterone-treated oophorectomized female rats. J Appl Physiol 88:2116–2122

Niklasson M, Holmäng A, Sjöstrand M, Strindberg L, Lönnroth P (2000b) Muscle glucose uptake is effectively activated by ischemia in type 2 diabetic subjects. Diabetes 49:1178–1185

Pachler C, Ikeoka D, Plank J, Weinhandl H et al (2007) Subcutaneous adipose tissue exerts proinflammatory cytokines after minimal trauma in humans. Am J Physiol 293:E690–E696

Qvisth V, Hagström-Toft E, Enoksson S, Sherwin R, Sjöberg S, Bolinder J (2004) Combined hyperinsulinemia and hyperglycemia, but not hyperinsulinemia alone, suppress human skeletal muscle lipolytic activity in vivo. J Clin Endocrinol Metab 89:4693–4700

Qvisth V, Enoksson S, Blaak E, Hagström-Toft E, Arner P, Bolinder J (2005) Major differences in noradrenaline action on lipolysis and blood flow rates in skeletal muscle and adipose tissue in vivo. Diabetologia 48:946–953

Rasio EA, Hampers CL, Soeldner JS, Cahill GF (1967) Diffusion of glucose, insulin, inulin, and Evans blue protein into thoracic duct lymph of man. J Clin Invest 46:903–910

Rattigan S, Clark MG, Barrett EJ (1997) Hemodynamic actions of insulin in rat skeletal muscle: evidence for capillary recruitment. Diabetes 46:1381–1388

Rice ME, Gerhardt GA, Hierl PM, Nagy G, Adams RN (1985) Diffusion coefficients of neurotransmitters and their metabolites in brain extracellular fluid space. Neuroscience 15:891–902

Rodbell M (1964) Metabolism of isolated fat cells. Effects of hormones on glucose metabolism and lipolysis. J Biol Chem 239:375–380

Rooyackers O, Thorell A, Nygren J, Ljungqvist O (2004) Microdialysis methods for measuring human metabolism. Curr Opin Clin Nutr Metab Care 7:515–521

Rooyackers O, Blixt C, Mattsson P, Wernerman J (2010) Continuous glucose monitoring by intravenous microdialysis. Acta Anaesthesiol Scand 54:841–847

Rosendal L, Kristiansen J, Gerdle B, Sögaard K, Peolsson M, Kjaer M, Sörensen J, Larsson B (2005) Increased levels of interstitial potassium but normal levels of of muscle IL-6 and LDH in patients with trapezius myalgia. Pain 119:201–209

Samuelsson A, Steinvall I, Sjöberg F (2006) Microdialysis shows metabolic effects in skin during fluid resuscitation in burn-injured patients. Crit Care 10:R172

Sandqvist M, Eriksson JW, Jansson P-A (2001) Increased lactate release per fat cell in normoglycemic first-degree relatives of individuals with type 2 diabetes. Diabetes 50: 2344–2348

Sandqvist M, Johanson EH, Ahrén B, Axelsen M, Schmelz M, Smith U, Jansson PA (2006) Postprandial interstitial insulin concentrations in type 2 diabetes relatives. Eur J Clin Invest 36:383–388

Sandqvist M, Strindberg L, Schmelz M, Lönnroth P, Jansson PA (2011) Impaired delivery of insulin to adipose tissue and skeletal muscle in obese women with postprandial hyperglycemia. J Clin Endocrinol Metab 96:1320–1324

Scheller D, Kolb J (1991) The internal reference technique in microdialysis: a practical approach to monitoring dialysis efficiency and to calculating tissue concentration from dialysate samples. J Neurosci Methods 40:31–38

Scholander PF, Hargens AR, Miller SL (1968) Negative pressure in the interstitial fluid of animals. Science 161:321–328

Shichiri M, Kawamori R, Hakui N, Yamasaki Y, Abe H (1984) Closed-loop glycemic control with a wearable artificial endocrine pancreas. Variations in daily insulin requirements to glycemic response. Diabetes 33:1200–1202

Simonsen L, Bülow J, Madsen J (1994) Adipose tissue metabolism in humans determined by vein catheterization and microdialysis techniques. Am J Physiol 266:E357–E365

Sjöstrand M, Holmäng A, Lönnroth P (1999) Measurements of of interstitial insulin in human muscle. Am J Physiol 276:E151–E154

Sjöstrand M, Holmäng A, Strindberg L, Lönnroth P (2000) Estimations of muscle interstitial insulin, glucose, and lactate in type 2 diabetic subjects. Am J Physiol 279:E1097–E1103

Smit J, Zeebregts CJ, Acosta R, Werker PM (2010) Advancements in free flap monitoring in the last decade: a critical review. Plast Reconstr Surg 125:177–185

Solligård E, Wahba A, Skogvoll E, Stenseth R, Grönbech JE, Aadahl P (2007) Rectal lactate levels in endoluminal microdialysate during routine coronary surgery. Anaesthesia 62: 250–258

Sopasakis VR, Sandqvist M, Gustafson B, Hammarstedt A, Schmelz M, Yang X, Jansson PA (2004) High local concentrations and effects on differentiation implicate interleukin-6 as a paracrine regulator. Obes Res 12:454–460

Stallknecht B, Larsen JJ, Mikines KJ, Simonsen L, Bülow J, Galbo H (2000) Effect of training on insulin sensitivity of glucose uptake and lipolysis in human adipose tissue. Am J Physiol 279:E376–E385

Stenken JA, Church MK, Gill CA, Clough GF (2010) How minimally invasive is microdialysis sampling? A cautionary note for cytokine collection in human skin and other clinical studies. AAPSJ 12:73–78

Strindberg L, Lönnroth P (2000) Validation of the endogenous reference technique for the calibration of microdialysis catheters. Scand J Clin Lab Invest 60:205–211

Summers LK, Arner P, Ilic V, Clark ML, Humphreys SM, Frayn K (1998) Adipose tissue metabolism in the postprandial period: microdialysis and arteriovenous techniques compared. Am J Physiol 274:E651–E655

Suzuki H, Lijima K, Moriya A, McElroy K, Scobie G, Fyfe V, McColl KE (2003) Conditions for acid catalysed luminal nitrosation are maximal at the gastric cardia. Gut 52:1095–1101

Toldt C (1870) Contribution to the histology and physiology of adipose tissue. Akad Wiss Wien Math Naturwiss Kl 62:445

Ungerstedt U (1984) Measurement of neurotransmitter release by intracranial dialysis. In: Marsden CA (ed) Measurement of neurotransmitter release in vivo. Wiley, London, pp 81–105

Ungerstedt U (1991) Microdialysis—principles and applications for studies in animals and man. J Intern Med 230:365–373

Ungerstedt U, Pycock C (1974) Functional correlates of dopamine neurotansmission. Bull Schweiz Akad Med Wiss 1278:1–13

Updike SJ, Hicks GP (1967) The enzyme electrode. Nature 214:986–988

Vestergaard ET, Gormsen LC, Jessen N, Lund S, Hansen TK, Moller N, Jorgensen JO (2008) Ghrelin infusion in humans induces acute insulin resistance and lipolysis independent of growth hormone signalling. Diabetes 57:3205–3210

Virtanen KA, Peltoniemi P, Marjamäki P, Asola M, Strindberg L, Parkkola R, Huupponen R, Knuuti J, Lönnroth P, Nuutila P (2001) Human adipose tissue glucose uptake determined using [^{18}F]-fluoro-deoxy-glucose ([^{18}F]FDG) and PETPET in combination with microdialysis. Diabetologia 44:2171–2179

Weinhandl H, Pachler C, Mader JK, Ikeoka D, Mautner A, Falk A, Suppan M, Pieber T, Ellmerer M (2007) Physiological hyperinsulinemia has no detectable effect on access of macromolecules to insulin-sensitive tissues in healthy humans. Diabetes 56:2213–2217

Wellhöner P, Rolle D, Lönnroth P, Strindberg L, Elam M, Dodt C (2006) Laser-doppler flowmetry reveals rapid perfusion changes in adipose tissue of lean and obese females. Am J Physiol 291:E1025–E1030

Wellhöner P, Welzel M, Rolle D, Dodt C (2007) In vivo effects of corticotrtropin-releasing hormone on femoral adipose tissue adipose tissue metabolism in women. Int J Obes 31: 718–722

Yuen DY, Dwyer RM, Matthews VB, Zhang L et al (2009) Interleukin-6 attenuates insulin-mediated increases in endothelial cell signaling but augments skeletal muscle insulin action via different effects on tumor necrosis factor-alpha expression. Diabetes 58:1086–1095

Zeitlinger M, Müller M, Joukhadar C (2005) Lung microdialysis—a powerful tool for the determination of exogenous and endogenous compounds in the lower respiratory tract (mini-review). AAPS J 7:E600–E608

Zetterström T, Vernet L, Ungerstedt U, Tossman U, Jonzon B, Fredholm BB (1982) Purine levels in the intact rat brain, studies with an implanted perfused hollow fibre. Neurosci Lett 29: 111–115

Chapter 13
High Molecular Weight Targets and Treatments Using Microdialysis

Geraldine F. Clough, Julie A. Stenken and Martin K. Church

Abstract Advances in proteomics, the quest for biomarker discovery and the desire to understand drug mechanisms at the cellular level all require the ability to obtain relevant biological samples, preferably from humans. Microdialysis sampling is a well-established and widely accepted method for the in vivo collection of solutes, including high molecular weight compounds, from a number of complex matrices, but principally from the extracellular fluid space. Here, we summarize the current status of microdialysis sampling of high molecular weight targets which, together with novel analytical methodologies, can lead to a better understanding of the healthy tissue proteome and how this may be changed in disease. Further, the potential for microdialysis to be used to explore the penetration of high molecular weight drugs into the extravascular compartment and the mechanisms by which they exert their beneficial effects is considered.

13.1 Introduction

Microdialysis sampling is a well-established and widely accepted method for the continuous collection of solutes from a number of complex matrices, but principally from the extracellular fluid space in vivo. The sampling method involves

---

G. F. Clough (✉)
Institute of Developmental Sciences, Faculty of Medicine, University of Southampton, Southampton General Hospital, MP 887, Southampton, SO16 6YD, UK
e-mail: G.F.Clough@soton.ac.uk

J. A. Stenken
Department of Chemistry and Biochemistry, University of Arkansas, Fayetteville, AR, USA

M. K. Church
Charité Universitätsmedizin, Berlin, Germany

M. Müller (ed.), *Microdialysis in Drug Development*, AAPS Advances in the Pharmaceutical Sciences Series, DOI: 10.1007/978-1-4614-4815-0_13,

implantation of a small porous hollow fiber dialysis membrane into the tissue such that the membrane is in contact with the surrounding fluid-filled extracellular tissue space. Perfusion of the fiber allows the exchange of fluid and dissolved solutes across the membrane, primarily by diffusion and/or ultrafiltration, to provide an analytically clean sample that requires little to no treatment before analysis. For this reason, microdialysis sampling has become a standard technique in many laboratories to explore the mediator mechanisms of physiological and pathological tissue events and to examine the uptake and distribution of xenobiotics in conscious and freely moving animals and in humans in a clinical research setting.

The dynamic range of protein concentrations within a biofluid such as serum or plasma covers 8–10 orders of magnitude, with a few abundant proteins making up over 90 % of the proteome (Anderson 2005). Obtaining plasma is a simple and minimally invasive method of sampling; however, as a systemic biofluid, it contains solutes sourced from many different tissues. Thus, while plasma provides a rich source of potential biomarkers, the origins of these biomarkers and their relevance to a (patho)physiological process is uncertain. The use of a localized sampling technique such as microdialysis provides a physiologically relevant biofluid sample for analysis that is less complex and therefore more amenable to biomarker analysis, as well as giving a greater insight into the biological process underlying the local rather than systemic (patho)physiological state.

The development of membranes suitable for the in vivo recovery of larger molecules has allowed the recovery of a wide range of high molecular weight compounds, particularly bioactive proteins from the interstitial space of many tissues including skin, brain, muscle, liver, and adipose tissue (Dostalova et al. 2009; Helmy et al. 2011a; Kendrick 1990; Riese et al. 2003; Salgo et al. 2011; Sjogren et al. 2002; Sopasakis et al. 2005; Takeda et al. 2011; Waelgaard et al. 2008; Winter et al. 2002). Recent studies have also successfully demonstrated in vivo sampling of free fatty acids using large-pore membrane in the rat (Cooley and Lunte 2011; Jensen et al. 2007). However, the ability to recover larger molecules is not without its challenges and several factors, including low tissue concentrations of many target analytes, low dialysis recovery rates, and non-specific adsorption to the probe membrane and tubing (Ao and Stenken 2006; Clough 2005; Maurer 2010), that together give rise to challengingly low dialysate concentrations.

Additionally, while microdialysis sampling is frequently referred to as a minimally invasive procedure and the probe as a passive artificial capillary, it is clear that there are both acute and chronic tissue responses to the insertion of the probe, its subsequent perfusion, and the process of dialysis (Stenken et al. 2010). Thus, the application of microdialysis systems to the measurement of in vivo levels of biologically active macromolecules has yet to be fully optimized. This chapter seeks to summarize the current status of microdialysis sampling of high molecular weight targets, particularly for collection of chemokines, cytokines, and growth factors, which together with novel analytical methodologies can lead to a better understanding of the healthy tissue proteome and the changes that occur during pathology. Further, we consider the potential for microdialysis in drug discovery and testing of high molecular mass compounds (biologicals/antibodies) and their effects.

13.2 Basic Issues of "Recovery" of Large Molecules

Large molecule microdialysis has been driven by the need to sample the interstitial fluid space for bioactive proteins and regulatory peptides as markers of tissue homeostasis and, more importantly in a clinical setting, tissue pathology and repair. Some of the first researchers to report the successful recovery of large molecules by microdialysis used a polycarbonate membrane extracted from a plasmapheresis capsule (Plasmaflo, OP-02 Asahi Medical Co Ltd, Kimal plc Bromsgrove, Worcs UK) configured as a linear, flow-through dialysis probe (Schmelz et al. 1997). These membranes had a molecular weight cutoff (MWCO) of 3,000,000 Da and a maximum pore size of $\sim$0.3 μm, as stated by the manufacturer. With these high-MWCO membranes it was possible to recover serum proteins, neuropeptides, cytokines, and growth factors (Angst et al. 2008; Lischetzki et al. 2001; Schmelz et al. 1997; Winter et al. 2002). Since then, commercially available polyether-sulfone microdialysis probes with a 100 kDa MWCO have been used to recover a wide range of molecular species ranging in size from 3,000 to 120,000 Da (Schutte et al. 2004; Waelgaard et al. 2006).

Theoretically, the probes at the higher end of the MWCO range may appear to be satisfactory for the dialysis of cytokines, growth factors, and even large serum proteins. Practically, however, the porous area capable of allowing the passage of such large molecules is small and results in an increased mass transport resistance and a low relative recovery (Helmy et al. 2009). The published dialysis extraction efficiency of a 10,000 Da protein across a commercial 100,000 Da MWCO membrane perfused at a flow rate of 1 μL/min is typically below 5 % (Clough 2005).

Net movement of high molecular weight analytes across the dialysis membrane is thus determined by tissue concentration, molecular size, aqueous solubility, and binding of the analyte to other molecules and the membrane itself (Ao and Stenken 2006; Chaurasia et al. 2007; Elmquist and Sawchuk 1997; Trickler and Miller 2003). It is also dependent on the nature of the physical characteristics of the membrane and whether membranes retain their ex vivo functionality in relation to passage of target molecules in vivo in the short- and longer term (e.g. biofouling and/or tissue reactions) and whether depletion of target molecules in the immediate area around the membrane occurs (Anderson et al. 2008; Norton et al. 2007; Stenken et al. 2010; Wisniewski et al. 2001).

13.2.1 Diffusivity of Proteins and Reduction in Recovery

A reduced microdialysate extraction of high molecular weight analytes such as proteins is in part due to their much lower diffusivity both within the interstitial space and across the membrane when compared to smaller hydrophilic solutes. Continuous perfusion of large-pore dialysis membranes for long periods results in an apparent fall in analyte recovery, with initial total protein levels significantly higher than those recovered in later samples (Clough 2005). One explanation for this is that the amount of material available for dialysis recovery falls with time. It is well recognized that the concentration gradient of any solute within the tissue space within the proximity of the probe will be determined by the rate at which it is supplied by, and diffuses through, the tissue and also by the rate at which it is removed in the dialysate. Unlike low molecular weight solutes that are delivered to the tissue space and/or cleared via the microcirculation, most protein biomarkers such as the cytokines are locally produced by different cell populations residing within the tissue space. The kinetics for the production processes may vary among different cytokines depending upon which cells (e.g., monocyte/macrophage, neutrophils, or T-cells) produce the proteins (Mascher et al. 1999). Proteases have been demonstrated to degrade cytokines (Zhao et al. 2005). In addition to cellular-bound receptors, cytokines can also be bound to their soluble receptors (Rose-John and Heinrich 1994). Additionally, chemokines, the subset of cytokines responsible for directing chemotaxis, bind to tissue glycosaminoglycans (Lortat-Jacob 2009; Rot 2010). This phenomenon may be particularly true for larger, more slowly diffusing molecules at low biological concentrations within the tissue space. Taken together these factors suggest that for high molecular mass solutes, it is unlikely that even at very low rates of perfusion a diffusion equilibrium will be reached and a sample representative of the biological concentration of the analyte will be recovered.

In order to recover large molecules, dialysis membranes with larger pores must be employed. However, larger pores allow probe perfusate to leak across the membrane into the surrounding tissue. This may result in disruption of the probe environment, local edema and/or dilution of the analyte(s) of interest. One solution has been to use a "push–pull" perfusion system that minimizes the hydrostatic pressure across the membrane and/or a probe design that allows rapid equalization of the pressure inside the probe with the atmospheric pressure through a "vent" hole (Takeda et al. 2011). This novel microdialysis system appears able to maintain stable recovery and track acute changes of inflammatory macromolecules in the brain following biological stimuli.

Fluid loss into the tissue and local tissue edema as well as having the potential to reduce further the often low concentrations of the target molecules analyte concentration within the micro-environment of the probe, may also compromise tissue function and stimulate changes in the tissue levels of the target molecules. Several groups have attempted to overcome fluid loss into the tissue and to improve total volume recovery by the addition of osmotic agents to the probe perfusate (Ernberg and Alstergren 2004; Hamrin et al. 2002; Trickler and Miller 2003). These agents have included serum albumin and modified dextrans. Some of the additives used have the added advantage of reducing non-specific adsorption of analyte onto the material of the probe or act as binding agents to stabilize and enhance recovery of hydrophobic or highly tissue-bound molecules.

Recovery of large molecules by dialysis may be further influenced by the ease with which the analyte diffuses within the extracellular space. As described by many (Goodman 2011) the brain's extracellular space, which comprises approximately

20–25 % of the total volume, has a complex geometry (Sykova 2004; Sykova and Nicholson 2008). The complex geometry of the extracellular space restricts diffusion of analytes to the probe. The degree of restriction of diffusion can be measured and expressed mathematically as the tissue tortuosity. The greater the tortuosity, the more diffusion is impeded and the lower the analyte recovery. Tortuosity can vary depending on tissue pathology. So, for example in tumors or in skin diseases such as psoriasis, an increased tortuosity may restrict diffusion and apparent analyte concentrations in dialysate will be lower than in healthy tissue. These types of changes in tissue extracellular volume have also been reported for the brain under various conditions (Vargova et al. 2011).

13.2.2 Ultrafiltration as a Driving Force to Advantage Dialysis Recovery of Large Molecules

In many cases, diffusion across the dialysis membrane may be the limiting factor for recovery of large molecules. To circumvent these problems and make it possible to sample larger molecules as well as obtain higher recoveries of small molecules, researchers have used ultrafiltration membranes and low-flow, push–pull perfusion systems to withdraw fluid and solutes from the extracellular space (Asai et al. 1996; Huang et al. 2006b; Huinink et al. 2010; Yang and Huang 2007). Flow rates using this method can be as low as 10–50 nL min^{-1} and analyte recovery approach up to 100 %.

13.2.3 Impact of Blood Flow on Recovery of Large Molecules

The impact of changes in local blood flow on the recovery of small, diffusible molecules has been extensively investigated and modeled (Benveniste and Huttemeier 1990; Boutsiouki et al. 2004, 2001; Bungay et al. 2011; Dancik et al. 2011; Sun et al. 2001) and normalization of data considered. However, few studies have explored the impact of a variable blood perfusion on macromolecular recovery from the extravascular tissue space by microdialysis, either theoretically or in vivo (Sjogren and Anderson 2010; Stenken et al. 2010). For larger molecules like cytokines, blood flow is probably less important, although the question as to whether changes in local blood flow influence the volume of distribution of the molecule of interest and the maintenance of a concentration gradient across the dialysis membrane is yet not fully elucidated.

13.2.4 Problems with Loss on Contact with Polymeric Material (Either Membrane or Tubing) and the Use of Recovery Enhancers

A significant challenge with protein collection via microdialysis sampling is that it is currently not possible to predict whether or not a particular protein will be recovered through the chosen dialysis membrane (or device). For example, cytokines with similar molecular weight and hydrodynamic radii have exhibited very different extraction efficiencies (Ao et al. 2004). Others have reported the in vitro extraction efficiencies of interleukin 2 (IL-2), IL-4, IL-5, interferon gamma (IFN-γ), and tumor necrosis factor alpha (TNF-α) using an polyethersulfone (10 mm) membrane all to be as low as 1 % (Dabrosin 2001). While for other cytokines, the in vitro efficiency of dialysis using a 300,000 Da membrane is somewhat higher: 45 ± 8 %, 28 ± 4 %, and 22 ± 8 % for IL-6, IL-1β, and nerve growth factor (NGF), respectively, compared with 17.5 ± 6 % for serum albumin (Helmy et al. 2011a; Schutte et al. 2004; Winter et al. 2002, 2004). The chemical mechanisms behind these large discrepancies are poorly understood, and no quantitative modeling mechanism has been provided to help researchers predict which protein targets may be recovered through dialysis membranes.

Some researchers claim that non-specific adsorption of the targeted analyte to the membrane and the microdialysis plastic outlet tubing is the primary cause for observations of decreased recovery for some proteins. Given the typical low concentrations (ng/mL to pg/mL) of proteins of interest in biomarker studies, this claim may have significant merit for many biomarkers. The use of albumin as a blocking agent at concentrations of 0–10 wt % in the dialysis perfusion fluid has been systematically studied by Trickler and Miller with fluorescein-labeled dextrans (FDX-3,10 and 19) as well as rat IL-1β and TNF-α shown to improve the recovery of proteins (Trickler and Miller 2003). However, the proteomics community has clearly shown that albumin can also act as a "sponge" for different proteins and allow the protein to be carried by the albumin. In other words, the albumin may actually serve to help solubilize these proteins that are poorly recovered. Yet, no studies of different protein "cocktails" that may be useful for blocking non-specific adsorption sites (e.g., casein or fetal bovine serum, FBS) have been performed. In an in vitro experiment, a perfusion fluid that contained the assay diluent used for an ELISA, 10 mM PBS and 10 % fetal bovine serum (FBS), pH 7.0, exhibited extraction efficiencies (*EE*) at 0.5 µL/min of roughly 19 % for human CCL2 (MCP-1) and 9 % using the Luminex wash buffer as the perfusion fluid (50 mM PBS with 25 mM EDTA, 0.08 % sodium azide, 0.05 % Tween-20 and 1 % bovine serum albumin (BSA), pH 7.4) (Wang and Stenken 2009).

An early recognition of difficulties incurred during microdialysis sampling within protein-containing solutions is from work performed in the Lo Gorton group at the University of Lund. Their work initially focused on sampling of carbohydrates from pulp and paper mill fluids (Torto et al. 1998). Issues of membrane fouling from these complex media have been described. To overcome some of the fouling issues, Torto et al. coated 5,000 Da (MWCO) polyethersulfone membranes with polyiminine as a means to prevent adsorption of enzymes (Torto et al. 2004).

Recently, Dahlin et al. passed through the probe FDA-approved triblock copolymer, Pluronic F-127, consisting of a hydrophobic block of polypropylene oxide (PPO), a polyethylene oxide block (PEO) followed by another PPO block (PPO_{98}-PEO_{67}-PPO_{98}) as a means to improve recovery of proteins collected in vitro from cerebral spinal fluid (CSF) (Dahlin et al. 2010). The use of this polymer also eliminated the need to use albumin as an agent to reduce non-specific adsorption since albumin would have to be removed prior to proteomics-type measurements. Six proteins, albumin (66 kDa), clusterin (50 kDa), complement C3 (185 kDa), transferrin (75 kDa), hemopexin (49 kDa), and hemoglobin-β (16 kDa) were tested under different conditions—treated vs. untreated CMA 71 catheters consisting of a 100 kDa cutoff membrane with a 0.3 μL/min flow rate. Interestingly, several observations were made for treated (contains Pluronic F-127)-vs. untreated catheters with five separate concentrations of different dextrans included in the perfusion fluid plus a control (no dextran). The *EE* values reported were approximately: albumin (17.5 %), clusterin (15–30 %), hemopexin (20–45 %), transferrin (20–40 %), complement C3 (20–50 %), hemoglobin-β (20–50 %). While some increased improvement in recovery with the F-127 treatment seemed apparent, these increases were only statistically significant for the transferrin collections.

The same group followed up on this work and performed a study of the proteins that were adsorbed onto the microdialysis membrane that was immersed into the CSF fluid (Dahlin et al. 2012). A solution of trypsin was added to dialysis membranes that had been in contact with the CSF fluid and the resulting peptides were analyzed in an LC–MS approach. Interestingly, while albumin was found in high concentrations in the dialysate, little albumin was found adsorbed to the membranes used in this study. A total of 19 proteins were found adsorbed to the membrane, with the most abundant appearing to be hemoglobin and clusterin.

The major driving force for mass transport of solutes into microdialysis probes is diffusion. Mass transport of any solute across a semipermeable membrane can be increased by inclusion of a binding agent that serves to trap the solute of interest. Albumin is known to bind many drugs as well as different proteins (Hage et al. 2011; Zolotarjova et al. 2005). Cyclodextrins have been used to improve recovery of hydrophobic leukotrienes (Kjellstrom et al. 1998; Sun and Stenken 2003).

The use of antibodies would be a logical choice for improving the recovery of proteins through microdialysis membranes. This has been demonstrated with anti-sera for studies in the brain (Pich et al. 1993). Inclusion of free antibodies in the perfusion fluid has been used to increase the recovery of Met-Enkephalin in vitro (Fletcher and Stenken 2008). In vivo, CCL2 (MCP-1) was collected from rat brain using antibodies as compared to standard perfusion fluids. The recovered amounts of CCL2 using the antibodies were more than double that of the control perfusion fluid without antibodies (Herbaugh and Stenken 2011).

The combination of microdialysis sampling with antibody-immobilized beads commonly used bead-based immunoassay has been demonstrated both in vitro and in vivo for collection of different cytokines. Increases in cytokine recovery in some cases exceeding 10-fold as compared to a perfusion fluid without the beads (Ao et al. 2004; Ao and Stenken 2006; Duo et al. 2006). The beads commonly used for these immunoassays are roughly 5–7 μm in diameter and can be passed through the microdialysis device. While it is reasonable to assume that beads may get lost or stuck in some of the dead spaces of the dialysis device, this has not been observed. The real culprit of bead loss is the syringe used to perfuse the fluid, and overall bead mass balance can be achieved by simply increasing the volumetric flow rate of the perfusion fluid (Duo and Stenken 2011).

A difficulty with antibody-immobilized beads is the possibility of saturation (Ao and Stenken 2006). This is a problem with bead-based immunoassays since recovering the beads after analysis is not possible on the standard instrumentation used for the analysis. Heparin and other glycosaminoglycans are known to bind to many different proteins including chemokines (chemoattractant cytokines) (Hamel et al. 2009). To decouple the capture step from the analysis step, heparin has been used as an alternative affinity agent in dialysis perfusion fluids to capture chemokines (Duo and Stenken 2011; Wang et al. 2009). During in vivo studies, the inclusion of heparin or heparin-immobilized beads gave measureable concentrations of CCL2 as compared to control perfusion fluids from subcutaneous tissue in rats (Duo and Stenken 2011). There are limited studies of whether these recovery enhancers are as effective in vivo as they appear to be in vitro. Additionally, whether the presence of these enhancers influence the immune response to probe placement or the micro-environment of the probe in other ways has yet to be determined.

13.2.5 Impact of Probe-Associated Tissue Injury on Microdialysis Sampling

Implantation of microdialysis probes into the majority of tissues involves either using a guide needle or removal of enough tissue fascias to allow probe insertion. There is now a growing body of evidence to show that implantation of microdialysis probes triggers a generalized injury response which is similar to the response to any foreign object (Anderson et al. 2008). This foreign body reaction involves not only humoral mechanisms, but also the influx and activation of inflammatory cells (neutrophils, monocytes, macrophages, etc.) resulting in cell activation and fibrin deposition on the probe. The chemical signaling processes, involving both acute phase inflammatory mediators and cytokines means that the tissue is in a dynamic rather than a static state.

Several recent publications using specific, targeted analyses of cellular components that are expected to be involved, in situ tissue imaging and post insertion tissue biopsy have reported on the acute trauma that is caused following the insertion of microdialysis probes; and that this can be long-lasting (Benveniste and Diemer 1987; Clapp-Lilly et al. 1999; Sjogren and Anderson 2010; Takeda et al. 2011; Wientjes et al. 2003). The damage occurs because microdialysis probes are generally much larger (≥200 μm) than the surrounding tissue elements and adjacent blood vessels since the distance between blood vessels is usually less than 50 μm and results in the perturbation of local blood flow and localized bleeding due to physical damage to blood vessels (Jaquins-Gerstl et al. 2011; Mitala et al. 2008). The result is disruption of the vascular architecture, ischemic and/or hyperemic events, perturbation of the blood tissue barrier and local tissue edema. Laser Doppler perfusion imaging has been used by many to determine the time course of the perturbation of localized blood flow post-microdialysis probe insertion into the forearm skin of human volunteers (Anderson et al. 1994; Stenken et al. 2010). These studies generally suggest that while there is an initial hyperemia, local blood flow returned to pre-insertion levels at approximately 60 min. However, while recent studies that have combined microdialysis with other end-point measures such as tissue biopsy and imaging by us and others (Gill et al. 2011; Helmy et al. 2009; Sjogren and Anderson 2010; Takeda et al. 2011) confirm the complex spatial and temporal nature of the host response to probe injury. They also illustrate that the tissue reactivity to probe insertion may be exploited in the study of the innate immune response and as Sjogren and Anderson (2010) suggest 'the phenotype of the individual under study'.

13.2.6 Encapsulation and Biofouling

As mentioned above, the implantation of any device into mammalian tissue will trigger the foreign body reaction that results in device encapsulation. Encapsulation of any device used to collect samples or even detect solutes (e.g., glucose sensors) will cause solute mass transport from the tissue to the device to be affected. The extent to which solute mass transport will be affected is dependent upon the solute chemical properties and how it is transported through the tissue. Another additional factor to consider is the time of microdialysis catheter implantation since the foreign body reaction is a dynamic reaction involving initial recruitment of macrophages within the first 5–7 days followed by fibroblast infiltration after five days.

The initial act of catheter implantation will cause recruitment of neutrophils as well as activation of resident tissue macrophages. Since the implantation of a microdialysis catheter is really a type of puncture wound, it would be expected to behave as such and release a cytokine profile equivalent to those observed in puncture wounds. In particular, inflammatory chemokines and cytokines would be expected to be present after a puncture wound. However, few studies that have focused on protein biomarkers have fully investigated the effect of merely implanting a dialysis probe into healthy tissue and the resulting cytokine profile. This is a somewhat serious issue since so many uses of microdialysis sampling in humans is focused toward investigating different disease states. Without appropriate controls for cytokines produced due only to the presence of the implanted microdialysis catheter, it is very difficult to ascribe the cytokine profiles and concentrations reported as solely due to the disease state. To complicate this even further, the alterations in tissue properties such as edema and changes in vasculature may also influence solute recovery thus making comparison of concentrations found between control and diseased tissue also very difficult.

The research that has focused on how solute recovery is affected by long-term implantation of microdialysis catheters has typically used small molecules to investigate the role of alterations in the vasculature (Clough et al. 2002; Wisniewski et al. 2001). Low molecular weight biomarkers such as glucose, pyruvate, and lactate have their recovery affected through implanted microdialysis probes during long-term implantations. The role of the alteration in vasculature due to implantation between an initial implant versus 7–10 days has been demonstrated by pharmacokinetic studies of the hydrophilic drug antipyrine for two separate membrane types, CMA 20 probes with 10 mm of either polycarbonate/polyether (PC) or polyethersulfone (PES) (Mou et al. 2010). The area under the curve was greatly reduced for the 7- and 10-day studies as compared to the initial implantation.

The effect of encapsulation on cytokine collection was demonstrated in an anesthetized rat subcutaneous tissue model. CMA 20 microdialysis probes with 10 mm PES membranes were implanted for three and seven days. IL-6 was collected as a baseline as well as with a lipopolysaccharide (LPS) induction (Wang et al. 2007). For probes implanted three and seven days, the collection of IL-6 produced due to the injection of LPS was significantly reduced as compared to the initial implant day. In awake and freely moving rat model, different cytokine levels (CCL2, IL-6 and IL-10) were observed over a range of 1 week (von Grote et al. 2011). These protein levels were compared to excised tissue that underwent qRT-PCR analysis to estimate the gene expression levels.

13.2.7 Equilibration Times

Since cells are broken upon implantation of the microdialysis probes, it is reasonable to expect that cellular content will leak into the extracellular fluid space and be collected during sampling. From the early studies focused on collection of neurotransmitters, the complexity of the obtained chromatograms changed as a function of post-implantation time. This was due to very hydrophilic (NADPH or dopamine and other neurotransmitters from vesicles that would be disrupted) contents being removed from the ECF due to the dialysis sampling process. Because of this cellular disruption process, researchers typically waited until the observable large peaks or additional peaks within the non-retained material (hydrophilic solutes) or within the run time on a reversed-phase LC separation decreased. Many of these hydrophilic materials seemed to clear rather rapidly both in the brain and other tissues. For example in the liver, it only took 45 min post implantation for chromatograms to resemble Ringer's solution blanks (Stenken et al. 1997).

Others have used Laser Doppler Scanning as a means to determine when blood flows surrounding the probe implant appear to return to normal levels (Clough et al. 2002; Groth et al. 1998). It appears that with these studies, blood flow was reported to return to control levels within a few hours. Based on this information, many current microdialysis sampling researchers appear to use this as a "standard." Although, in some neurochemical studies in awake and freely moving animals, dialysis probes may often be implanted and left perfusing overnight prior to experimentation the next morning. For small hydrophilic molecules that are rapidly released (vesicular release) into the tissue space, transported into the tissue space via the vasculature and rapidly removed from the tissue space, differences in "equilibration" or "run-in" times may not critically affect the data quality obtained. However, for larger molecular weight biomarkers, this may severely affect the data quality especially since cytokines are known to be produced upon implantation.

The only way for the microdialysis sampling community to properly evaluate equilibration times for different markers, tissues, and animal or human models will require us not to throw out those first few samples that are collected. Collections of the initial dialysates obtained immediately after implantation followed by quantitation of the targeted analytes for the study will greatly improve our understanding of the dynamics and homeostasis of different protein biomarkers.

13.2.8 Use of Reference Markers

A common concern during microdialysis sampling is that properties of either the membrane or the tissue will change in such a way as to affect the recovery of the intended solute. Across all applications of microdialysis sampling, the use of some type of internal standard has been limited. The use of such an internal standard not only provides information on the functioning of the microdialysis system but can also inform the interpretation of the data where microdialysis efficiency may vary and/or an intervention affect the dialysis probe environment.

Potential reference markers for a collection of large molecules include antipyrine (Yokel et al. 1992), urea (Strindberg and Lonnroth 2000), inulin (Sopasakis et al. 2005) and albumin (Takeda et al. 2011). Tyvold and colleagues have used FITC-dextran 4000 as an infusion marker for several bronchial studies (Tyvold et al. 2007). FITC-dextran 4000 could certainly be used as a reference marker for larger molecular weight solute collections. However, the possibility of these reference molecules influencing recovery of the target molecule or the microenvironment of the probe cannot be ignored.

More recently Takeda et al. (Takeda et al. 2011) have argued that changes (or lack of them) in the concentrations of simultaneously collected endogenous high molecular weight analytes can be used as an internal standard of dialysis performance. In their study in which they measured IL-6 and beta-amyloid (Aβ) peptide in the mouse brain, local treatment with LPS increased the concentration of IL-6 while Aβ peptide remained unchanged and inhibition of Aβ peptide left IL-6 levels unaffected.

There is increasing awareness of the wide range of factors that may influence dialysis recovery of large molecules and ways in which these may be addressed and in some cases overcome. Nonetheless, low and inconsistent recovery rates, due to the slower diffusion of large molecules, fluctuations in membrane properties and changes in tissue structure and function, remain a major obstacle in the microdialysis of macromolecules particularly in free-moving animals and in pathophysiological states. In summary, the consequence of these various physical restraints alone or in combination is that the amount of high molecular mass material recovered in dialysate, even under optimal conditions, is frequently very low and due to detection inability not observed. Furthermore, the concentrations recovered in dialysate may be unrepresentative of tissue levels. Evidence for changes in macromolecular analyte levels from baseline may thus be more effectively sought than absolute values.

13.3 Analysis Platforms for Large Molecules in Dialysis Samples

The current and future application of microdialysis to the monitoring of large molecules as markers of tissue homeostasis and as prognostic or diagnostic markers of disease depends very much on an ability to detect and assay them with sufficient sensitivity and reproducibility. The analytical methods used to detect abundant low molecular mass solutes in dialysate are reviewed elsewhere (Nandi and Lunte 2009). There are several challenges with the analysis of larger biomolecules from dialysate samples. Cytokines are typically observed in dialysate samples in concentrations approaching the pg/mL range. The volume-limited nature (μL) of microdialysis sampling adds another challenge. Temporal resolution is additionally challenged with respect to assay volume requirements and the necessity to use low flow rates to achieve improved recovery. Standard immunoassays can have sample volume requirements in the 50–100 μL range. If protein biomarkers are being collected using 0.3 or 1.0 μL/min perfusion rates through the microdialysis catheter, then the temporal resolution for the experiment is diminished.

13.3.1 Targeted Molecules e.g., Cytokines, Neuropeptides, etc. Multiplexed Cytokine Arrays and ELISA

13.3.1.1 Immunoassay-Based Detection Methods

Early microdialysis sampling studies of larger molecular weight solutes such as cytokines (IL-1β, IL-6, TNF-α) (Shaw and Britt 1995; Woodroofe et al. 1991), ghrelin (de la Cour et al. 2007), oxytocin (Shaw and Britt 2000) and neuropeptide Y (Mertes et al. 1993; Thompson et al. 1995) commonly used radioimmunoassay or enzyme-linked immunosorbent assays (ELISA) for quantitation of the protein concentrations in the dialysate. Commercial immunoassay kits are available for many different human, mouse and rat proteins from major vendors such as BD Biosciences. Depending on the manufacturer, enough reagents (standards, antibodies, and other materials) for twenty 96-well plate assays will be approximately ~$600. Wash buffers are typically sold separately. This type of approach is economical if the research is attempting to determine if a certain biomarker is collected.

The low sample volumes obtained from microdialysis experiments are quite amenable with newer platforms used to measure cytokines. These new platforms include bead-based immunoassays and planar surface arrays. Many researchers have reported use of the bead-based assays for measurement of cytokine content in microdialysis samples (Angst et al. 2008; Helmy et al. 2011a; Tyvold et al. 2010). Reichert's group has developed cytokine arrays (Li et al. 2003, 2005). Although this group has only used their protein arrays for exudate samples (Schutte et al. 2009), there is nothing that precludes the use of these arrays for microdialysis samples.

Bead-based immunoassays include antibodies immobilized onto polystyrene beads (5–7 μm, i.d.) with different concentrations of a marker fluorophore embedded into the bead (Vignali 2000). The instrumentation then reads both the fluorophore intensity within the bead as well as the intensity from the fluorescent-labeled detection antibody. Different manufacturers claim the ability to measure up to 100 different solutes in a single sample. However, measurement of 15 different solutes is the practical analytical limit (Elshal and McCoy 2006). Volumes needed for analysis are variable and depend on the manufacturer of the bead-based assay, but typically range between 5 and 25 μL of sample. The two most widely available multiplex bead assay platforms are the Luminex system, requiring its own analysis instrumentation, and the cytometric bead array produced by BD Biosciences, which is compatible with most BD flow cytometer instruments. BD also markets an instrument for bead-based assays only. For BD instruments, only BD bead assays can be used and the number of analytes available for study across the human, mouse, and rat cytokine needs are fewer than for the Luminex-based system. Many different vendors sell kits for the Luminex system with Millipore being the largest. These assays are not inexpensive. Typically, a 5-plex kit that measures five cytokines with a single 96-well plate will cost ~$1,000.

13.3.1.2 Separation-Based Methods

Phillips and colleagues have described immunoaffinity capillary electrophoresis as a means to separate and detect different cytokines or any targeted analyte to which an antibody can be made (Phillips 2001; Phillips and Dickens 1998). In this work, the protein targeted for analysis is labeled with a fluorophore. This mixture then is added to a capillary to which the antibody has been immobilized. The targeted protein binds to the antibody while other non-retained materials are removed from the capillary. Finally, the protein is released from the antibody by use of an acid or pH shift and the resultant electrophoretic band is measured by laser-induced fluorescence. Dialysate samples have been analyzed for cytokine content using this approach (Kalish and Phillips 2009; Summy-Long et al. 2008).

For peptide analysis, LC–MS approaches have been described. Andren and Caprioli were the first to demonstrate the use of LC–MS to quantify neuropeptides from dialysate samples (Andren and Caprioli 1995; Emmett et al. 1995). Although, it should to be noted that in that original work the perfusion fluids contained water and not a typical Ringer's solution. More recently, work from the Kennedy and Smolder's research groups has described different aspects of using LC–MS for microdialysis samples particularly with respect to neuropeptides (Lanckmans et al. 2007a, b, 2008; Li et al. 2009). In a completely different application, Wang et al. (2009) infused different matrix metalloproteinase substrates through microdialysis probes and quantified the resultant products.

13.3.2 Proteomic Studies

While targeted approaches, such as multiplexed cytokine arrays and ELISA studies are particularly sensitive, they can only be used to quantify known components. With the increase in availability of proteomics technologies it has now become possible to measure multiple proteins that have not been specifically targeted, e.g. for biomarker discovery, and to reveal proteins involved in specific processes. Proteomic analysis of dialysate from human brain has revealed a small number of proteins, the majority of which were acute phase proteins (Haskins et al. 2004; Maurer et al. 2003, 2007). In skin, proteomic analysis of tissue fluid extracted from biopsy samples or suction blister fluid has predominantly identified proteins that are abundant in plasma, including members of the albumin and apolipoprotein families and other proteins involved associated with the acute phase of inflammation, as well as structural proteins such as keratins and collagen.

Maurer et al. (2003, 2007) and Haskins et al. (2004) were among the first groups to publish on the proteome of human brain dialysate in stroke victims in order to assess changes during ischemia. Dialysate was collected over 1–2 h intervals from the non-diseased, contralateral hemisphere, 18 h after probe insertion. This study demonstrated for the first time that proteins could be recovered and identified using microdialysis coupled with gel-based separation and mass spectrometric analysis. From 95 spots obtained by 2D gel electrophoresis, 27 individual proteins were identified by mass spectroscopy. Of these, the majority were acute phase proteins, with considerable overlap with those seen in the suction blister skin studies (Maurer et al. 2003). They included malate dehydrogenase, creatine kinase, tubulin, and several common plasma proteins (e.g. serum albumin, transferrin, haptoglobin, and immunoglobulins). The range of molecular mass and pI values for the identified proteins was between 13–190 kDa and 4.5–7.8, respectively. This is a broad range of size and a pI range similar to that of the skin (Marro et al. 2001). Ten of the proteins were apparently specific to brain dialysate and had not been previously identified in cerebrospinal fluid. These data, together with more recent studies on brain dialysate (Behrens et al. 2008; Behrens and Li 2010) and other interstitial fluid samples (Haslene-Hox et al. 2011), suggest that locally produced bioactive macromolecules can be recovered from the microenvironment of the probe distinct from those of systemic origin, using these techniques.

Recently, Gill et al. (2011) provided the first report of the proteome of healthy human dermal dialysate using a "shotgun" proteomics approach which allows identification of both known and unknown peptides for profiling and biomarker discover. This analysis confirmed that highly abundant plasma proteins comprise approximately 85 % of dermal dialysate fluid. Once these were removed, a further 25 proteins were revealed by GeLC-MS/MS, many with functions associated with the response to injury. A total of 89 proteins were assigned using iTRAQ analysis, although many were variants of those identified in the GeLC-MS/MS experiments. Analysis of the changes in protein ratios between the early and late samples, 60–150 and 300–390 min after probe insertion respectively, showed changes in the recovery of several proteins. The high proportion of plasma proteins identified suggests a considerable extravasation of protein during probe insertion and injury (Figs. 13.1 and 13.2).

While the proteomics of skin has been described, the majority of this work used murine skin or human skin cancer samples (Haslene-Hox et al. 2011; Huang et al. 2006a), and comparison of these proteomes to that of dialysate from healthy human tissue should be made with some caution. The number of proteins identified in dermal dialysate was low compared to 401 in suction blister fluid (Kool et al. 2007), of which 202 were not found in plasma. While four of the proteins (apolipoprotein A1, ceruloplasmin, complement factor B and vitamin D binding protein), identified in blister fluid from one individual (Macdonald et al. 2006) are shared with dermal dialysate (Gill et al. 2011), several of the most abundant blister fluid proteins (plasma proteins and key systemic inflammatory markers such as serum amyloid, C-Reactive Protein) were not detected in dermal dialysate. The absence of such proteins from dialysate may be explained by their large size (C-Reactive Protein forms a pentamer of 105 kDa) which may cause them to be retained by the dialysis membrane. Alternatively, as protein identities are inferred according to the matching of sequenced peptides obtained from tandem MS analyses, complications in protein identification may arise where peptide sequences are not unique to a single protein. This is particularly problematic when trying to distinguish between different isoforms of a protein or different proteins that share identical regions of sequence (Nesvizhskii and Aebersold 2005). Finally, one over-riding concern with the global profiling approach is that with the limited capacity of the mass spectrometer, the abundance of several plasma proteins, particularly immunoglobulins and serum albumin, may reduce the detection of less abundant components. This is particularly problematic with profiling of plasma/serum and may similarly be a problem in

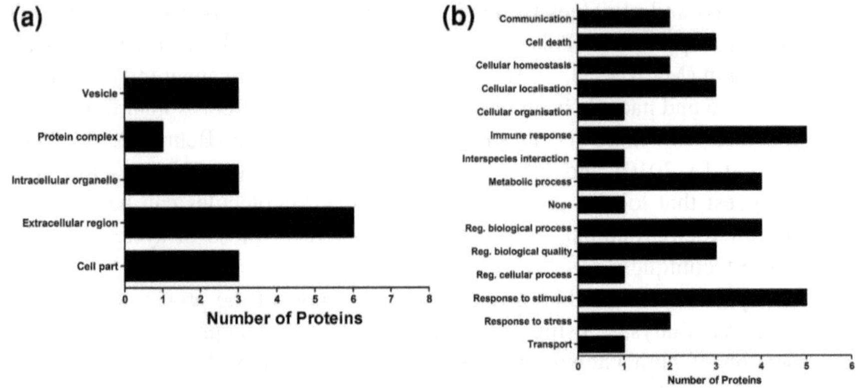

Fig. 13.1 Proteomic analysis using GeLC-MS/MS of non-depleted dialysate from healthy skin. Dialysate was collected using a 3,000 kDa MWCO plasmaphoresis membrane inserted for a length of 20 mm and perfused at a rate of 3 μl/min with buffered saline solution. Characterization of identified proteins was undertaken using AmiGO database classification. **a** The bar graph shows a summary the different cellular compartments from which the proteins identified in the non-depleted dialysate sample (collected 30–90 min after injury from a single volunteer) are associated. Each protein may be active within more than one component, in which case, all associated components were included. Components correspond to tree level 3 terms and are grouped to avoid repetition of very similar terms, e.g., membrane-bound organelle and intracellular organelle. **b** The bar graph shows the different biological processes for the proteins identified in non-depleted dialysate (collected 30–90 min after injury from a single volunteer), according to the Gene Ontology classification. Processes correspond to level 3 terms within the Gene Ontology classification, with similar terms e.g., regulation of the response to stimulus and response to external stimulus grouped appropriately. Regulation is abbreviated to "Reg." for esthetic purposes. Redrawn from Gill (Gill 2009)

cerebral and dermal dialysate given that serum albumin is estimated to comprise approximately 70 % of the total protein in these fluids (Gill et al. 2011; Winter et al. 2002).

Interestingly and perhaps unsurprisingly, the published protein profiles of dialysate obtained from the skin and from the brain differ. Eight of the proteins identified in cerebral microdialysate using gel-based separation and mass spectrometric analysis (Maurer et al. 2003) were not present in dermal dialysate, while 15 proteins were found in dermal dialysate by GeLC-MS/MS but not reported in cerebral dialysate. Common to both were albumin, immunoglobulin (IgG) light and heavy chain, and various binding proteins. Specific to the dialysate were proteins that were brain-specific, such as the creatinine kinase B subunit and tubulin alpha-1, and others that might indicate disease severity or progression. These differences may reflect 'real' differences in composition of the dialysate from the two tissues (brain and dermis). Alternatively, they may arise from differences in instrumentation and/or analysis protocols or are due to a differing response to the probe insertion trauma.

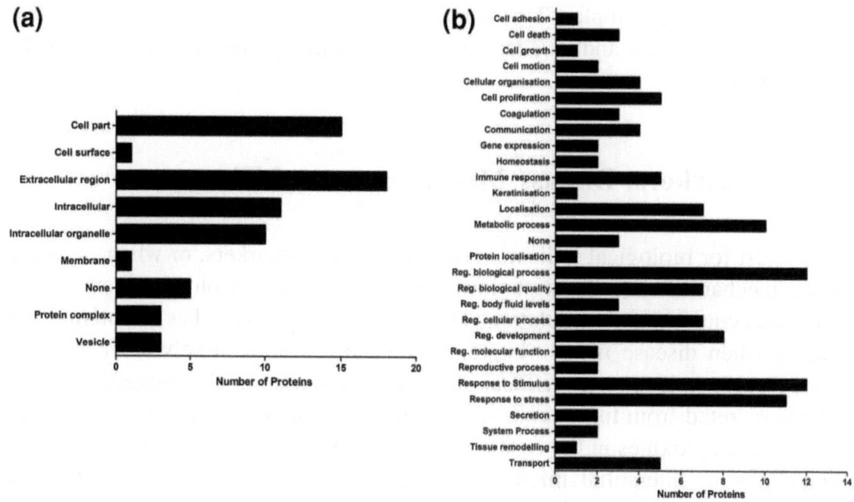

Fig. 13.2 Proteomic analysis using GeLC-MS/MS of depleted dialysate from healthy skin. Dialysate was collected using a 3,000 kDa MWCO plasmaphoresis membrane inserted for a length of 20 mm and perfused at a rate of 3 μl/min with buffered saline solution. Depletion was performed using by high performance liquid chromatography using a multi-affinity removal system (MARS) column. Characterization of identified proteins was undertaken using AmiGO database classification. **a** Bar graph showing the different cellular compartments from which the proteins identified in the depleted dialysate sample (collected 30–120 min after injury and pooled from 2 volunteers) originate. Each protein may be derived from more than one component, in which case, all components were included. Components correspond to level 3 terms within the Gene Ontology classification. From Gill PhD thesis. **b** Bar graph showing the different biological processes in which the proteins identified in dialysate (collected 30–150 min after injury from 2 volunteers) are involved, according to the Gene Ontology classification. Each protein may perform more than process, in which case, all level 3 process terms were included. Redrawn from Gill (Gill 2009)

Taken together, these initial studies of the dialysate proteome provide substantive evidence that such proteomic approaches can be used to investigate the protein content of dialysate with a view to identifying novel biomarkers of tissue (patho)physiology. To date, however, such studies are restricted by technological limitations which need to be overcome if comprehensive information on the tissue microenvironment is to be obtained. While the combination of microdialysis and proteomics appears to be compatible, the proteins identified represent only a small fraction of the total protein complement that might be anticipated in dialysate from the interstitial space. This is perhaps to be expected, given the experimental conditions imposed by the microdialysis sampling method. While depletion of the most highly abundant proteins significantly increased the number of less abundant proteins identified, the dynamic range of concentrations remains too large for comprehensive characterization. For example, no cytokines have been reported using these proteomic approaches, despite their detection using more targeted assays in equivalent samples. This is most likely due to a suboptimal sensitivity of the mass spectrometer and its analysis capacity, although this will improve as the technology matures.

13.4 Biomarkers, Disease Mechanisms, and Therapy

In the quest for biological indicators of disease, or biomarkers, or when exploring disease mechanisms, the most obvious starting point is to explore the presence of target molecules in the circulation. However, this approach has problems, particularly when disease is not systemic but localized to relatively small areas of tissue; for example tumors or local sites of inflammation. The concentrations of proteins secreted from tumor cells or shed from tumor cell membranes in the case of tumors, or cytokines in the case of inflammation, may be high in the immediate extracellular or interstitial microenvironment compared to the plasma. Furthermore, the wide diversity and high concentrations of proteins in plasma are a major obstacle in the discovery of biomarkers using proteomics.

In easily accessible solid tumors, isolation of interstitial fluid from fresh biopsies has been suggested as the most appropriate method to obtain fluid for proteomic analysis of carcinoma tissue (Haslene-Hox et al. 2011). However, in inaccessible tumors, such as those in the brain where this technique cannot be applied, the use of microdialysis for the recovery tumor-associated macromolecules in the brain is being explored. For example, microdialysis of brain tumors has been suggested for the identification and assessment of the activity of cysteine protease and cathepsin S, known to facilitate astrocytoma invasion (Flannery et al. 2007). More recently, microdialysis has been used to assess the response of brain tumors to therapy (Goodman 2011) and the concentrations of growth factors, cytokines and other proteins involved in the pathogenesis of high-grade gliomas in vivo, in order to understand more of their metabolism (Marcus et al. 2010). In addition to tumors, microdialysis is also being used to explore the multiplicity of cytokines generated in the brain as a result of severe brain injury (Cederberg and Siesjo 2010; Helmy et al. 2011b). These are described in detail in the chapter on microdialysis and the brain.

The second example we will use in this section is that of microdialysis of cytokines and their relationship to inflammation in the skin. Much of our knowledge cytokines molecules has been gained using isolated cells, cell lines and, more recently, genetically engineered models in mice or flies. However, to really understand the control of their generation and function in disease, it is necessary to study them in clinical environments, focusing, wherever possible, on the interstitial micro-environment of the skin. To do this, either biopsy analysis or microdialysis may be used. Skin biopsy has the advantage of being able to examine histological aspects to study cellular mRNA levels, in addition to assessing cytokine levels but has the drawback of giving only single snapshots in time. In contrast, dermal microdialysis allows continuous in vivo sampling of the interstitial fluid in the intact skin with minimal trauma. However, microdialysis, as highlighted earlier, suffers from the drawback of inducing the production of IL-1β, IL-6, and IL-8 in an innate response to minimal trauma (Clough et al. 2007; Sjogren and Anderson 2010; Sjogren et al. 2002). While several groups have used microdialysis to explore changes in cytokine generation in dermal inflammation (Angst et al. 2008) and its treatment (Salgo et al. 2011), a particular example of the use of continuous sampling is the study of Averbeck and co-workers (Averbeck et al. 2006) who followed the cytokine response to ultraviolet B-induced inflammation over a 48 h period.

When investigating disease mechanisms, the combination of microdialysis and microdialysis may provide insights which the use of either in isolation cannot. In our study of the dermal late-phase response to allergen challenge of allergic individuals (Clough et al. 2007), two microdialysis probes were inserted, one close to the point of allergen injection (injection site) and the other 1 cm away (distal site) within the area of the late-phase induration. Skin biopsies taken at both sites were stained immunocytochemically to investigate possible relationships between cytokine generation, expression of adhesion molecules, and recruitment of neutrophils and eosinophils during the late-phase allergic response. The cytokine response to probe insertion was assessed using a single 'control' probe in the opposite arm. The results showed significant increases above control in IL-6 and IL-8 at 3 and 6 h at both injection and distal sites. However, adhesion molecule expression and leukocyte infiltration were elevated only at the allergen injection site, suggesting a more complex relationship between cytokine generation and the cellular events in allergic inflammation than can be predicted from in vitro experiments.

13.5 Conclusions and Perspectives for the Future

The recovery of high molecular weight molecules is more challenging than recovery of analytes with low molecular mass, and large-pore high-MWCO membranes are harder to use than those with small pores with recoveries that are more subject to changes in the local environment, particularly biofouling. However, such approaches allow the recovery of larger bioactive molecules and protein-bound substrates that could otherwise not be recovered in vivo. Furthermore, recovery of large molecules by microdialysis allows the identification of patterns of protein expression in clinical environments, thereby giving the potential of identifying clinically useful biomarkers of disease, investigating disease mechanisms, and developing appropriate therapies.

But what of the future? Up to now, the use of microdialysis techniques to investigate disease has been hampered by lack of sensitivity of assays to cope with the small volumes generated. Advances in proteomics and analysis of candidate molecules, especially cytokines, will surely make quantitative microdialysis of large molecules less problematic. Also, development of 'fit for purpose' membranes and protocols that optimize stable macromolecule sampling will allow researchers to monitor the changes in interstitial levels of large molecules with excellent temporal resolution in a wide range of physiological and pathological processes in humans and in relevant animal models. Furthermore, development of membranes which are less subject to rejection in vivo would allow for more prolonged sampling necessary in the investigation of more chronic conditions.

Another area of microdialysis, mentioned only as a post script to this chapter, is the delivery of molecules for therapeutic purposes. Administration of therapies such as interferon-γ and interleukin-2 for the treatment of viral and neoplastic diseases is dogged by systemic toxicity and poor local efficacy (Shaker and Younes 2009; Younes and Amsden 2002). Could microdialysis be a suitable method to deliver such chemotherapies (Bungay et al. 2010)? Although their studies are in their infancy, Martins and colleagues (Martins and de Oliveira 2009; Martins et al. 2012) believe drug delivery by microdialysis could be the way forward.

References

Anderson C, Andersson T, Wårdell K (1994) Changes in skin circulation after insertion of a microdialysis probe visualised by laser Doppler perfusion imaging. J Invest Dermatol 102:807–811

Anderson JM, Rodriguez A, Chang DT (2008) Foreign body reaction to biomaterials. Semin Immunol 20(2):86–100

Anderson L (2005) Candidate-based proteomics in the search for biomarkers of cardiovascular disease. J Physiol 563(Pt 1):23–60

Andren PE, Caprioli RM (1995) In vivo metabolism of substance P in rat striatum utilizing microdialysis/liquid chromatography/micro-electrospray mass spectrometry. J Mass Spectrom 30:817–824

Angst MS, Clark JD, Carvalho B, Tingle M, Schmelz M, Yeomans DC (2008) Cytokine profile in human skin in response to experimental inflammation, noxious stimulation, and administration of a COX-inhibitor: a microdialysis study. Pain 139(1):15–27

Ao X, Sellati TJ, Stenken JA (2004) Enhanced microdialysis relative recovery of inflammatory cytokines using antibody-coated microspheres analyzed by flow cytometry. Anal Chem 76(13):3777–3784

Ao X, Stenken JA (2006) Microdialysis sampling of cytokines. Methods 38(4):331–341

Asai S, Kohno T, Ishii Y, Ishikawa K (1996) A newly developed procedure for monitoring of extracellular proteins using a push-pull microdialysis. Anal Biochem 237(2):182–187

Averbeck M, Beilharz S, Bauer M, Gebhardt C, Hartmann A, Hochleitner K et al (2006) In situ profiling and quantification of cytokines released during ultraviolet B-induced inflammation by combining dermal microdialysis and protein microarrays. Exp Dermatol 15(6):447–454

Behrens HL, Chen R, Li L (2008) Combining microdialysis, NanoLC-MS, and MALDI-TOF/TOF to detect neuropeptides secreted in the crab, Cancer borealis. Anal Chem 80(18):6949–6958

Behrens HL, Li L (2010) Monitoring neuropeptides in vivo via microdialysis and mass spectrometry. Methods Mol Biol 615:57–73

Benveniste H, Diemer NH (1987) Cellular reactions to implantation of a microdialysis tube in the rat hippocampus. Acta Neuropathol 74(3):234–238

Benveniste H, Huttemeier PC (1990) Microdialysis–theory and application. Prog Neurobiol 35(3):195–215

Boutsiouki P, Georgiou S, Clough GF (2004) Recovery of nitric oxide from acetylcholine-mediated vasodilatation in human skin in vivo. Microcirculation 11(3):249–259

Boutsiouki P, Thompson JP, Clough GF (2001) Effects of local blood flow on the percutaneous absorption of the organophosphorus compound Malathion: a microdialysis study in man. Arch Toxicol 75(6):321–328

Bungay PM, Sumbria RK, Bickel U (2011) Unifying the mathematical modeling of in vivo and in vitro microdialysis. J Pharm Biomed Anal 55(1):54–63

Bungay PM, Wang T, Yang H, Elmquist WF (2010) Utilizing transmembrane convection to enhance solute sampling and delivery by microdialysis: theory and in vitro validation. J Membr Sci 348(1–2):131–149

Cederberg D, Siesjo P (2010) What has inflammation to do with traumatic brain injury? Childs Nerv Syst 26(2):221–226

Chaurasia CS, Muller M, Bashaw ED, Benfeldt E, Bolinder J, Bullock R et al (2007) AAPS-FDAFDA workshop white paper: microdialysis principles, application and regulatory perspectives. Pharm Res 24(5):1014–1025

Clapp-Lilly KL, Roberts RC, Duffy LK, Irons KP, Hu Y, Drew KL (1999) An ultrastructural analysis of tissue surrounding a microdialysis probe. J Neurosci Methods 90(2):129–142

Clough GF (2005) Microdialysis of large molecules. AAPS J 7(3):E686–E692

Clough GF, Boutsiouki P, Church MK, Michel CC (2002) Effects of blood flow on the in vivo recovery of a small diffusible molecule by microdialysis in human skin. J Pharmacol Exp Ther 302(2):681–686

Clough GF, Jackson CL, Lee JJ, Jamal SC, Church MK (2007) What can microdialysis tell us about the temporal and spatial generation of cytokines in allergen-induced responses in human skin in vivo? J Invest Dermatol 127(12):2799–2806

Cooley JC, Lunte CE (2011) Detection of malondialdehyde in vivo using microdialysis sampling with CE-fluorescence. Electrophoresis 32(21):2994–2999

Dabrosin C (2001) Technical aspects of microdialysis of human breast. Scand J Clin Lab Invest 61(4):269–272

Dahlin AP, Hjort K, Hillered L, Sjodin MO, Bergquist J, Wetterhall M (2012) Multiplexed quantification of proteins adsorbed to surface-modified and non-modified microdialysis membranes. Anal Bioanal Chem 402(6):2057–2067

Dahlin AP, Wetterhall M, Caldwell KD, Larsson A, Bergquist J, Hillered L et al (2010) Methodological aspects on microdialysis protein sampling and quantification in biological fluids: an in vitro study on human ventricular CSF. Anal Chem 82(11):4376–4385

Dancik Y, Anissimov YG, Jepps OG, Roberts MS (2012) Convective transport of highly plasma protein bound drugs facilitates direct penetration into deep tissues after topical application. Br J Clin Pharmacol 73(4):564–578

de la Cour CD, Norlen P, Hakanson R (2007) Secretion of ghrelin from rat stomach ghrelin cells in response to local microinfusion of candidate messenger compounds: a microdialysis study. Regul Pept 143(1–3):118–126

Dostalova I, Kavalkova P, Haluzikova D, Housova J, Matoulek M, Haluzik M (2009) The use of microdialysis to characterize the endocrine production of human subcutaneous adipose tissue in vivo. Regul Pept 155(1–3):156–162

Duo J, Fletcher H, Stenken JA (2006) Natural and synthetic affinity agents as microdialysis sampling mass transport enhancers: current progress and future perspectives. Biosens Bioelectron 22(3):449–457

Duo J, Stenken JA (2011) In vitro and in vivo affinity microdialysis sampling of cytokines using heparin-immobilized microspheres. Anal Bioanal Chem 399(2):783–793

Elmquist WF, Sawchuk RJ (1997) Application of microdialysis in pharmacokinetic studies. Pharm Res 14:267–288

Elshal MF, McCoy JP (2006) Multiplex bead array assays: performance evaluation and comparison of sensitivity to ELISA. Methods 38(4):317–323

Emmett MR, Andren PE, Caprioli RM (1995) Specific molecular mass detection of endogenously released neuropeptides using in vivo microdialysis/mass spectrometry. J Neurosci Methods 62:141–147

Ernberg MM, Alstergren PJ (2004) Microdialysis of neuropeptide Y in human muscle tissue. J Neurosci Methods 132(2):185–190

Flannery T, McConnell RS, McQuaid S, McGregor G, Mirakhur M, Martin L et al (2007) Detection of cathepsin S cysteine protease in human brain tumour microdialysates in vivo. Br J Neurosurg 21(2):204–209

Fletcher HJ, Stenken JA (2008) An in vitro comparison of microdialysis relative recovery of Met- and Leu-enkephalin using cyclodextrins and antibodies as affinity agents. Anal Chim Acta 620(1–2):170–175

Gill C, Parkinson E, Church MK, Skipp P, Scott D, White AJ et al (2011) A qualitative and quantitative proteomic study of human microdialysate and the cutaneous response to injury. AAPS J 13(2):309–317

Gill CA (2009) Proteomic analysis of interstitial fluid for novel markers of the cutaneous response to Injury University of Southampton 2009

Goodman JC (2011) Clinical microdialysis in neuro-oncology: principles and applications. Chin J Cancer 30(3):173–181

Groth L, Jorgensen A, Serup J (1998) Cutaneous microdialysis in the rat: insertion trauma and effect of anaesthesia studied by laser Doppler perfusion imaging and histamine release. Skin Pharmacol Appl Skin Physiol 11(3):125–132

Hage DS, Anguizola J, Barnaby O, Jackson A, Yoo MJ, Papastavros E et al (2011) Characterization of drug interactions with serum proteins by using high-performance affinity chromatography. Curr Drug Metab 12(4):313–328

Hamel DJ, Sielaff I, Proudfoot AE, Handel TM (2009) Chapter 4. Interactions of chemokines with glycosaminoglycans. Methods Enzymol 461:71–102

Hamrin K, Rosdahl H, Ungerstedt U, Henriksson J (2002) Microdialysis in human skeletal muscle: effects of adding a colloid to the perfusate. J Appl Physiol 92(1):385–393

Haskins WE, Watson CJ, Cellar NA, Powell DH, Kennedy RT (2004) Discovery and neurochemical screening of peptides in brain extracellular fluid by chemical analysis of in vivo microdialysis samples. Anal Chem 76(18):5523–5533

Haslene-Hox H, Oveland E, Berg KC, Kolmannskog O, Woie K, Salvesen HB et al (2011) A new method for isolation of interstitial fluid from human solid tumors applied to proteomic analysis of ovarian carcinoma tissue. PLoS One 6(4):e19217

Helmy A, Carpenter KL, Menon DK, Pickard JD, Hutchinson PJ (2011a) The cytokine response to human traumatic brain injury: temporal profiles and evidence for cerebral parenchymal production. J Cereb Blood Flow Metab 31(2):658–670

Helmy A, Carpenter KL, Skepper JN, Kirkpatrick PJ, Pickard JD, Hutchinson PJ (2009) Microdialysis of cytokines: methodological considerations, scanning electron microscopy, and determination of relative recovery. J Neurotrauma 26(4):549–561

Helmy A, De Simoni MG, Guilfoyle MR, Carpenter KL, Hutchinson PJ (2011b) Cytokines and innate inflammation in the pathogenesis of human traumatic brain injury. Prog Neurobiol 95(3):352–372

Herbaugh AW, Stenken JA (2011) Antibody-enhanced microdialysis collection of CCL2 from rat brain. J Neurosci Methods 202(2):124–127

Huang CM, Ananthaswamy HN, Barnes S, Ma Y, Kawai M, Elmets CA (2006a) Mass spectrometric proteomics profiles of in vivo tumor secretomes: capillary ultrafiltration sampling of regressive tumor masses. Proteomics 6(22):6107–6116

Huang CM, Wang CC, Kawai M, Barnes S, Elmets CA (2006b) In vivo protein sampling using capillary ultrafiltration semi-permeable hollow fiber and protein identification via mass spectrometry-based proteomics. J Chromatogr A 1109(2):144–151

Huinink KD, Lambooij B, Jansen-van ZK, Cremers TI, van OW, Bakker PL et al. (2010) Microfiltration sampling in rats and in cows: toward a portable device for continuous glucocorticoid hormone sampling. Analyst 135(2):390–396

Jaquins-Gerstl A, Shu Z, Zhang J, Liu Y, Weber SG, Michael AC (2011) Effect of dexamethasone on gliosis, ischemia, and dopamine extraction during microdialysis sampling in brain tissue. Anal Chem 83(20):7662–7667

Jensen SM, Hansen HS, Johansen T, Malmlof K (2007) In vivo and in vitro microdialysis sampling of free fatty acids. J Pharm Biomed Anal 43(5):1751–1756

Kalish H, Phillips TM (2009) Application of immunoaffinity capillary electrophoresis to the measurements of secreted cytokines by cultured astrocytes. J Sep Sci 32(10):1605–1612

Kendrick KM (1990) Microdialysis measurement of in vivo neuropeptide release. J Neurosci Methods 34(1–3):35–46

Kjellstrom S, Emneus J, Laurell T, Heintz L, Marko-Varga G (1998) On-line coupling of microdialysis sampling with liquid chromatography for the determination of peptide and non-peptide leukotrienes. J Chromatogr A 823(1–2):489–496

Kool J, Reubsaet L, Wesseldijk F, Maravilha RT, Pinkse MW, D'Santos CS et al (2007) Suction blister fluid as potential body fluid for biomarker proteins. Proteomics 7(20):3638–3650

Lanckmans K, Sarre S, Smolders I, Michotte Y (2007a) Use of a structural analogue versus a stable isotope labeled internal standard for the quantification of angiotensin IV in rat brain dialysates using nano-liquid chromatography/tandem mass spectrometry. Rapid Commun Mass Spectrom 21(7):1187–1195

Lanckmans K, Sarre S, Smolders I, Michotte Y (2008) Quantitative liquid chromatography/mass spectrometry for the analysis of microdialysates. Talanta 74(4):458–469

Lanckmans K, Stragier B, Sarre S, Smolders I, Michotte Y (2007b) Nano-LC-MS/MS for the monitoring of angiotensin IV in rat brain microdialysates: limitations and possibilities. J Sep Sci 30(14):2217–2224

Li Q, Zubieta JK, Kennedy RT (2009) Practical aspects of in vivo detection of neuropeptides by microdialysis coupled off-line to capillary LC with multistage MS. Anal Chem 81(6):2242–2250

Li Y, Nath N, Reichert WM (2003) Parallel comparison of sandwich and direct label assay protocols on cytokine detection protein arrays. Anal Chem 75:5274–5281

Li Y, Schutte RJ, Abu-Shakra A, Reichert WM (2005) Protein array method for assessing in vitro biomaterial-induced cytokine expression. Biomaterials 26(10):1081–1085

Lischetzki G, Rukwied R, Handwerker HO, Schmelz M (2001) Nociceptor activation and protein extravasation induced by inflammatory mediators in human skin. Eur J PainPain 5(1):49–57

Lortat-Jacob H (2009) The molecular basis and functional implications of chemokine interactions with heparan sulphate. Curr Opin Struct Biol 19(5):543–548

Macdonald N, Cumberbatch M, Singh M, Moggs JG, Orphanides G, Dearman RJ et al (2006) Proteomic analysis of suction blister fluid isolated from human skin. Clin Exp Dermatol 31(3):445–448

Marcus HJ, Carpenter KL, Price SJ, Hutchinson PJ (2010) In vivo assessment of high-grade glioma biochemistry using microdialysis: a study of energy-related molecules, growth factors and cytokines. J Neurooncol 97(1):11–23

Marro D, Guy RH, Delgado-Charro MB (2001) Characterization of the iontophoretic permselectivity properties of human and pig skin. J Controlled Release 70(1–2):213–217

Martins FC, de Oliveira CF (2009) Chemotherapy and the future: microdialysis as a local administration technique. Eur J Gynaecol Oncol 30(1):5–8

Martins FC, Santos JL, Oliveira CF (2012) Microdialysis: improving local chemotherapy in cancer using a mathematical model. Front Biosci (Elite Ed) 4:401–409

Mascher B, Schlenke P, Seyfarth M (1999) Expression and kinetics of cytokines determined by intracellular staining using flow cytometry. J Immunol Methods 223(1):115–121

Maurer MH (2010) Proteomics of brain extracellular fluid (ECF) and cerebrospinal fluid (CSF). Mass Spectrom Rev 29(1):17–28

Maurer MH, Berger C, Wolf M, Futterer CD, Feldmann RE Jr, Schwab S et al (2003) The proteome of human brain microdialysate. Proteome Sci 1(1):7

Maurer MH, Haux D, Sakowitz OW, Unterberg AW, Kuschinsky W (2007) Identification of early markers for symptomatic vasospasm in human cerebral microdialysate after subarachnoid hemorrhage: preliminary results of a proteome-wide screening. J Cereb Blood Flow Metab 27(10):1675–1683

Mertes PM, Beck B, Jaboin Y, Stricker A, Carteaux JP, Pinelli G et al (1993) Microdialysis in the estimation of interstitial myocardial neuropeptide Y release. Regul Pept 49(1):81–90

Mitala CM, Wang Y, Borland LM, Jung M, Shand S, Watkins S et al (2008) Impact of microdialysis probes on vasculature and dopamine in the rat striatum: a combined fluorescence and voltammetric study. J Neurosci Methods 174(2):177–185

Mou X, Lennartz MR, Loegering DJ, Stenken JA (2010) Long-term calibration considerations during subcutaneous microdialysis sampling in mobile rats. Biomaterials 31(16):4530–4539

Nandi P, Lunte SM (2009) Recent trends in microdialysis sampling integrated with conventional and microanalytical systems for monitoring biological events: a review. Anal Chim Acta 651(1):1–14

Nesvizhskii AI, Aebersold R (2005) Interpretation of shotgun proteomic data: the protein inference problem. Mol Cell Proteomics 4(10):1419–1440

Norton LW, Koschwanez HE, Wisniewski NA, Klitzman B, Reichert WM (2007) Vascular endothelial growth factor and dexamethasone release from nonfouling sensor coatings affect the foreign body response. J Biomed Mater Res A 81(4):858–869

Phillips TM (2001) Multi-analyte analysis of biological fluids with a recycling immunoaffinity column array. J Biochem Biophys Meth 49(1–3):253–262

Phillips TM, Dickens BF (1998) Analysis of recombinant cytokines in human body fluids by immunoaffinity capillary electrophoresis. Electrophoresis 19(16–17):2991–2996

Pich EM, Koob GF, Heilig M, Menzaghi F, Vale W, Weiss F (1993) Corticotropin-releasing factor release from the mediobasal hypothalamus of the rat as measured by microdialysis. Neuroscience 55(3):695–707

Riese J, Boecker S, Hohenberger W, Klein P, Haupt W (2003) Microdialysis: a new technique to monitor perioperative human peritoneal mediator production. Surg Infect (Larchmt) 4(1):11–15

Rose-John S, Heinrich PC (1994) Soluble receptors for cytokines and growth factors: generation and biological function. Biochem J 300(Pt 2):281–290

Rot A (2010) Chemokine patterning by glycosaminoglycans and interceptors. Front Biosci 15:645–660

Salgo R, Thaci D, Boehncke S, Diehl S, Hofmann M, Boehncke WH (2011) Microdialysis documents changes in the micromilieu of psoriatic plaques under continuous systemic therapy. Exp Dermatol 20(2):130–133

Schmelz M, Luz O, Averbeck B, Bickel A (1997) Plasma extravasation and neuropeptide release in human skin as measured by intradermal microdialysis. Neurosci Let 230:117–120

Schutte RJ, Oshodi SA, Reichert WM (2004) In vitro characterization of microdialysis sampling of macromolecules. Anal Chem 76(20):6058–6063

Schutte RJ, Xie L, Klitzman B, Reichert WM (2009) In vivo cytokine-associated responses to biomaterials. Biomaterials 30(2):160–168

Shaker MA, Younes HM (2009) Interleukin-2: evaluation of routes of administration and current delivery systems in cancer therapy. J Pharm Sci 98(7):2268–2298

Shaw DW, Britt JH (1995) Concentrations of tumor necrosis factor alpha and progesterone within the bovine corpus luteum sampled by continuous-flow microdialysis during luteolysis in vivo. Biol Reprod 53(4):847–854

Shaw DW, Britt JH (2000) In vivo oxytocin release from microdialyzed bovine corpora lutea during spontaneous and prostaglandin-induced regression. Biol Reprod 62(3):726–730

Sjogren F, Anderson CD (2010) Are cutaneous microdialysis cytokine findings supported by end point biopsy immunohistochemistry findings? AAPS J 12(4):741–749

Sjogren F, Svensson C, Anderson C (2002) Technical prerequisites for in vivo microdialysis determination of interleukin-6 in human dermis. Br J Dermatol 146(3):375–382

Sopasakis VR, Nagaev I, Smith U (2005) Cytokine release from adipose tissue of nonobese individuals. Int J Obes (Lond) 29(9):1144–1147

Stenken JA, Church MK, Gill CA, Clough GF (2010) How minimally invasive is microdialysis sampling? A cautionary note for cytokine collection in human skin and other clinical studies. AAPS J 12(1):73–78

Stenken JA, Lunte CE, Southard MZ, Stahle L (1997) Factors that influence microdialysis recovery. Comparison of experimental and theoretical microdialysis recoveries in rat liver. J Pharm Sci 86(8):958–966

Strindberg L, Lonnroth P (2000) Validation of an endogenous reference technique for the calibration of microdialysis catheters. Scand J Clin Lab Invest 60(3):205–211

Summy-Long JY, Hu S, Long A, Phillips TM (2008) Interleukin-1beta release in the supraoptic nucleus area during osmotic stimulation requires neural function. J Neuroendocrinol 20(11):1224–1232

Sun H, Bungay PM, Elmquist WF (2001) Effect of capillary efflux transport inhibition on the determination of probe recovery during in vivo microdialysis in the brain. J Pharmacol Exp Ther 297(3):991–1000

Sun L, Stenken JA (2003) Improving microdialysis extraction efficiency of lipophilic eicosanoids. J Pharm Biomed Anal 33(5):1059–1071

Sykova E (2004) Extrasynaptic volume transmission and diffusion parameters of the extracellular space. Neuroscience 129(4):861–876

Sykova E, Nicholson C (2008) Diffusion in brain extracellular space. Physiol Rev 88(4):1277–1340

Takeda S, Sato N, Ikimura K, Nishino H, Rakugi H, Morishita R (2011) Novel microdialysis method to assess neuropeptides and large molecules in free-moving mouse. Neuroscience 14(186):110–119

Thompson AC, Justice JB Jr, McDonald JK (1995) Quantitative microdialysis of neuropeptide Y. J Neurosci Methods 60(1–2):189–198

Torto N, Bang J, Richardson S, Nilsson GS, Gorton L, Laurell T et al (1998) Optimal membrane choice for microdialysis sampling of oligosaccharides. J Chromatogr A 806(2):265–278

Torto N, Ohlrogge M, Gorton L, Van AJM, Laurell T, Marko-Varga G (2004) In situ poly(ethylene imine) coating of hollow fiber membranes used for microdialysis sampling. Pure Appl Chem 76:879–888

Trickler WJ, Miller DW (2003) Use of osmotic agents in microdialysis studies to improve the recovery of macromolecules. J Pharm Sci 92(7):1419–1427

Tyvold SS, Solligard E, Gunnes S, Lyng O, Johannisson A, Gronbech JE et al (2010) Bronchial microdialysis of cytokines in the epithelial lining fluid in experimental intestinal ischemia and reperfusion before onset of manifest lung injury. Shock 34(5):517–524

Tyvold SS, Solligard E, Lyng O, Steinshamn SL, Gunnes S, Aadahl P (2007) Continuous monitoring of the bronchial epithelial lining fluid by microdialysis. Respir Res 8:78

Vargova L, Homola A, Cicanic M, Kuncova K, Krsek P, Marusic P et al (2011) The diffusion parameters of the extracellular space are altered in focal cortical dysplasias. Neurosci Lett 499(1):19–23

Vignali DA (2000) Multiplexed particle-based flow cytometric assays. J Immunol Methods 243(1–2):243–255

von Grote EC, Venkatakrishnan V, Duo J, Stenken JA (2011) Long-term subcutaneous microdialysis sampling and qRT-PCR of MCP-1, IL-6 and IL-10 in freely-moving rats. Mol Biosyst 7(1):150–161

Waelgaard L, Pharo A, Tonnessen TI, Mollnes TE (2006) Microdialysis for monitoring inflammation: efficient recovery of cytokines and anaphylotoxins provided optimal catheter pore size and fluid velocity conditions. Scand J Immunol 64(3):345–352

Waelgaard L, Thorgersen EB, Line PD, Foss A, Mollnes TE, Tonnessen TI (2008) Microdialysis monitoring of liver grafts by metabolic parameters, cytokine production, and complement activation. Transplantation 86(8):1096–1103

Wang X, Lennartz MR, Loegering DJ, Stenken JA (2007) Interleukin-6 collection through long-term implanted microdialysis sampling probes in rat subcutaneous space. Anal Chem 79(5):1816–1824

Wang Y, Stenken JA (2009) Affinity-based microdialysis sampling using heparin for in vitro collection of human cytokines. Anal Chim Acta 651(1):105–111

Wang Y, Zagorevski DV, Lennartz MR, Loegering DJ, Stenken JA (2009) Detection of in vivo matrix metalloproteinase activity using microdialysis sampling and liquid chromatography/mass spectrometry. Anal Chem 81(24):9961–9971

Wientjes KJ, Grob U, Hattemer A, Hoogenberg K, Jungheim K, Kapitza C et al (2003) Effects of microdialysis catheter insertion into the subcutaneous adipose tissue assessed by the SCGM1 system. Diabetes Technol Ther 5(4):615–620

Winter CD, Iannotti F, Pringle AK, Trikkas C, Clough GF, Church MK (2002) A microdialysis method for the recovery of IL-1beta, IL-6 and nerve growth factor from human brain in vivo. J Neurosci Methods 119(1):45–50

Winter CD, Pringle AK, Clough GF, Church MK (2004) Raised parenchymal interleukin-6 levels correlate with improved outcome after traumatic brain injury. Brain 127(Pt 2):315–320

Wisniewski N, Klitzman B, Miller B, Reichert WM (2001) Decreased analyte transport through implanted membranes: differentiation of biofouling from tissue effects. J Biomed Mater Res 57(4):513–521

Woodroofe MN, Sarna GS, Wadhwa M, Hayes GM, Loughlin AJ, Tinker A et al (1991) Detection of interleukin-1 and interleukin-6 in adult rat brain, following mechanical injury, by in vivo microdialysis: evidence of a role for microglia in cytokine production. J Neuroimmunol 33(3):227–236

Yang S, Huang CM (2007) Recent advances in protein profiling of tissues and tissue fluids. Expert Rev Proteomics 4(4):515–529

Yokel RA, Allen DD, Burgio DE, McNamara PJ (1992) Antipyrine as a dialyzable reference to correct differences in efficiency among and within sampling devices during in vivo microdialysis. J Pharmacol Toxicol Methods 27(3):135–142

Younes HM, Amsden BG (2002) Interferon-gamma therapy: evaluation of routes of administration and delivery systems. J Pharm Sci 91(1):2–17

Zhao W, Oskeritzian CA, Pozez AL, Schwartz LB (2005) Cytokine production by skin-derived mast cells: endogenous proteases are responsible for degradation of cytokines. J Immunol 175(4):2635–2642

Zolotarjova N, Martosella J, Nicol G, Bailey J, Boyes BE, Barrett WC (2005) Differences among techniques for high-abundant protein depletion. Proteomics 5(13):3304–3313

Chapter 14
Complementary Techniques: Positron Emission Tomography

Oliver Langer

Abstract Positron emission tomography (PET) is a non-invasive nuclear imaging technique which can be used to assess the tissue distribution and pharmacokinetics of drugs labeled with short-lived positron-emitting radionuclides, such as carbon-11 (^{11}C, half-life: 20.4 min) or fluorine-18 (^{18}F, half-life: 109.8 min). In addition, the availability of long-lived positron-emitters, such as bromine-76 (^{76}Br, half-life: 16 h), iodine-124 (^{124}I, half-life: 100.2 h), copper-64 (^{64}Cu, half-life: 12.7 h), or zirconium-89 (^{89}Zr, half-life: 78.4 h), permits PET imaging of macromolecules, such as antibodies or peptides. As the amount of unlabeled drug associated with a PET tracer is usually in the range of a few micrograms, PET with radiolabeled drugs satisfies the definition of a microdose study. Consequently, regulatory authorities require reduced preclinical safety testing as compared to conventional phase 1 studies. PET imaging can be considered as an interesting complementary technique to clinical microdialysis as it can be applied in cases in which clinical microdialysis has its limitations (e.g. assessment of blood–brain barrier penetration of drugs in humans, assessment of tissue distribution of radiolabeled antibodies). The availability of small-animal PET systems makes PET a very powerful translational research tool, which enables a seaming less transition between preclinical and clinical study protocols.

O. Langer (✉)
Department of Clinical Pharmacology, Medical University of Vienna, Vienna, Austria
e-mail: oliver.langer@meduniwien.ac.at

O. Langer
Health and Environment Department, AIT Austrian Institute of Technology GmbH, Seibersdorf, Austria

M. Müller (ed.), *Microdialysis in Drug Development*, AAPS Advances in the Pharmaceutical Sciences Series, DOI: 10.1007/978-1-4614-4815-0_14,

14.1 Introduction

Previous chapters have demonstrated that microdialysis is a powerful experimental tool to assess unbound concentrations of drugs in the interstitial space fluid of different tissues and organs, both in preclinical species and in human subjects. In the present chapter, the non-invasive nuclear imaging method positron emission tomography (PET) will be discussed as a complementary technique to study drug distribution and pharmacokinetics (PK) in the living body. The concept of using PET in conjunction with radiolabeled drugs to assess drug disposition in vivo has been termed “PET microdosing” (Bergström et al. 2003; Wagner et al. 2008; Wagner and Langer 2011).

14.2 Principle of PET

PET is a nuclear imaging technique which relies on the use of molecules labeled with positron-emitting radioisotopes (so-called radiotracers) which are usually administered by intravenous (i.v.) injection. Inside the body, the emitted positrons travel a few millimeters until they combine with an electron (“annihilation event”), which generates two collinear photons with an energy of 511 keV each. In PET cameras, which consist of several stacked detector rings containing scintillation crystals, these photons are recorded simultaneously by detector pairs oriented at 180° to each other (“coincidence detection”). These unique decay characteristics obviate the need for physical collimators and result in a considerable increase in sensitivity as compared to other nuclear imaging techniques. Moreover, correction for tissue attenuation is straightforward, thus affording quantitative radioactivity distribution data (e.g. in kBq per cm^3 or mL tissue). From the recorded decay events, 3D pictures of the distribution of the positron-emitting molecules in the body are reconstructed using either a filtered back projection technique or an iterative reconstruction algorithm. The spatial resolution obtained with modern PET cameras ranges from 1 to 5 mm. PET cameras can perform dynamic imaging and monitor radioactivity distribution as a function of time with a maximum temporal resolution in the range of a few seconds to minutes.

14.3 Radiolabeling with PET Radionuclides

Commonly employed PET radionuclides are oxygen-15 (^{15}O, half-life: 2.07 min), nitrogen-13 (^{13}N, half-life: 9.96 min), carbon-11 (^{11}C, half-life: 20.4 min), gallium-68 (^{68}Ga, half-life: 67.6 min), and fluorine-18 (^{18}F, half-life: 109.7 min). The first three radionuclides are isotopes of elements occurring in virtually all organic molecules, thus giving the possibility of obtaining tracer molecules which retain the structure of the unlabeled drug molecule. However, the radioactive half-lives of ^{15}O and ^{13}N are very short, which does not allow monitoring tissue concentrations of radiolabeled drugs over several elimination half-lives as usually desired in PK measurements. With regard to radioactive half-life, ^{18}F is the most attractive PET radioisotope, since it allows for imaging durations of up to 10 h. A considerable drawback, however, stems from the fact that relatively few drug molecules contain fluorine in their native structure. Consequently, despite the rather impracticable half-life of ^{11}C (20.4 min), the majority of PET PK experiments have relied on ^{11}C-labeled tracer molecules. A commonly employed method for the synthesis of ^{11}C-labeled drugs is the methylation of amino, phenolic hydroxyl or carboxyl groups with [^{11}C]methyl iodide or [^{11}C]methyl triflate. However, the field of PET radiochemistry is continuously expanding and new and innovative radiolabeling approaches are being developed, which rely on the use of a wide array of [^{11}C]synthons, such as [^{11}C]carbon monoxide, [^{11}C]phosgene, or [^{11}C]cyanide, thus giving the possibility of labeling a range of different functional groups and structures with ^{11}C (Miller et al. 2008). PET radiochemistry usually proceeds at high specific activity (i.e. the ratio of radioactivity to molar mass, typically given in GBq/μmol). Consequently, the amount of unlabeled drug associated with a PET tracer is usually in the range of a few micrograms. Particular challenges in the synthesis of PET radiotracers include time constraints due to the short radioactive half-lives of the employed radioisotopes, the need for automatization of procedures to protect the chemist from radiation exposure and the necessity to produce a radiotracer which meets the quality criteria of a drug for i.v. injection into humans.

Macromolecules, such as therapeutic antibodies or peptides, generally possess slow in vivo kinetics which requires the use of longer lived positron-emitters in PET measurements, such as the radiohalogens bromine-76 (^{76}Br, half-life: 16 h) and iodine-124 (^{124}I, half-life: 100.2 h) or the radiometals copper-64 (^{64}Cu, half-life: 12.7 h) and zirconium-89 (^{89}Zr, half-life: 78.4 h) (Nayak and Brechbiel 2009). Challenges which are often associated with the use of these radionuclides are their restricted availability, high costs, inferior imaging characteristics, and higher dosimetry as compared to ^{11}C- or ^{18}F-labeled radiotracers, and unsatisfactory in vivo stability of macromolecules labeled with such radionuclides (Nayak and Brechbiel 2009). In addition, the introduction of prosthetic groups or chelating moieties is often required to facilitate radiolabeling of macromolecules with long-lived positron-emitters which may affect the bioactivity of these molecules.

14.4 Radiation Exposure in PET Scans

Owing to the short radioactive half-lives of PET radionuclides the radiation exposure of a PET examination is lower than that of other nuclear medicine procedures, such as a computer tomography (CT) investigation. A typically administered activity of a ^{18}F-tracer of 400 MBq given i.v. corresponds to a total effective dose of about 5 mSv, the same amount of ^{11}C-tracer corresponds to about 2 mSv. Therefore, the radiation exposure of one PET scan is approximately in the same order as the level of natural background irradiation in Europe (1–2 mSv/year). The relatively low radiation exposure of PET scans, in particular with ^{11}C-tracers, usually permits the performance of repeated PET scans in one single subject (e.g. before and after administration of an unlabeled drug in different doses). Due to the short half-life of ^{11}C, such repeated scans are usually possible within one scan session on a single study day. It has been argued that dosimetry measurements of new ^{11}C-radiotracers in humans make no sense as the dosimetry is in general governed by the radioactive half-life of ^{11}C and not by the biological half-life of the drug molecule thereby affording very similar total effective doses for most ^{11}C-radiotracers (Zanotti-Fregonara and Innis 2012).

14.5 Limitations of PET

One limitation of PET imaging is that parent drug cannot be distinguished from radiolabeled metabolites in tissue, because both give the same signal. Consequently, for drugs which are extensively metabolized in vivo, the interpretation of drug tissue PK may be confounded by the presence of radiolabeled metabolites. For providing a quantitative description of PET data, such as the rate constants for transfer of radiolabeled drug between plasma and different tissue compartments, the concentration–time profile of the unmetabolized radiolabeled drug in arterial plasma is required ("arterial input function"). Although approaches have been described to obtain this measure directly from the PET data ("image-derived input function") (Zanotti-Fregonara et al. 2011), the gold standard remains arterial blood sampling, which is an invasive procedure, the applicability of which might be restricted in certain subject groups, such as elderly patients. Arterial blood samples are usually counted for total radioactivity using conventional gamma counting and then processed by assays involving chromatographic separation of parent drug from radiolabeled metabolites. Due to the short radioactive half-lives of most PET radionuclides the determination of unchanged parent might be inaccurate, particularly at late time points of the scan session, which might have a considerable impact on the accuracy of quantitative parameter estimates derived from PET data. In fact, the arterial input function determination can be considered as the weakest link in a PET experiment, as this part of the measurement is most likely prone to errors.

As PET is a functional imaging technique, the amount of anatomic information obtained in a PET scan is limited and usually governed by the distribution characteristics of the particular radiotracer. For highly target-specific molecules (e.g. radiolabeled antibodies), almost no anatomic information can be obtained from a PET scan. This limitation can be overcome by combining PET with CT or

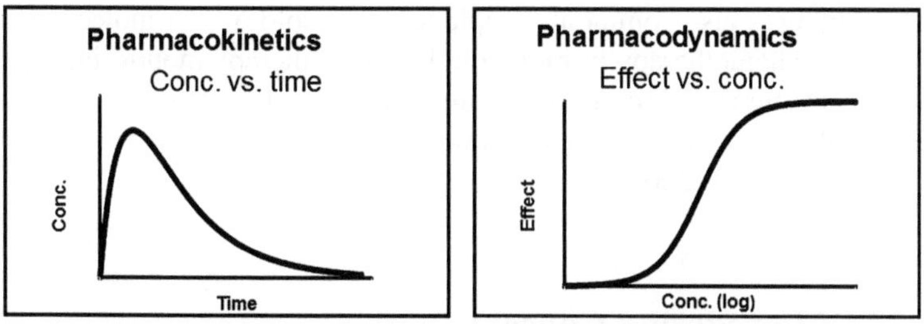

Fig. 14.1 Two different approaches how PET can be used in drug development

magnetic resonance (MR) measurements, either by using separate scanners or by employing dedicated PET/CT (Beyer et al. 2000) or PET/MR scanners (Judenhofer et al. 2007).

14.6 Use of PET in Drug Development

In principle, there are two different ways how PET can be employed in drug development (Fig. 14.1). The first one will be dealt with in this chapter and involves direct radiolabeling of drug molecules for the sake of assessing tissue distribution and PK ("PET microdosing") (Bergström et al. 2003; Wagner et al. 2008; Wagner and Langer 2011). The second approach studies the interaction of a drug molecule with its pharmacodynamic (PD) target (e.g. receptor and transporter proteins, enzymes etc.) (Wong et al. 2009). This is done by employing a validated PET radiotracer which is usually different from the drug under investigation and which allows for accurate quantification of parameters related to binding site density. The investigated (unlabeled) drug is then given at different doses and the level of binding site occupancy induced by the drug is monitored with PET. This indirect approach is more commonly employed than direct drug labeling and nowadays a broad range of radiotracers for various drug targets, particularly those implicated in neuropsychiatric disorders, has been validated (Halldin et al. 2001). Receptor occupancy studies have proven very valuable in dose finding studies in central nervous system (CNS) drug development [see for instance (Bergström et al. 2004)]. In many instances, however, a distinction between the two approaches to use PET in drug development may not be clearcut as a radiolabeled drug candidate may turn out to be a suitable radioligand for visualization of a PD target site. In other words, the PK of the radiolabeled drug in target tissue may provide information on target site density.

PET imaging is also commonly employed in oncology as a molecular readout of response to chemotherapy or radiation therapy ("therapy monitoring"), e.g., by assessment of tumor glucose metabolism using 2-[^{18}F]fluoro-2-deoxyglucose [^{18}F]FDG (Weber et al. 2001) or tumor proliferation rate using [^{18}F]fluorothymidine (Chen et al. 2007).

14.7 The Microdosing Concept

In order to obtain absorption, distribution, metabolism, and excretion (ADME) data of an investigational drug in humans at an early stage of the development process, ideally prior to conventional phase 1 trials, thus providing the possibility of quickly identifying drug candidates with unfavorable PK properties, the concept of microdose studies has been proposed (Lappin and Garner 2003). The European Medicines Agency (EMA) and the Food and Drug Administration (FDA) have provided the legal framework for such kinds of studies by issuing guidance documents (European Medicines Agency, Committee for Medicinal Products for Human Use (CHMP) 2004; Food and Drug Administration, Center for Drug Evaluation and Research (CDER) 2006). Herein, the term microdose is defined as "less than 1/100th of the dose calculated to yield a pharmacological effect of the test substance based on primary PD data obtained in vitro or in vivo and at a maximum dose of ≤100 microgram". For low-dose clinical studies in humans that satisfy the microdose definition, an abbreviated non-clinical safety testing package has been proposed (European Medicines Agency, Committee for Medicinal Products for Human Use (CHMP) 2004; Food and Drug Administration, Center for Drug Evaluation and Research (CDER) 2006). The feasibility for performing such kinds of studies depends on the availability of ultrasensitive analytical methods which are capable of detecting minute drug amounts in human body tissues. When high specific-activity radiotracers are employed, the sensitivity of PET for the detection of mass is in the lower femtomole (10^{-15} mol) range (i.e. concentrations as low as 10^{-12} mol/L can be detected), which makes this method attractive for first-time-in-human studies of new drug candidates for which only limited toxicological information is available (Bergström et al. 2003). Another analytical technique which has proven suitable for detection of ultralow drug concentrations is accelerator mass spectrometry (AMS), which is capable of quantifying ^{14}C-labeled drugs in biological fluids in the attomole to zeptomole range (10^{-18}–10^{-21} mol) (Lappin and Garner 2003). An important issue in this context is the ability of microdose PK parameters to predict therapeutic-dose data. Many enzyme systems and carrier-mediated transport systems can be saturated at therapeutic drug concentrations and drug distribution at very low doses can be governed by high-affinity binding sites. To what extent the principle of "dose linearity" is fulfilled in individual cases has been investigated in Europe in the so-called Consortium for Resourcing and Evaluating AMS Microdosing (CREAM) trial (Lappin et al. 2006) and the "European Union Microdose AMS Partnership

Programme" (EUMAPP) (Lappin et al. 2011) and in Japan in a New Energy and Industrial Technology Development Organization (NEDO) project entitled "Establishment of Evolutional Drug Development with the Use of Microdosing Clinical Trial: Based on the Quantitative Prediction Technology of ADME" (Maeda et al. 2011).

14.8 PET Studies with Radiolabeled Drugs

14.8.1 Assessment of Blood–Brain Barrier Penetration of Drugs

Pharmaceuticals targeted to the CNS often fail to reach therapeutically effective concentration levels inside the brain. This is mainly due to the action of adenosine triphosphate-binding cassette (ABC) transporters, such as P-glycoprotein (Pgp) or breast cancer resistance protein (BCRP), which are located in the luminal membrane of vascular endothelial cells forming the blood–brain barrier (BBB) (Löscher and Potschka 2005). These multidrug efflux pumps actively transport a broad range of drugs against concentration gradients from brain into the blood compartment. As distribution of drugs to the human brain is often difficult to predict from preclinical studies, PET imaging with radiolabeled CNS drugs has proven useful to assess BBB penetration in humans. However, it has now become widely accepted that the unbound and not the total concentration of drug in brain parenchyma is of therapeutic relevance as non-specifically bound drug is usually not able to bind to the drug target (Hammarlund-Udenaes et al. 2008). PET measures the sum of bound and unbound drug concentration in tissue and may therefore not allow judging per se if drug tissue exposure is high enough to be therapeutically effective. Therefore, a combination of PET imaging with equilibrium dialysis assays to estimate the extent of non-specific drug binding in brain tissue has been proposed (Gunn et al. 2012). We have recently conducted a PET microdosing study to assess the local brain tissue PK of an investigational ^{11}C-labeled antiamyloid drug (ST1859) (Bauer et al. 2006). This study showed that ST1859 displayed good BBB penetration and sustained brain tissue retention, which constitutes an important prerequisite for the therapeutic applicability of this agent.

PET imaging with radiolabeled CNS agents may also provide information if the drug binds to its intended target site in cases when the PD target has a distinct anatomic distribution pattern inside the brain. This is because drug tissue concentration levels achieved with radiotracers with high specific activity may be in the same range as a drug's affinity constant (K_i, K_d) for its pharmacological target—in particular for novel therapeutic drugs with high target site binding affinities (in the nanomolar or subnanomolar range)—, which might cause drug tissue distribution to be governed by binding to its pharmacological target. One such example is the characteristic cerebral distribution pattern of a tracer dose of the ^{11}C-labeled antipsychotic drug raclopride, which is dominated by high-affinity

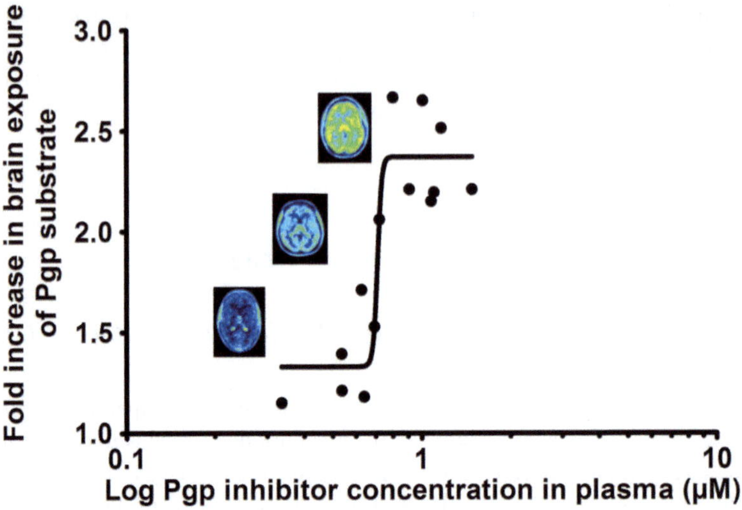

Fig. 14.2 Use of PET to study the interaction between the model Pgp substrate (*R*)-[^{11}C]verapamil and the third-generation Pgp inhibitor tariquidar at the human BBB (Bauer et al. 2012). Cold tariquidar was given to healthy male volunteers (n = 3 per dose group) at doses of 3, 4, 6, and 8 mg/kg at 1 h before a 60 min (*R*)-[^{11}C]verapamil PET scan. A sigmoidal relationship was found between tariquidar plasma concentration at time of the PET scan (μM) and increase in brain exposure of (*R*)-[^{11}C]verapamil with a half-maximum effect concentration of tariquidar of 0.71 μM (95 % confidence interval: 0.68–0.73 μM). The maximum increase in (*R*)-[^{11}C]verapamil brain exposure was 2.4-fold relative to baseline scans without tariquidar administration (95 % confidence interval: 2.2–2.6)

binding to dopamine D_2 receptors in the basal ganglia, the brain region with the highest concentration of these receptors (Farde 1996).

PET imaging has also proven very useful to assess transporter-mediated drug–drug interactions at the BBB (Eyal et al. 2009). Such drug–drug interactions are of great concern in drug development as they may lead to pronounced changes in drug–brain exposure and potentially cause severe side effects (Giacomini et al. 2010). We have recently assessed the Pgp-mediated interaction between the model Pgp substrate (*R*)-[^{11}C]verapamil and the potent third-generation Pgp/BCRP inhibitor tariquidar at the human BBB (Fig. 14.2) (Bauer et al. 2012). We were able to show that tariquidar caused at the highest investigated dose (8 mg/kg) a 2.4-fold increase in brain exposure of (*R*)-[^{11}C]verapamil as compared to baseline PET scans without tariquidar administration.

14.8.2 Assessment of Tumor Distribution of Radiolabeled Anticancer Agents

In clinical oncology, PET microdosing has been successfully used to determine the PK and intratumoral distribution of radiolabeled cytotoxic agents, such as

5-[^{18}F]fluorouracil (Saleem et al. 2000), [^{18}F]fluoropaclitaxel (Kurdziel et al. 2011) or [^{11}C]docetaxel (van der Veldt et al. 2011), in cancer patients. Many anticancer agents are substrates of multidrug efflux transporters (Pgp, BCRP) which are often overexpressed in chemoresistant cancer (Sarkadi et al. 2006). In an ideal scenario, PET imaging with tracer amounts of radiolabeled chemotherapeutics may be used to predict tumor response to treatment, before therapeutic doses of chemotherapeutic drug are administered, based on individual patterns of drug resistance (personalized medicine). Proof-of-concept preclinical studies have provided evidence that such an approach may be indeed feasible in the clinic (Memon et al. 2009). Recently, a series of tyrosine kinase inhibitors has been labeled with ^{11}C (Memon et al. 2009; Asakawa et al. 2011; Zhang et al. 2010) and might find future application in PET studies in cancer patients.

14.8.3 Macromolecules

To date, the drugs studied with PET have mainly been small molecules. However, an interesting application of PET microdosing would be to evaluate the distribution and PK of macromolecules, such as therapeutic antibodies or peptides. Promising results have been already obtained with ^{89}Zr-labeled antibodies ("immuno-PET") which might potentially find application for selection of cancer patients who will benefit from radio immunotherapy or for individual treatment planning (estimation of radiation-absorbed dose) (Börjesson et al. 2006).

14.9 Small-Animal PET

Whereas PET was initially developed for an application in humans, dedicated high-resolution and high-sensitivity PET scanners have been developed which allow for conducting PET experiments in small laboratory animals (Chatziioannou 2002). Such experiments are of particular interest since diverse animal models of human disease (e.g. transgenic mice, tumor-xenograft mice etc.) are commonly used in drug research. The PET technology therefore takes a key position at the interface between preclinical and clinical—that is translational—research. For the purpose of drug development, imaging with PET bears significant potential in translational medicine as it allows the same methodology to be employed in animal experiments and human studies.

Small-animal PET is increasingly used to study drug PK and PD, before conducting human PET studies. The interaction between a radiolabeled drug with its PD target can best be studied by determining radiotracer distribution at baseline, as well as after pretreatment with one or more pharmacological agents that specifically bind at the same target. Unlike in human PET studies, tissue biopsy samples can be obtained in preclinical studies to determine if radiolabeled drug metabolites contribute to the PET signal in tissue. Because radiolabeled drug can be administered in vivo together with increasing doses of unlabeled drug, preclinical PET enables assessment of dose linearity of the PK of investigational drugs (Balani et al. 2006). After validating preclinical assumptions with small-animal PET, they can be transferred to human PET studies in a translational approach. Species differences in drug PK and PD which might be observed may in turn permit to develop improved preclinical models for prediction of human data.

14.10 Comparison of PET and Microdialysis

As apparent from the examples given above and in previous chapters, PET and microdialysis differ with respect to individual strengths and shortcomings (see Table 14.1). Clinical microdialysis is probably more widely applied than PET imaging for drug distribution monitoring; the required equipment is relatively inexpensive and time for method development is comparatively short. Provided the availability of a suitable analytical assay to quantify the drug of interest, an appropriate ethical setting and that appropriate attention is given to recovery measurements, clinical microdialysis studies can be performed for virtually any small drug molecule. However, there exists a cutoff in molecular weight so that macromolecules (e.g. antibodies, peptides) can usually not straightforwardly be subjected to microdialysis experiments. Issues like radioactive waste handling, radiation exposure of patients/volunteers and clinical staff are not relevant for microdialysis experiments, as radiolabeled compounds are usually not employed. The major disadvantage of microdialysis lies in the restricted applicability of the technique to certain tissues of interest, such as brain, lung, liver, etc., owing to its invasive nature. PET, on the other hand, is fully non-invasive and can measure drug concentrations in virtually any organ. PET additionally offers excellent spatial resolution in the order of a few millimeters, so that regionally different drug concentrations in a given organ can be monitored. Microdialysis is a focal sampling method and measures substances in a small volume of tissue, which is defined by the position of the microdialysis probe. What PET cannot provide is chemical resolution, i.e., bound and unbound drug or parent drug and metabolites give the same signal. Therefore, PET imaging should preferably be applied to those compounds which are metabolically stable or whose metabolites are not taken up and/or retained in the tissue being investigated. Microdialysis, when being combined with a separation-based analytical method, is also well suited for metabolite monitoring. PET is not able to discern drug concentrations in different compartments in a given volume of tissue (e.g. intracellular and extracellular), whereas microdialysis measures selectively the unbound drug fraction in one clearly defined compartment, i.e., the interstitial space fluid. PET is highly sensitive and is therefore suited for studying the disposition of novel drug candidates, for which the toxicity is not yet fully characterized, at very low doses (microdose studies). Sensitivity of microdialysis-based protocols depends on the employed

Table 14.1 Comparative characteristics of positron emission tomography (PET) and clinical microdialysis

	PET	Microdialysis
Detection method	Radioactive decay (radiolabeling required)	Absorption, fluorescence, electrochemical, etc.
Employed radionuclides	^{18}F ($t_{1/2}$: 109.7 min) ^{11}C ($t_{1/2}$: 20.4 min)	Not applicable
Invasiveness	Non-invasive	Invasive
Administration mode of drug	Mostly intravenous	Any mode
Measured compartment	Total tissue concentration (bound + unbound)	Extracellular tissue concentration (unbound)
Chemical resolution (metabolite monitoring)	No	Yes
Temporal resolution	Seconds-minutes	Minutes
Spatial resolution	1–5 mm	Focal sampling
Sensitivity	10^{-12} mol/L	10^{-9}–10^{-3} mol/L
Cost	High	Low

analytical assay. In conjunction with ultrasensitive analytical procedures, such as AMS, the microdialysis technique in principle lends itself to performance of microdose studies even though no such examples have appeared in the literature yet. Performance of PET experiments is certainly more complex with regard to the required infrastructure (cyclotron, radiochemistry laboratory, PET camera) than performing a clinical microdialysis study. In addition, in microdialysis experiments samples can be collected at one site and analyzed at another site, which is not possible in PET experiments.

A limitation of PET for direct drug labeling lies in the fact that often time- and labor-intensive development of radiolabeling chemistry is required. Therefore, the prospective use of PET should be integrated as early as possible in the preclincial drug development process. Synthesis of the required radiolabeling precursors is often not associated with extra work when being embedded in the chemical synthesis program of the drug candidates. In addition, the increasing need to produce PET radiotracers for human use under good manufacturing practice (GMP) conditions further increases the costs of PET studies.

A limitation of PET stems from the short radioactive half-lives of most available radioisotopes. For instance, maximum possible imaging times with ^{11}C-labeled drugs do not exceed 2 h. Microdialysis, on the other hand, can perform continuous sampling without removal of body fluid and therefore lends itself to longer sampling periods (some clinical applications report up to 12 h). PET imaging can in principle also be applied to macromolecules. However, in this case the chemical structure of the molecule needs to be modified to introduce long-lived radionuclides (^{124}I, ^{76}Br, ^{64}Cu, ^{89}Zr), which may change the bioactivity of the molecule of interest.

14.11 Conclusion

PET imaging with radiolabeled drugs is a powerful and highly innovative tool along the critical path of pharmaceutical development which allows assessing drug–tissue distribution in humans. PET studies can be included at an early stage of the drug development process as one means for selection or rejection of compounds based on their in vivo performance in humans. PET imaging can be considered as an interesting complementary technique to clinical microdialysis as it can be applied in cases in which clinical microdialysis has its limitations (e.g. assessment of BBB penetration of drugs in humans; assessment of tissue distribution of radiolabeled antibodies). Finally, the availability of small-animal PET systems provides a translational approach to PET imaging.

References

Asakawa C, Ogawa M, Kumata K, Fujinaga M, Kato K, Yamasaki T et al (2011) [11C]sorafenib: radiosynthesis and preliminary PET study of brain uptake in P-gp/Bcrp knockout mice. Bioorg Med Chem Lett 21(8):2220–2223

Balani SK, Nagaraja NV, Qian MG, Costa AO, Daniels JS, Yang H et al (2006) Evaluation of microdosing to assess pharmacokinetic linearity in rats using liquid chromatography-tandem mass spectrometry. Drug Metabol Dispos 34(3):384–388

Bauer M, Langer O, Dal-Bianco P, Karch R, Brunner M, Abrahim A et al (2006) A positron emission tomography microdosing study with a potential antiamyloid drug in healthy volunteers and patients with Alzheimer's disease. Clin Pharmacol Ther 80(3):216–227

Bauer M, Zeitlinger M, Karch R, Matzneller P, Stanek J, Jager W et al (2012) Pgp-mediated interaction between (*R*)-[^{11}C]verapamil and tariquidar at the human blood-brain barrier: a comparison with rat data. Clin Pharmacol Ther 91(2):227–233

Bergström M, Grahnen A, Langström B (2003) Positron emission tomography microdosing: a new concept with application in tracer and early clinical drug development. Eur J Clin Pharmacol 59(5–6):357–366

Bergström M, Hargreaves RJ, Burns HD, Goldberg MR, Sciberras D, Reines SA et al (2004) Human positron emission tomography studies of brain neurokinin 1 receptor occupancy by aprepitant. Biol Psychiatry 55(10):1007–1012

Beyer T, Townsend DW, Brun T, Kinahan PE, Charron M, Roddy R et al (2000) A combined PET/CT scanner for clinical oncology. J Nucl Med 41(8):1369–1379

Börjesson PK, Jauw YW, Boellaard R, de Bree R, Comans EF, Roos JC et al (2006) Performance of immuno-positron emission tomography with zirconium-89-labeled chimeric monoclonal antibody U36 in the detection of lymph node metastases in head and neck cancer patients. Clin Cancer Res 12(7 Pt 1):2133–2140

Chatziioannou AF (2002) PET scanners dedicated to molecular imaging of small animal models. Mol Imaging Biol 4(1):47–63

Chen W, Delaloye S, Silverman DH, Geist C, Czernin J, Sayre J et al (2007) Predicting treatment response of malignant gliomas to bevacizumab and irinotecan by imaging proliferation with [18F] fluorothymidine positron emission tomography: a pilot study. J Clin Oncol 25(30): 4714–4721

European medicines agency, committee for medicinal products for human use (CHMP) (2004) Position paper on non-clinical safety studies to support clinical trials with a single microdose. CPMP/SWP/2599/02

Eyal S, Hsiao P, Unadkat JD (2009) Drug interactions at the blood-brain barrier: fact or fantasy? Pharmacol Ther 123(1):80–104

Farde L (1996) The advantage of using positron emission tomography in drug research. Trends Neurosci 19(6):211–214

Food and drug administration, center for drug evaluation and research (CDER) (2006) Guidance for industry, investigators, and reviewers. Exploratory IND studies

Giacomini KM, Huang SM, Tweedie DJ, Benet LZ, Brouwer KL, Chu X et al (2010) Membrane transporters in drug development. Nat Rev Drug Discov 9(3):215–236

Gunn RN, Summerfield SG, Salinas CA, Read KD, Guo Q, Searle GE et al (2012) Combining PET biodistribution and equilibrium dialysis assays to assess the free brain concentration and BBB transport of CNS drugs. J Cereb Blood Flow Metab 32(5):874–883

Halldin C, Gulyas B, Farde L (2001) PET studies with carbon-11 radioligands in neuropsychopharmacological drug development. Curr Pharm Des 7(18):1907–1929

Hammarlund-Udenaes M, Friden M, Syvänen S, Gupta A (2008) On the rate and extent of drug delivery to the brain. Pharm Res 25(8):1737–1750

Judenhofer MS, Catana C, Swann BK, Siegel SB, Jung WI, Nutt RE et al (2007) PET/MR images acquired with a compact MR-compatible PET detector in a 7-T magnet. Radiology 244(3):807–814

Kurdziel KA, Kalen JD, Hirsch JI, Wilson JD, Bear HD, Logan J et al (2011) Human dosimetry and preliminary tumor distribution of 18F-fluoropaclitaxel in healthy volunteers and newly diagnosed breast cancer patients using PET/CT. J Nucl Med 52(9):1339–1345

Lappin G, Garner RC (2003) Big physics, small doses: the use of AMS and PET in human microdosing of development drugs. Nat Rev Drug Discov 2(3):233–240

Lappin G, Kuhnz W, Jochemsen R, Kneer J, Chaudhary A, Oosterhuis B et al (2006) Use of microdosing to predict pharmacokinetics at the therapeutic dose: experience with five drugs. Clin Pharmacol Ther 80(3):203–215

Lappin G, Shishikura Y, Jochemsen R, Weaver RJ, Gesson C, Brian Houston J et al (2011) Comparative pharmacokinetics between a microdose and therapeutic dose for clarithromycin, sumatriptan, propafenone, paracetamol (acetaminophen), and phenobarbital in human volunteers. Eur J Pharm Sci 43(3):141–150

Löscher W, Potschka H (2005) Role of drug efflux transporters in the brain for drug disposition and treatment of brain diseases. Prog Neurobiol 76(1):22–76

Maeda K, Takano J, Ikeda Y, Fujita T, Oyama Y, Nozawa K et al (2011) Nonlinear pharmacokinetics of oral quinidine and verapamil in healthy subjects: a clinical microdosing study. Clin Pharmacol Ther 90(2):263–270

Memon AA, Jakobsen S, Dagnaes-Hansen F, Sorensen BS, Keiding S, Nexo E (2009) Positron emission tomography (PET) imaging with [11C]-labeled erlotinib: a micro-PET study on mice with lung tumor xenografts. Cancer Res 69(3):873–878

Miller PW, Long NJ, Vilar R, Gee AD (2008) Synthesis of 11C, 18F, 15O, and 13N radiolabels for positron emission tomography. Angew Chem Int Ed 47(47):8998–9033

Nayak TK, Brechbiel MW (2009) Radioimmunoimaging with longer-lived positron-emitting radionuclides: potentials and challenges. Bioconjug Chem 20(5):825–841

Saleem A, Yap J, Osman S, Brady F, Suttle B, Lucas SV et al (2000) Modulation of fluorouracil tissue pharmacokinetics by eniluracil: in vivo imaging of drug action. Lancet 355(9221): 2125–2131

Sarkadi B, Homolya L, Szakacs G, Varadi A (2006) Human multidrug resistance ABCB and ABCG transporters: participation in a chemoimmunity defense system. Physiol Rev 86(4):1179–1236

van der Veldt AA, Lubberink M, Greuter HN, Comans EF, Herder GJ, Yaqub M et al (2011) Absolute quantification of [(11)C]docetaxel kinetics in lung cancer patients using positron emission tomography. Clin Cancer Res 17(14):4814–4824

Wagner CC, Langer O (2011) Approaches using molecular imaging technology—use of PET in clinical microdose studies. Adv Drug Deliv Rev 63:539–546

Wagner CC, Müller M, Lappin G, Langer O (2008) Positron emission tomography for use in microdosing studies. Curr Opin Drug Discov Devel 11(1):104–110

Weber WA, Ott K, Becker K, Dittler HJ, Helmberger H, Avril NE et al (2001) Prediction of response to preoperative chemotherapy in adenocarcinomas of the esophagogastric junction by metabolic imaging. J Clin Oncol 19(12):3058–3065

Wong DF, Tauscher J, Gründer G (2009) The role of imaging in proof of concept for CNS drug discovery and development. Neuropsychopharmacology 34(1):187–203

Zanotti-Fregonara P, Innis RB (2012) Suggested pathway to assess radiation safety of (11)C-labeled PET tracers for first-in-human studies. Eur J Nucl Med Mol Imaging 39(3):544–547

Zanotti-Fregonara P, Chen K, Liow JS, Fujita M, Innis RB (2011) Image-derived input function for brain PET studies: many challenges and few opportunities. J Cereb Blood Flow Metab 31(10):1986–1998

Zhang MR, Kumata K, Hatori A, Takai N, Toyohara J, Yamasaki T et al (2010) [11C]Gefitinib ([11c]Iressa): radiosynthesis, in vitro uptake, and in vivo imaging of intact murine fibrosarcoma. Mol Imaging Biol 12(2):181–191

Chapter 15
Open Flow Microperfusion: An Alternative Method to Microdialysis?

Thomas Pieber, Thomas Birngruber, Manfred Bodenlenz, Christian Höfferer, Selma Mautner, Katrin Tiffner and Frank Sinner

Abstract Membrane-based sampling systems encounter problems when sampling high molecular weight or highly lipophilic substances in the interstitial fluid. Open flow microperfusion (OFM) overcomes these problems by replacing the membrane with a steel mesh featuring macroscopic openings in combination with a peristaltic OFM pump in push/pull mode to achieve stable recovery of OFM samples. Unfiltered sampling results in a complete representation of the ISF for relative and absolute quantification in the target tissue. Current applications in adipose subcutaneous tissue (aOFM) and dermal tissue (dOFM) range from preclinical studies to clinical trials, and cover a wide range of substances from small ions to lipophilic topical drugs to large antibodies. The latest development in OFM has been designed for use in cerebral tissue (cOFM). Currently used in preclinical research, cOFM allows effective sampling in brain tissue with an intact blood–brain barrier. Future work will combine OFM with metabolomics for a more complete assessment of metabolic pathways.

15.1 OFM General Concept

Open flow microperfusion (OFM) is an alternative in vivo sampling technique that builds on the strengths of continuous, minimally invasive interstitial sampling methods such as microdialysis (MD). But instead of a membrane, OFM uses probes with macroscopic openings to exchange substances in a liquid pathway that is based on convection rather than diffusion so that the exchange can occur

T. Pieber · T. Birngruber · M. Bodenlenz · C. Höfferer ·
S. Mautner · K. Tiffner · F. Sinner (✉)
Health, Institute for Biomedicine and Health Sciences,
Joanneum Research, Graz, Austria
e-mail: frank.sinner@joanneum.at

M. Müller (ed.), *Microdialysis in Drug Development*, AAPS Advances in the Pharmaceutical Sciences Series, DOI: 10.1007/978-1-4614-4815-0_15,

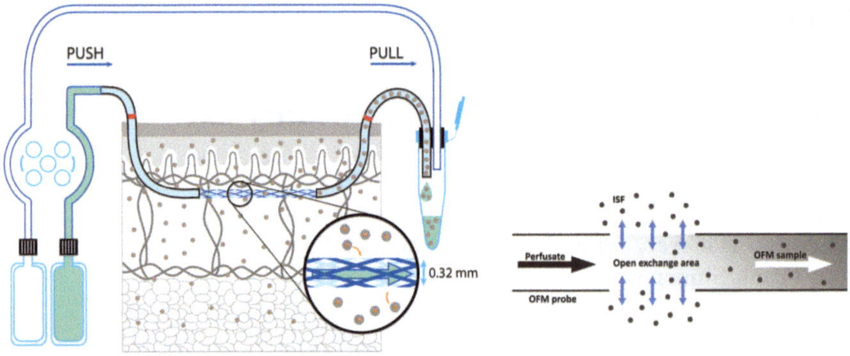

Fig. 15.1 Schematic figure of the OFM system with a linear membrane-free OFM probe. The inserted OFM probe is connected to a peristaltic pump via push–pull tubing. The OFM pump simultaneously pushes the perfusate into the OFM probe and pulls the OFM sample into an easily exchangeable vial. At the exchange area, substances are freely exchanged between the ISF and the perfusate

nonselectively in either direction (Fig. 15.1). The direct liquid pathway between the probe's perfusate and the surrounding interstitial fluid (ISF) provides a diluted but otherwise unchanged ISF sample, since no dialysis or filtering occurs. The membrane-free OFM design also avoids problems such as membrane fouling (Rosenbloom et al. 2005) or clotting or other limitations that can arise when sampling high molecular weight or lipophilic substances. In contrast to MD, OFM provides unfiltered, merely diluted ISF from the investigated target tissues. Endogenous and exogenous substances present in the ISF of targeted tissues can therefore be quantified without the risk of investigating an altered sample.

Therefore, OFM represents a universal, continuous, tissue-specific sampling method with no nominal cutoff value, which thus provides the opportunity to investigate all substances regardless of their physico-chemical properties. All substances ranging from ions up to entire immune competent cells are present in the collected OFM sample, thus closely reflecting the actual ISF composition around the probe. But as proteins and peptides are present that may not be the analytes of interest, the OFM sample requires specifically adapted analytical methods.

To avoid perfusate loss into the tissue and ensure stable sampling volumes, OFM systems actively push the perfusate into the probe lumen, and actively pull the OFM sample into a sampling vial using push–pull pumps similar to those that would be required for high molecular weight cutoff MD systems (Sjögren et al. 2002; Jensen et al. 2007). The technical implementation of the push–pull mode in OFM systems has already been shown to work effectively and to ensure stable, drift-free recovery and sampling for at least 24 h in clinical studies (Fig. 15.2). In these studies the pump was realized as a small, wearable, multi-channel, push–pull peristaltic pump (OFM pump) which increases the mobility of study subjects and allows longer study periods. In addition, the current OFM probes have been optimized for application in different target tissues and for minimal trauma during insertion.

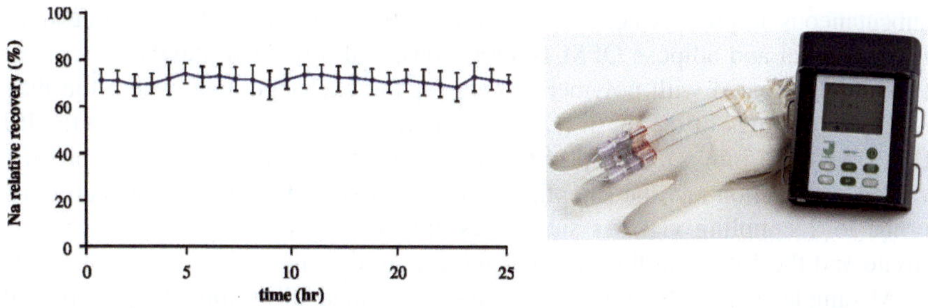

Fig. 15.2 A small, wearable peristaltic OFM pump in push–pull mode enables stable sampling of analytes over long periods (Bodenlenz, unpublished data)

Table 15.1 Key features of OFM and MD sampling systems

Component	OFM	MD
Probe	membrane-free no nominal cutoff 0.5 mm insertion needle	semi-permeable membrane different cutoffs available <100 kDa 0.8 mm insertion needle
Sample	unfiltered, diluted long-term sampling low–high molecular weight low–high lipophilicity high protein load	filtered, diluted membrane fouling or clotting can limit sampling time low–moderate molecular weight low–moderate lipophilicity no–low protein load
Pump	push–pull mode	push suitable for low molecular cutoff
Tubing	no to low adsorption due to direct coupling of sampling vial to OFM probe	possible adsorption in outflow tubing
Analysis	sample pretreatment required	sample pretreatment optional

The OFM system facilitates reliable quantification of analytes including lipophilic and high molecular weight substances. OFM and its CE-certified devices can be used for basic and pharmaceutical research not only in a clinical in vivo setting but also in preclinical in vivo and ex vivo settings. Both MD and OFM are efficient sampling techniques for substances in the ISF. While OFM offers advantages regarding the range of sampled substances, it also faces challenges when analyzing the more complex OFM samples. Table 15.1 gives a brief summary of differences between the two sampling systems.

15.2 OFM Design

OFM is based on the use of a membrane-free probe with macroscopic openings. The probe design can vary considerably, depending on the specific target tissue properties. For superficial peripheral tissues, such as the dermis of the skin or subcutaneous adipose tissue, a very thin probe (Ø 0.36 mm) with a linear design is used. Dermal and adipose OFM systems consist of a braided, flexible steel mesh that is fully coated with polymer, except for the exchange area, where the macroscopic openings of the mesh (Ø 100 μm) provide unrestricted access to the ISF (Fig. 15.1). To minimize unspecific adsorption, the inner lumen of the probe is coated with Teflon and the sampling path is kept short by directly connecting the probe to a sampling vial. As substances are exchanged freely between the perfusate and the ISF regardless of their molecular size, lipophilicity, or charge, the OFM sample represents an unfiltered, merely diluted ISF sample (Schaupp et al. 1999; Ellmerer et al. 2000; Bodenlenz et al. 2005; Bodenlenz et al. 2012).

To facilitate minimally invasive insertion into superficial tissues, a thin needle (not much thicker than the probe itself), is attached to the OFM probe while a wire inside the probe stabilizes the exchange area during handling and insertion. Markings on the OFM probe ensure exact positioning of the exchange area inside the tissue. After probe insertion the needle and the wire are removed. One end of the OFM probe is connected to the sampling vial and the other end is connected to the pump. Using the peristaltic OFM pump in push–pull mode ensures a stable volume and analyte recovery for sampling periods of up to two days.

The sterile linear OFM probes used in adipose OFM (aOFM) and dermal OFM (dOFM), along with the OFM pump, sterile tubings and the perfusate reservoirs, are all CE-certified for clinical use for up to 48 h (class I-IIa medical device). OFM use in the brain (cerebral OFM, cOFM) requires a different, concentric design to allow stereotactical insertion with minimal trauma, as described later in Sect. 15.4.3. cOFM probes are currently realized as research tools that can be used in preclinical trials.

15.3 Quantification and Analytical Aspects

Quantification: Although the membrane-free OFM design allows barrier-free transport of substances from the ISF into the OFM probe, the OFM sample represents a diluted ISF because, at common flow rates of ≥0.5 μl/min, the equilibration between the ISF and the perfusate is not complete. Therefore, the concentration of a substance in the OFM sample represents a dilution rather than its absolute interstitial concentration. To be able to determine the absolute substance concentrations, the degree of dilution must be known. The relationship between the substance concentration in the ISF and the concentration in the OFM sample is called the 'relative recovery' (RR). The RR primarily depends on the flow rate, the length of the exchange area, substance-specific parameters, such as lipophilicity, molecular weight and solubility, and on the physiological parameters of the examined tissue. Absolute quantification of all sampled substances can be achieved by several different calibration methods which can also be used with MD.

In combination with OFM, the *zero flow rate* method has been used to assess ISF concentrations of potassium, sodium, and glucose (Schaupp et al. 1999), while the *no net flux* method has been frequently implemented to obtain the absolute lactate concentration in the adipose ISF (Ellmerer et al. 1998a), glucose concentrations in the adipose ISF (Schaupp et al. 1999), the absolute ISF albumin concentration in adipose tissue and skeletal muscle (Ellmerer et al. 2000), absolute ISF inulin concentration after continuous intravenous infusion (Bodenlenz, unpublished data), and the interstitial insulin concentration during moderate hyperinsulinemic conditions (Bodenlenz et al. 2005). In order to achieve complete equilibrium between the ISF and the perfusate by using the *recirculation method*, the inflow and the outflow tubing of the probe are connected and the perfusate is recirculated several times. The substance concentration in the OFM sample is then an approximation of the substance concentration in the ISF. In combination with OFM, the recirculation method has been used to estimate the absolute ISF concentration of analytes during steady-state conditions to obtain the adipose ISF concentrations of potassium, sodium, and glucose (Schaupp et al. 1999).

In contrast to the above described methods, which require stable analyte concentrations in the ISF, the *ionic reference technique* allows a calibration for each sample and thus can also be used to investigate ISF under dynamic conditions. The ionic reference technique uses sodium as an endogenous reference following the assumptions that the ISF sodium concentration is constant and known during the experiment, and that the relative recoveries of sodium and the substance to calibrate are comparable. The method was initially used to measure the absolute lactate concentration (Ellmerer et al. 1998a) and the absolute glucose concentration (Schaupp et al. 1999). Similar to the ionic reference technique, the calibration can be based on inulin after a stable ISF inulin level is established by primed continuous intravenous inulin infusion. Inulin has been validated as an exogenous reference substance for insulin and successfully used to quantify the levels of human insulin and an insulin analog (Bodenlenz, unpublished data).

Analytical Aspects: Membrane-free OFM systems generate complex matrix samples, mainly due to the presence of proteins in OFM samples. Some analytical methods consequently require OFM sample pretreatment to remove the proteins without losing the analyte of interest. The unfiltered OFM sample might also include substances that could lead to an enzymatic degradation of the analytes. Similar to MD, dilution of the samples and the small sample volume make the analysis challenging because many of the analytes of interest, which already occur in small quantities, are diluted even further during the sampling process.

15.4 OFM Applications

Stable sampling and quantification of lipophilic and high molecular weight substances in the ISF can now be performed in three tissues: adipose (aOFM, since 1995), dermal (dOFM, since 2006), and cerebral (cOFM, since 2008). To date a CE-certified linear OFM probe is used in clinical studies in the dermis (dOFM) and adipose tissue (aOFM), while a concentric OFM probe in brain tissue (cOFM) is currently being used in preclinical studies. The OFM concept has the potential for use in many different tissues.

15.4.1 Adipose Tissue OFM (aOFM)

Subcutaneous adipose tissue (SAT) has traditionally been seen as a passive storage site for excess lipids, but it is currently receiving increased attention as an important endocrine organ that modulates metabolism, immunity, and satiety (Ferris and Crowther 2011). Many biological mediators are produced in the SAT, such as adipose-derived hormones (e.g. leptin, resistin, adiponectin), adipokines (e.g. chemerin, PAI-1, RBP4, visatin), and inflammatory cytokines (e.g., IL-1α, IL-6, IL-8, TNF-β). These mediators may play an important role in various physiological and pathophysiological processes. Successful sampling and quantification of substances in the SAT depends on the specific physico-chemical properties of the analytes of interest. In particular, macromolecules or highly lipophilic substances pose a challenge for sampling and quantification. aOFM has been optimized to enable stable sampling and quantification of a wide range of analytes in preclinical and clinical studies. aOFM has been used to quantify lactate (Ellmerer et al. 1998a, b), peptides such as insulin and cytokines (Bodenlenz et al. 2005; Pachler et al. 2007; Ikeoka et al. 2010a, b, 2011) and proteins (Ellmerer et al. 2000) in SAT. But aOFM can also be used to sample analytes in standard procedures such as continuous glucose monitoring, where glucose levels are assessed in the ISF rather than in the blood (Skrabal et al. 1995; Trajanoski et al. 1997; Schaupp et al. 1999; Ellmerer et al. 2000; Lindpointner et al. 2010a, b), and in combination with sensors for ex vivo online monitoring.

aOFM can also be used to deliver drugs directly to the target tissue and to simultaneously measure their effects. Drugs are mixed into the OFM perfusate, giving them direct access to target cells in known concentrations. Local effects of the drugs can be assessed by the biomarkers in the OFM sample.

15.4.1.1 Materials and Design

The current version of the aOFM probe uses the linear design in combination with the OFM pump (Fig. 15.1). aOFM is CE certified for up to 48 h use in SAT. The wearable OFM pump enables clinical studies of up to two days without interruption of the sampling phase. Clinical studies over longer periods are possible because volunteers tolerate the procedure more readily with the wearable pump, as it allows some mobility. Moreover, because of the closed push–pull sampling system, it is possible to implant the probe in any position or direction in the tissue regardless of the position between probe outflow and sampling vial. This flexibility allows volunteers to lie sit, or even walk around.

15.4.1.2 Stable Sampling and Quantification

aOFM has been used to quantify absolute substance concentrations in the ISF: for example, glucose ISF concentrations have been quantified during basal and hyperglycemic conditions by using various calibration methods, such as zero flow rate, no net flux, recirculation, and the ionic reference technique (Schaupp et al. 1999). Results showed significantly lower glucose concentration in adipose tissue than in plasma, which is similar to results from earlier studies (Schmidt et al. 1993; Maggs et al. 1995, 1997; Sternberg et al. 1995; Müller et al. 1996). It has been shown that the mass-transfer resistance for ions and glucose is similar, which makes continuous estimation of the recovery with the ionic reference technique possible. Thus time-consuming calibration procedures and exposure to labeled markers can be avoided. By combining aOFM with a glucose sensor and conductivity measurements, aOFM can be used to monitor the glucose concentration online in the SAT, extra-corporeally and continuously without any in vivo calibration, providing accurate measurements during hyper- and hypoglycemia (Trajanoski et al. 1997).

15.4.1.3 Sampling of Clinically Relevant Substances

Clinically relevant substances such as insulin are often readily adsorbed by the semi-permeable membranes used in many tissue sampling systems. With its membrane-free design, aOFM has been used to quantify insulin levels in the ISF during euglycemic hyperinsulinemic clamps with the no net flux calibration (Bodenlenz et al. 2005). Under steady-state conditions of moderate hyperinsulinemia after a euglycemic insulin clamp, the estimated ISF insulin concentration was about 25 % of the serum insulin concentration, which confirms reports of low peripheral insulin levels under these distinct insulinemic conditions (Fig. 15.3a).

aOFM has also been used to sample cytokines and assess the local cytokine response in SAT after probe insertion. A reproducible and highly consistent pro-inflammatory response for IL-1β, IL6, IL8, and TNF-α was found after aOFM probe insertion, most likely due to the insertion trauma, which results in a pathophysiological condition in the tissue surrounding the probe (Pachler et al. 2007). These results may influence the interpretation of previous studies using similar probe techniques that have addressed the adipose tissue physiology of small (e.g., glucose, lactate) or large (e.g. cytokines) compounds. aOFM also represents a novel in vivo model to further investigate acute inflammation at the level of ISF cytokines and determine the modulated response to the local delivery of active agents in humans.

15.4.1.4 Sampling of High Molecular Weight Substances

The membrane-free aOFM design has also been used to measure the absolute concentration of albumin in the ISF of SAT in combination with no net flux calibration (Fig. 15.3b, Ellmerer et al. 2000). The results showed stable sampling of macromolecules from the interstitial subcutaneous tissue with aOFM probes. Moreover, when sampling albumin it is also possible to sample albumin-bound substances such as insulin detemir (Bodenlenz, unpublished data). This provides the opportunity for preclinical investigations of various insulin formulations and insulin analogs in early drug development

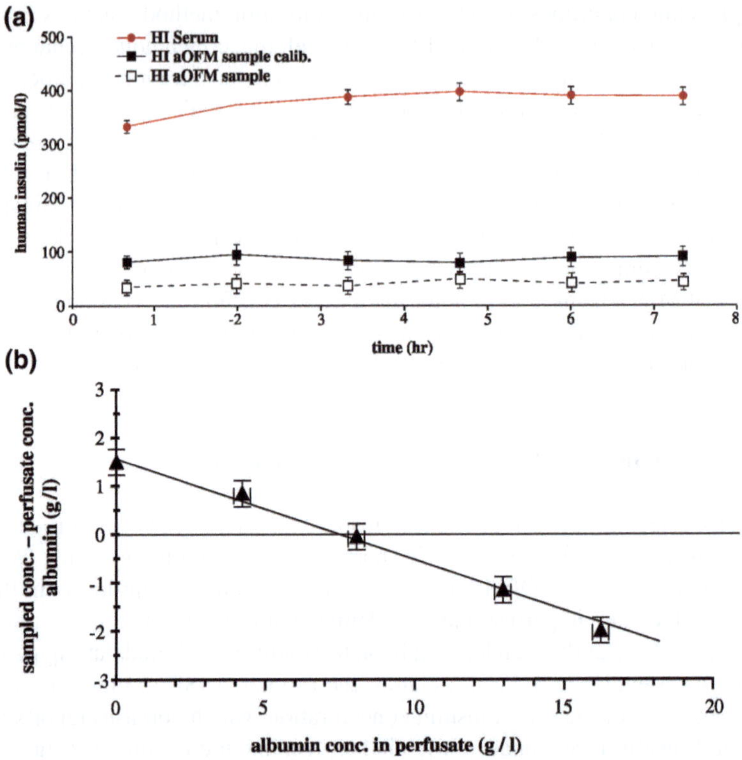

Fig. 15.3 **a** Insulin concentration (mean ± SE) in 11 healthy volunteers (reprinted with permission from Bodenlenz et al. 2005). **b** Albumin concentration (mean ± SE) in subcutaneous tissue of six healthy volunteers. Albumin concentrations are expressed as the difference between sampled concentration (sampled conc.) and perfusate fluid (perfusate conc.) on the y-axis and are plotted against albumin concentrations in perfusate (albumin conc. in perfusate) on the x-axis (reprinted with permission from Ellmerer et al. 2000)

followed by early clinical drug research. The measurement of high molecular weight drugs and protein bound drugs in their target tissues (pharmacokinetics; PK) and their action (pharmacodynamics; PD) is now possible.

15.4.1.5 Conclusions

aOFM systems are:

- used to perform preclinical studies in rats and pigs
- CE-certified and used for clinical studies of up to two days with mobile volunteers
- used for absolute quantification of all ISF analytes

aOFM future work will focus on:

- further quantification of an even wider range of substances
- the investigation of adipokines and cytokines in subcutaneous tissue
- sampling entire cells in sufficient amounts
- preclinical studies with transgenic mouse models and
- clinical studies in patients and healthy people to investigate various metabolic diseases.

15.4.2 Dermal OFM (dOFM)

The current CE-certified dOFM design has been tailored to the specific requirements of dermal applications, triggered by the urgent need for novel tools to efficiently test at target tissues in dermatology and pharmaceutical science (Chaurasia et al. 2007). The dOFM system and all procedures are optimized for the particular properties of the skin and its dermal layer. Skin anatomy and physiology pose certain challenges for sampling applications, both for dOFM and dermal MD. The skin is a large but thin organ (dermis ~ 1.5 mm) that is also flexible and tough, which affects probe insertion. Skin determines our individual appearance and is always visible, so extra care must be taken to ensure that volunteers in studies are never at risk of acquiring permanent scars or pigment changes. Thus, a safe probe structure and a thin needle are necessary. Chronically inflamed or diseased skin shows an increase in capillaries and blood flow with an increased risk of local bleeding around the probe that needs to be identified in the samples to omit them from further analyses.

Topical drugs have physico-chemical properties that make them suitable for skin penetration (Bos 2003), but their hydrophobic properties increase surface adsorption and make them challenging for PK sampling. Mediators of inflammation (e.g., cytokines, potential biomarkers) are also attracted to surfaces, are larger, and require proper controls, i.e., untreated or healthy control sites (Anderson 2006; Stenken et al. 2010), when used for PD investigations. Multiple probes and test sites are required to monitor variability in bioequivalence and bioavailability for generic products (Benfeldt et al. 2007), or when comparing penetration of chemicals into healthy, healthy disrupted, or diseased skin sites in occupational science.

15.4.2.1 Materials and Design

dOFM uses probes of linear design, 0.5 mm insertion needles for minimal invasiveness and a wearable pump for push–pull mode (Fig. 15.1 and 15.4a). In addition, dOFM uses a self-adhesive plastic ring to stabilize the selected skin zone (Ø 60 mm) for prolonged sampling. Thus, the skin cannot be stretched and the dOFM probes are prevented from sliding within the skin during sampling in

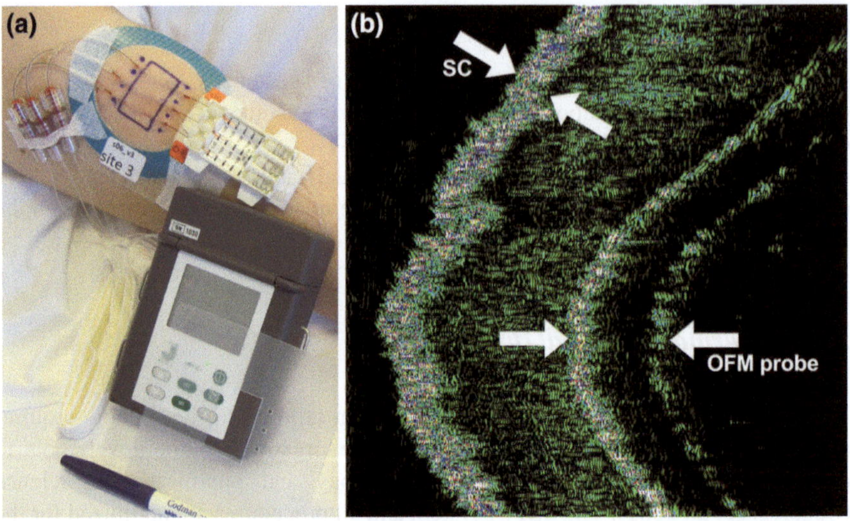

Fig. 15.4 **a** Skin site on forearm investigated by dOFM. The site is stabilized for a 25 h investigation using a self-adhesive ring, and a rectangular zone within the site is marked with the insertion pattern. Three probes are inserted per sampling site. Dermal ISF is sampled continuously using the wearable OFM pump, a push–pull tubing set and a sampling unit with glass capillaries as sampling vials. Straps are used to attach the pumps to the arm, thus enabling mobility and normal bathroom visits without interrupting the sampling protocol. **b** High-frequency ultrasound image (50 MHz) showing the stratum corneum (sc) and the probe (exchange area between *arrows*) within the dermis at a depth of ~0.6 mm

mobile subjects. With the help of a predefined insertion pattern, dOFM ensures precise application of probes and topical drugs, thus helping to standardize procedures to allow comparisons between skin sites within subjects as well as between subjects and study periods.

The dOFM system has been evaluated for applicability, tolerability, safety, principal performance and stable relative recovery over 25 h in a clinical setting (Bodenlenz, unpublished data). Healthy and psoriatic volunteers readily tolerated the application of 9 dOFM probes over a 25 h sampling period, and insertion pain was low on the visual analog scale ($VAS = 3$). Ultrasound imaging confirmed precise dOFM probe application and comparable probe depths when the insertion was performed by trained personnel (Fig. 15.4b).

15.4.2.2 Stable Sampling

The stability of dOFM sampling over long periods was investigated in the above-mentioned 25 h sampling study by monitoring the drift and variability of the relative recovery of sodium. The sampling recovery showed no drift for dOFM probes over 25 h (Fig. 15.2) using the OFM pump, which tightly controls the fluidic path in the OFM system. Stable dOFM recovery over a 12 h sampling period has also been confirmed for inulin (MW = 5 kDa), which can be used at a stable concentration as an exogenous reference substance in dOFM sampling (Bodenlenz, unpublished data).

15.4.2.3 Sampling of Highly Lipophilic Substances

In dermal sampling, the membrane-free design and anti-adsorptive inner coating of dOFM probes are of utmost importance. All topically applied drugs are amphiphilic and thus have some lipophilic character, with a $\log P \sim 3$, which enables them to penetrate the stratum corneum and then enter the aqueous milieu of the target site, the viable epidermis and dermis. Some topicals that should not be systemically available due to their potent immunosuppressive action, even feature a $\log P = 6$ (Tacrolimus) or $\log P = 7$ (Pimecrolimus).

dOFM has been used to sample an investigational topical antipsoriatic drug continuously in lesional and non-lesional skin. Intradermal drug concentration profiles were generated and compared after the first application and again after one week of daily treatment. The study demonstrated a significant increase of the drug in the dermis of lesional and non-lesional skin but, unexpectedly, the drug concentration in the dermis did not differ between lesional and non-lesional skin (Bodenlenz et al. 2012). A recent study attributed similar results to a thickened stratum corneum in lesional skin, which traps the applied topical drug (Rony and Scholler, personal communication).

Likewise, a highly potent class IV corticoid clobetasol-17-propionate ($\log P = 4$) has been investigated using dOFM. The number of application sites was increased to four and the number of probes to 12 per subject in order to compare both drug and vehicle in lesional and non-lesional skin for PK and PD from baseline up to 24 h post dose. The study design included proper controls for the active drug and the lesional skin site, and enabled the identification of cytokines as psoriasis-specific biomarkers (Fig. 15.5a Bodenlenz, unpublished data).

15.4.2.4 Sampling of High Molecular Weight Substances

Sampling of high molecular weight substances is a domain of OFM that can be best exploited by using dOFM in dermal studies of cytokines as biomarkers, and in studies assessing antibody distribution in the dermal target tissue. Cytokines are frequently sampled and quantified as mediators of inflammation and potential biomarkers in both MD and OFM studies. To quantify a rather large set of cytokines of interest, no more than 20–50 μl sample volume is required for multiplex bead assays. Cytokines range in size from 10 to 51 kDa (TNFα-trimer), posing a challenge for membrane-based sampling methods (Holmgaard et al. 2010).

dOFM probes and device design have been optimized to facilitate cytokine sampling, which is carried out in clinical research on inflammatory skin diseases to identify biomarkers of the disease, and in PK-PD studies of anti-inflammatory drugs to investigate whether the drug suppresses the release of the specific

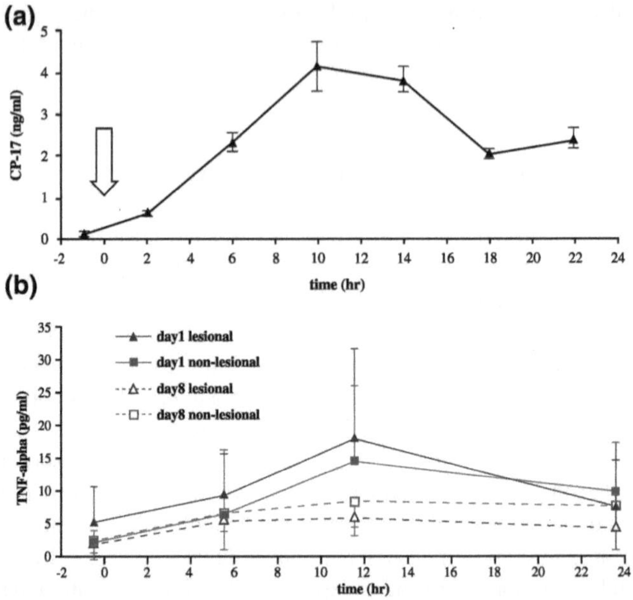

Fig. 15.5 **a** Dermal PK profiles of clobetasol-17-propionate (LogP = 4) at one sampling site on day 14 of once-daily treatments (*arrow*). **b** Dermal TNFα levels in four 60-min samples (mean ± SD)

cytokine at target site as intended (Fig. 15.5b). In a clinical trial, TNFα levels increased to a maximum on day 1, whereas on day 8 of treatment, the cytokine increase was clearly reduced. TNFα levels on day 8 at 6 and 12 h were much lower than on day 1 ($p = 0.039$, Bodenlenz et al. 2012).

Antibodies with a molecular weight of more than 100 kDa are out of range for clinical MD. The most permeable MD probe approved for clinical use features a molecular weight cutoff of 100 kDa (CMA66, CMA MD, Solna, Stockholm, Sweden), and recovers entities of up to ~30 kDa efficiently (Rosenbloom et al. 2005). Higher molecular weights can only be sampled with the use of an adequate push–pull pump to avoid perfusate loss to the surrounding tissue. The wearable OFM pump (multichannel, push–pull) is essential for membrane-free OFM probes, but could also be used for high molecular weight cutoff MD probes. Currently, dOFM studies on the PK-PD of a 160 kDa antibody in the dermis are being conducted, and suitable calibration methods are being evaluated. Preliminary data demonstrate the feasibility of antibody sampling with dOFM.

15.4.2.5 Conclusions

dOFM systems are:

- CE-certified
- membrane-free, super-permeable, robust

- universal in terms of sampling capacity; i.e., including lipophilic and large entities
- versatile and wearable, enabling multi-probe/site set-ups with proper controls
- facilitating informative PK-PD and bioequivalence studies

dOFM future work will focus on:

- investigating the influence of probe depth variation on PK/PD
- the characterization of entire sampled cells
- ex vivo human skin (EVA-cell) as a model for topical penetration PK in place of clinical trials
- the absolute quantification of antibodies in skin disease.

15.4.3 Cerebral OFM (cOFM)

cOFM is the most recently developed OFM technique. It utilizes the strengths of continuous membrane-free sampling in the brain. The blood–brain barrier (BBB) forms a protective shield for the brain, restricting the exchange between blood and brain tissue. The BBB features tight junctions that consist of occlusion proteins such as occludin, claudins and junctional adhesion molecules, which seal the gap between two capillary endothelial cells.

In 1913, Paul Ehrlich and Edwin Goldmann first described the special characteristics of the central nervous system (CNS) and demonstrated limited marker penetration into the CNS, which they attributed to the presence of the BBB (Goldmann 1913). The BBB probably evolved to restrict access to the complex electrical and chemical network of the brain, which contains diverse endogenous transmitters and modulators as well as hundreds of channels, synapses, and receptors. As interference of exogenous substances could easily result in the disruption of essential brain functions, some kind of BBB is present in all vertebrates and even some insect orders. Active and passive transport systems across the BBB ensure the supply of essential substances such as glucose and amino acids. But its protective function also limits access of potential therapeutic agents to the brain (Miller 2002; Pardridge 2002), leading to difficulties in the development of drugs that target the CNS. Overcoming the active and passive defense mechanisms of the BBB during drug development has caused neuroactive drugs to be underrepresented in terms of commercial availability, given the prevalence of CNS diseases (Pardridge 2005; Upton 2007), which affect ~127 M Europeans (Olesen and Leonardi 2003). In 2004, treatment costs for CNS diseases were estimated to be 368 bn €; 135 bn € in direct medical costs and 13 bn € in direct drug costs (Andlin-Sobocki and Rössler 2005). The global market for CNS pharmaceutical products is the second-fastest growing, although the probability of a phase I candidate product reaching the market is about three times lower than for drugs in comparable pharmaceutical fields (IMS Health Retail Drug Monitor 2004; Kola and Landis 2004; The Innovative Medicine Initiative Reseach Agenda 2008).

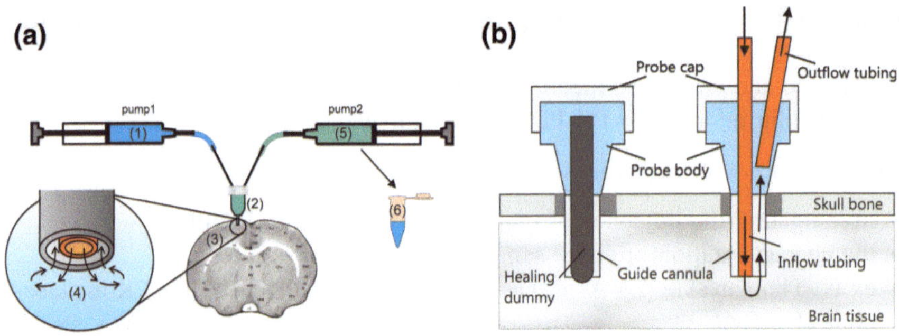

Fig. 15.6 **a** cOFM working principle during sampling: Pump1 pushes perfusate (*1*) through the cOFM probe (*2*) which is implanted in a rat brain (*3*). At the tip of the probe (*4*) substances are exchanged between ISF and perfusate. This mixture is withdrawn by pump2 (*5*) and collected at regular intervals in vials (*6*). The open structure of cOFM puts perfusate in direct contact with brain tissue and brain ISF, which enables sampling of lipophilic and high molecular weight substances. **b** Schematic drawing of the cOFM probe with inserted healing dummy on the left side and during sampling

This illustrates the need to develop new methods to determine the chemical and metabolic composition of fluids and tissues in the CNS without limitations in terms of lipophilicity and size of the analytes. cOFM is designed to measure substance transport across the intact BBB and shows the potential to bridge the gap in neuro-active drug development. Like all invasive sampling techniques, cOFM probe implantation causes trauma, destruction of capillaries and disruption of the BBB. But the cOFM design allows measurement after the BBB has been re-established. cOFM features no membrane, instead bringing the perfusate into direct contact with brain tissue, offering the benefit of constant measurement over a long period with no membrane occlusion due to protein clotting or biofouling (Rosenbloom et al. 2005; Rosenbloom et al. 2006). cOFM also enables the sampling of lipophilic substances, which cross the BBB, and are therefore of special interest in neuro-pharmacology.

15.4.3.1 cOFM Working Principle

The cOFM probe features a concentric design that allows implantation into different brain regions with minimal trauma. In order to provide mechanical stability during implantation, the healing dummy is inserted into the inner lumen of the probe and guide cannula (Fig. 15.6). The healing dummy allows tissue regeneration, prevents tissue migration into the guide tubing, and remains in place until re-establishment of the BBB. During sampling, the healing dummy is replaced by inflow/outflow tubing and perfusion fluid is pushed into brain tissue (via inflow tubing) and withdrawn at the same speed. The inflow/outflow tubing is connected to syringe pumps that allow operation under stringently controlled conditions regarding flow-rate precision and the balance between delivered and withdrawn fluid.

15.4.3.2 Surgery

The cOFM probe is implanted in anesthetized animals using a stereotactic frame in order to accurately address the targeted brain area. After exposing the skull, a hole for probe insertion is drilled and the cOFM probe is slowly inserted. The probe is fixed to the skull bone by using two anchor screws and biocompatible dental cement. For three days after surgery a daily dose of antibiotics and pain treatment is administered subcutaneously.

15.4.3.3 BBB Re-Establishment After cOFM Implantation

Implantation of the cOFM probe into brain tissue causes capillary rupture and therefore BBB damage. But as the measurement of substance transport across the BBB requires an intact BBB, a healing period between probe implantation and sampling is required. BBB re-establishment can be assessed using albumin-bound Evans Blue (EB), which has a MW of $\sim$66 kDa. EB is an established marker for BBB intactness because it does not cross the intact BBB but will stain brain tissue when the BBB is damaged (Manaenko et al. 2011). In our experiments, EB was injected into the vascular system 5–15 days after cOFM probe implantation. The intravascular EB fraction was removed by performing a transcardial flushing procedure. The amount of EB that crossed the BBB was assessed by fluorescence measurements in the extracted brain (Fig. 15.7a). Compared to a negative control group with an intact BBB, EB fluorescence after cOFM probe implantation indicated that the BBB was disrupted for nine days. After 11 days, measured values were comparable with the negative control group, indicating an intact BBB (Birngruber, unpublished data).

15.4.3.4 Continuous cOFM Monitoring of BBB Permeability

Sodium fluorescein (Naf) is a low molecular weight marker that is able to cross the intact BBB to a small degree. Therefore, Naf can be used to monitor BBB intactness throughout the whole cOFM sampling procedure. In order to highlight the difference between the two states of BBB intactness, hyperosmolar mannitol was added to the perfusate in one group because it increases BBB permeability. Naf measurements in cOFM samples showed a substantial difference between standard perfusate and perfusate with mannitol, clearly demonstrating the potential of cOFM for continuous monitoring of BBB permeability (Fig. 15.7b). Additionally, this experiment demonstrated the feasibility of substance application directly into brain tissue via the cOFM probe.

15.4.3.5 Constant Performance for Several Weeks of Repeated Sampling

The extracellular sodium concentration is maintained at a highly constant level throughout the whole organism. Therefore, the recovery rate can be determined by

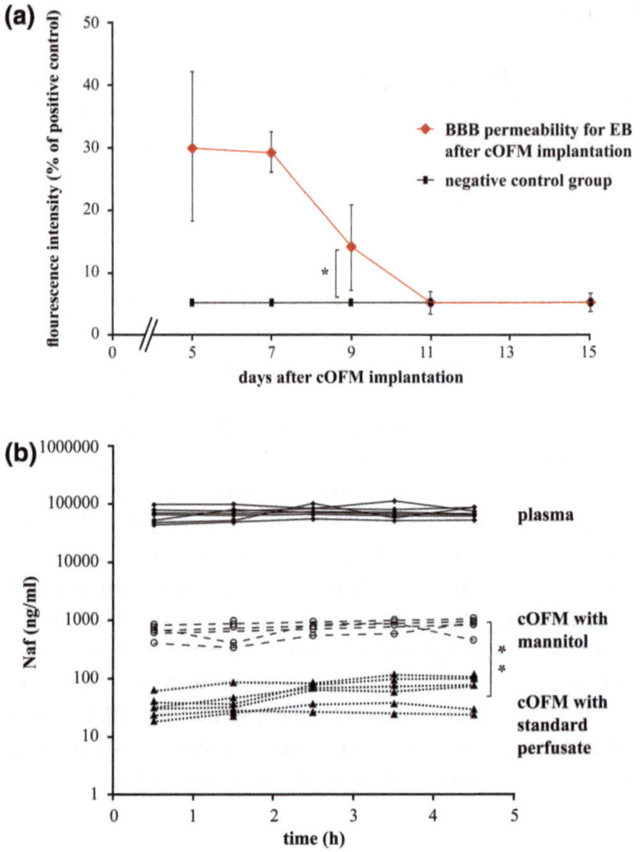

Fig. 15.7 **a** BBB re-establishment after cOFM implantation. **b** Continuous monitoring of BBB permeability using Naf (Birngruber, unpublished data)

measuring ion concentrations in cOFM samples. Using ion-free water with 5 % glucose as perfusate, stable recovery values of ~30 % for a flow rate of 1 μl/min and ~40 % for a flow rate of 0.5 μl/min were found. Repeated measurements showed no significant decrease in recovery rates over three weeks, indicating stable conditions that allow dynamic monitoring of PK/PD.

15.4.3.6 Changes in Tissue Surrounding the Probe

The techniques of MD and cOFM are comparable, as they are both based on continuous fluidic perfusion after an invasive implantation. It is well documented in MD studies, that implantation trauma combined with reactive probe surface causes a so-called "trauma layer" to appear, which surrounds and encapsulates the MD probe. The consequences of the trauma layer are not completely understood, but there is strong evidence that the presence and the extent of a trauma layer can influence results from MD measurements (Benveniste et al. 1987; Benveniste and Diemer 1987; Westergren et al. 1995; de Lange et al. 1997; Grabb et al. 1998; Groothuis et al. 1998; Clapp-Lilly et al. 1999). The influence of a trauma layer has also been estimated using mathematical modelling. (Bungay et al. 2003). In contrast to MD, preliminary histological investigations of brain tissue after cOFM probe implantation show that the trauma layer surrounding the cOFM probe is very small even after two weeks of probe implantation. The cOFM design provides increased sampling time without the hindrance of a trauma layer, and allows complete healing of the BBB. The low degree of tissue changes can be mainly attributed to the non-reactive materials used with the cOFM probe.

15.4.3.7 Conclusions

cOFM probes:

- can measure transport across the BBB
- can deliver substances into brain tissue, bypassing the BBB
- can measure changes in BBB permeability in addition to PK/PD as quality control in each animal
- can sample over weeks with stable recovery rates
- cause minimal tissue reaction.

cOFM future work will focus on:

- investigating PK and PD of neuroactive substances
- examining histological changes surrounding the probe
- using cOFM with neurodegenerative animal models
- the role of BBB permeability in neurodegenerative disease.

15.5 Outlook

Current research and development aims to further standardize all components and processes involved in the OFM concept. Standard operating procedures will be defined for all standard applications available for different tissues in order to minimize variability. To facilitate the use of OFM systems in therapeutic applications the system will be miniaturized even further. The absolute quantification of analytes will be extended to larger entities, including cells. OFM can achieve a more complete picture of the whole metabolism when OFM samples are analyzed using metabolomics instead of focusing on particular analytes. We plan to use metabolomics with OFM to compare different tissues and their respective metabolic pathways in both healthy and diseased subjects.

Future OFM applications will include preclinical trials with improved anesthesia to prolong experiment time from 8 to 12–14 hours. The use of the OFM system in awake animals will allow a trial time of several weeks, and enables a combination of OFM with behavioral experiments. Furthermore, OFM systems will be tested for use in knock-out mice to study a wide variety of diseases. Clinical experiments will be extended to diseases such as diabetes, cancer, auto-immune diseases and wound healing.

References

Anderson CD (2006) Cutaneous microdialysis: is it worth the sweat? J Invest Dermatol 126:1207–1209

Andlin-Sobocki P, Rössler W (2005) Cost of psychotic disorders in Europe. Eur J Neurol 12(Suppl 1):74–77

Benfeldt E, Hansen SH, Vølund A et al (2007) Bioequivalence of topical formulations in humans: evaluation by dermal microdialysis sampling and the dermatopharmacokinetic method. J Invest Dermatol 127:170–178

Benveniste H, Diemer NH (1987) Cellular reactions to implantation of a microdialysis tube in the rat hippocampus. Acta Neuropathol 74:234–238

Benveniste H, Drejer J, Schousboe A, Diemer NH (1987) Regional cerebral glucose phosphorylation and blood flow after insertion of a microdialysis fiber through the dorsal hippocampus in the rat. J Neurochem 49:729–734

Bodenlenz M, Höfferer C, Magnes C et al (2012) Dermal PK/PD of a lipophilic topical drug in psoriatic patients by continuous intradermal membrane-free sampling. Eur J Pharm Biopharm (in press)

Bodenlenz M, Schaupp L, Druml T et al (2005) Measurement of interstitial insulin in human adipose and muscle tissue under moderate hyperinsulinemia by means of direct interstitial access. Am J Physiol Endocrinol Metab 289:296–300

Bos JD (2003) Non-steroidal topical immunomodulators provide skin-selective, self-limiting treatment in atopic dermatitis. Eur J Dermatol 13:455–461

Bungay PM, Newton-Vinson P, Isele W et al (2003) Microdialysis of dopamine interpreted with quantitative model incorporating probe implantation trauma. J Neurochem 86:932–946

Chaurasia CS, Müller M, Bashaw ED et al (2007) AAPS-FDAFDA workshop white paper: microdialysis principles, application and regulatory perspectives. Pharm Res 24:1014–1025

Clapp-Lilly KL, Roberts RC, Duffy LK et al (1999) An ultrastructural analysis of tissue surrounding a microdialysis probe. J Neurosci Methods 90:129–142

Ellmerer M, Schaupp L, Brunner GA et al (2000) Measurement of interstitial albumin in human skeletal muscle and adipose tissue by open-flow microperfusion. Am J Physiol Endocrinol Metab 278:352–356

Ellmerer M, Schaupp L, Sendlhofer G et al (1998a) Lactate metabolism of subcutaneous adipose tissue studied by open flow microperfusion. J Clin Endocrinol Metab 83:4394–4401

Ellmerer M, Schaupp L, Trajanoski Z et al (1998b) Continuous measurement of subcutaneous lactate concentration during exercise by combining open-flow microperfusion and thin-film lactate sensors. Biosens Bioelectron 13:1007–1013

Ferris WF, Crowther NJ (2011) Once fat was fat and that was that: our changing perspectives on adipose tissue. Cardiovasc J Afr 22:147–154

Goldmann EE (1913) Vitalfärbung am Zentralnervensystem. In: Königliche Preußische Akademie der Wissenschaften. Verlag der Königlichen Akademie der Wissenschaften, Berlin, pp 1–13

Grabb MC, Sciotti VM, Gidday JM et al (1998) Neurochemical and morphological responses to acutely and chronically implanted brain microdialysis probes. J Neurosci Methods 82:25–34

Groothuis DR, Ward S, Schlageter KE et al (1998) Changes in blood-brain barrier permeability associated with insertion of brain cannulas and microdialysis probes. Brain Res 803:218–230

Holmgaard R, Nielsen JB, Benfeldt E (2010) Microdialysis sampling for investigations of bioavailability and bioequivalence of topically administered drugs: current state and future perspectives. Skin Pharmacol Physiol 23:225–243

Ikeoka D, Mader JK, Pieber TR (2010a) Adipose tissue, inflammation and cardiovascular disease. Rev Assoc Med Bras 56:116–121

Ikeoka D, Pachler C, Korsatko S et al (2010b) Interleukin-6 produced in subcutaneous adipose tissue is linked to blood pressure control in septic patients. Cytokine 50:284–291

Ikeoka DT, Pachler C, Mader JK et al (2011) Lipid-heparin infusion suppresses the IL-10 response to trauma in subcutaneous adipose tissue in humans. Obesity (Silver Spring) 19:715–721

IMS Retail Drug Monitor (2004) Health, (San Francisco), p 1–4

Innovative Medicines Initiative (IMI) (2008) Creating Biomedical R&D Leadership for Europe to Benefit Patients and Society. 2008:118

Jensen SM, Hansen HS, Johansen T, Malmlöf K (2007) In vivo and in vitro microdialysis sampling of free fatty acids. J Pharm Biomed Anal 43:1751–1756

Kola I, Landis J (2004) Can the pharmaceutical industry reduce attrition rates? Nat Rev Drug Discov 3:711–715

de Lange EC, Danhof M, de Boer AG, Breimer DD (1997) Methodological considerations of intracerebral microdialysis in pharmacokinetic studies on drug transport across the blood-brain barrier. Brain Res Rev 25:27–49

Lindpointner S, Korsatko S, Köhler G et al (2010a) Use of the site of subcutaneous insulin administration for the measurement of glucose in patients with type 1 diabetes. Diabetes Care 33:595–601

Lindpointner S, Korsatko S, Köhler G et al (2010b) Glucose levels at the site of subcutaneous insulin administration and their relationship to plasma levels. Diabetes Care 33:833–838

Maggs DG, Jacob R, Rife F et al (1997) Counterregulation in peripheral tissues: effect of systemic hypoglycemia on levels of substrates and catecholamines in human skeletal muscle and adipose tissue. Diabetes 46:70–76

Maggs DG, Jacob R, Rife F et al (1995) Interstitial fluid concentrations of glycerol, glucose, and amino acids in human quadricep muscle and adipose tissue. Evidence for significant lipolysis in skeletal muscle. J Clin Invest 96:370–377

Manaenko A, Chen H, Kammer J et al (2011) Comparison Evans Blue injection routes: Intravenous versus intraperitoneal, for measurement of blood-brain barrier in a mice hemorrhage model. J Neurosci Methods 195:206–210

Miller G (2002) Drug targeting. Breaking down barriers. Science (New York) 297:1116–1118

Müller M, Holmäng A, Andersson OK et al (1996) Measurement of interstitial muscle glucose and lactate concentrations during an oral glucose tolerance test. Am J Physiol 271:1003–1007

Olesen J, Leonardi M (2003) The burden of brain diseases in Europe. Eur J Neurol 10:471–477

Pachler C, Ikeoka D, Plank J et al (2007) Subcutaneous adipose tissue exerts proinflammatory cytokines after minimal trauma in humans. Am J Physiol Endocrinol Metab 293:690–696

Pardridge WM (2002) Drug and gene delivery to the brain: the vascular route. Neuron 36:555–558

Pardridge WM (2005) The blood-brain barrier: bottleneck in brain drug development. Neuron 2:3–14

Rosenbloom AJ, Ferris RL, Ferris R et al (2006) In vitro and in vivo protein sampling by combined microdialysis and ultrafiltration. J Immunol Methods 309:55–68

Rosenbloom AJ, Sipe DM, Weedn VW (2005) Microdialysis of proteins: performance of the CMA/20 probe. J Neurosci Methods 148:147–153

Schaupp L, Ellmerer M, Brunner GA et al (1999) Direct access to interstitial fluid in adipose tissue in humans by use of open-flow microperfusion. Am J Physiol 276:401–408

Schmidt FJ, Sluiter WJ, Schoonen AJ (1993) Glucose concentration in subcutaneous extracellular space. Diabetes Care 16:695–700

Sjögren F, Svensson C, Anderson C (2002) Technical prerequisites for in vivo microdialysis determination of interleukin-6 in human dermis. Br J Dermatol 146:375–382

Skrabal F, Trajanoski Z, Kontscheider H et al (1995) Portable system for on-line continuous ex vivo monitoring of subcutaneous tissue glucose using open tissue perfusion. Med Biol Eng 33:116–118

Stenken JA, Church MK, Gill CA, Clough GF (2010) How minimally invasive is microdialysis sampling? A cautionary note for cytokine collection in human skin and other clinical studies. The AAPS Journal 12:73–78

Sternberg F, Meyerhoff C, Mennel FJ et al (1995) Subcutaneous glucose concentration in humans. Real estimation and continuous monitoring. Diabetes Care 18:1266–1269

Trajanoski Z, Brunner GA, Schaupp L et al (1997) Open-flow microperfusion of subcutaneous adipose tissue for on-line continuous ex vivo measurement of glucose concentration. Diabetes Care 20:1114–1121

Upton RN (2007) Cerebral uptake of drugs in humans. Clin Exp Pharmacol Physiol 34:695–701

Westergren I, Nyström B, Hamberger A, Johansson BB (1995) Intracerebral dialysis and the blood-brain barrier. J Neurochem 64:229–234

Chapter 16
Microdialysis in Internal Organs and Tumors

Robert Sauermann and Markus Zeitlinger

Abstract Microdialysis in internal organs and tumors is recognized as a valuable tool in drug development because it provides the concentrations of unbound compounds in interstitial tissue fluid over time. Microdialysis has been used in internal organs and tumors to measure local pharmacokinetic profiles of various drugs, but also to study metabolites and organ-specific pathophysiological pathways. Measuring pharmacokinetics and quantifying pharmacodynamic effects at the target site instead of plasma helps proving or dismissing therapeutic concepts during early stages of drug development. Finally, microdialysis has been evaluated as a tool for direct local delivery of drugs to organs and tumors. Because insertion of microdialysis probes into internal organs is relatively invasive, in humans essentially the need for clinically required operations offers the possibility of performing microdialysis studies in internal organs. This chapter provides an overview of relevant microdialysis applications in various internal organs and tumors of humans and animals, discussing their benefits and limitations.

16.1 General Considerations

A major strength of microdialysis is its applicability in virtually every tissue or bodily compartment of solid or liquid consistency. In contrast to other methods like tissue biopsy or positron emission tomography, it offers the possibility of continuously measuring unbound drug concentrations in the interstitial tissue fluid of living organisms. Therefore, scientists have frequently used the microdialysis

R. Sauermann (✉) · M. Zeitlinger
Department of Clinical Pharmacology, Medical University of Vienna, Vienna, Austria
e-mail: Robert.sauermann@meduniwien.ac.at

M. Müller (ed.), *Microdialysis in Drug Development*, AAPS Advances in the Pharmaceutical Sciences Series, DOI: 10.1007/978-1-4614-4815-0_16,

technique also in internal organs and tumors to investigate various scientific questions and for drug development.

However, the practical application of microdialysis in internal organs bears considerable ethical and safety concerns as well as other limitations. While microdialysis probes can be inserted easily into superficial compartments like skin or subcutaneous adipose tissue without notable risks to healthy subjects, this is not the case in deeper compartments. Thus, in general microdialysis in internal organs is not possible in healthy volunteers. Because insertion of microdialysis probes into internal organs is more invasive, in strictly experimental studies the technique has generally been limited to animals. In some rather rare cases, explanted organs like parts of the urinary bladder have been subject to microdialysis studies ex vivo. In humans, essentially the need for clinically required operations offers and justifies the possibility of performing microdialysis in internal organs. During surgery probe placement can be done quite safely without separate invasive intervention or major additional risks for the subjects. Then, however, the microdialysis can provide unique and extremely valuable data from the analyzed organs of humans, and indeed a range of studies has been performed in organs and tumors of humans.

Usually, at the end of surgery the microdialysis probe tip is inserted in the target organ. The shaft of the probes should be long enough to pass the outer bodily layers including the incised skin. After suturing the wound the tip of the probe should remain in the target tissue while the end of the probe shaft must be situated outside the body. This allows connecting the probe tubes to a perfusion pump and sampling of dialysates, as well as safe removal of the entire probe after the end of the experimental period. Some practical challenges of microdialysis studies in organs of animals or humans have to be specified. First, probe insertion in vulnerable organs situated in protected anatomic positions might cause adverse events like hemorrhages. In case of hematoma in the target area, the probe should be withdrawn because direct diffusion of substances from blood to the probe membrane could falsify the study results. Second, the size of the investigated tissue must be capacious enough to incorporate the entire membranous part of the microdialysis probe. Otherwise, the microdialysis measurements are performed not entirely in the tissue of interest, but partly in neighboring compartments yielding potentially misleading results. The risk of wrongly placed probe membranes is naturally higher in tissues with a small diameter or in small animals. If the tissue of interest or the entire subject is exposed to movements or contractions during the investigational period, proper probe placement and fixation are particularly critical to avoid dislocation. Due to such difficulties, it is often recommendable to make use of imaging techniques like ultrasound to ensure exact and correct placement of the probe tips in the target compartment. Ultimately, probe insertion must be performed under strictly sterile conditions to avoid severe infections like peritonitis.

Before starting a microdialysis study in organs or tumors the shaft length, length and cutoff of the semipermeable membranes and flow rates have to be chosen cautiously case-by-case depending on species, organ and agent of interest. For example, relatively short microdialysis probes with short membranous tips will be employed in small animals while in mobile or contracting anatomic areas the flexibility of the microdialysis probes will be of high importance.

When conducting a microdialysis study attention should also be drawn to the question whether the measurement is really obtained from the interstitial space fluid as generally assumed. Microdialysis measurements generally reflect the conditions in the interstitium when performed in compact and homogenous tissues. However, for organs housing several functional compartments apart of tissue cells, like epithelial lining fluid (ELF) in the lung or bile cannliculi in the liver, an effect of these additional compartments on the composition of dialysates has to be considered if they are in contact with the membrane of the probe. This may be of less importance when trends or existence of certain effects are studied, but may be more relevant if a project aims at exactly quantifying absolute interstitial concentrations.

Evidently, the focuses of scientific interest have been different in every organ according to its respective function. Nevertheless, in simple terms, four main areas of employing microdialysis in internal organs might be distinguished. First, strictly metabolic processes have been studied under different conditions, e.g., metabolites like lactate have been quantified to monitor organ ischemia. Second, researchers have aimed at elucidating organ-specific physiological or pathophysiological pathways. For instance, cytokine cascades and their interactions have been identified in internal organs in addition to metabolites. Third, a considerable number of pharmacokinetic studies have been conducted in organs with different drugs. Finally, microdialysis has been evaluated as a tool permitting local delivery of drugs to specific organs or tumors.

During the process of drug development all of the named areas of microdialysis may be relevant and can be coupled in one experiment. For example, during a basically pharmacokinetic study changing levels of metabolites may be monitored simultaneously over time, serving as indicator of drug efficacy. Thus, more sophisticated microdialysis study designs can contribute to revealing pharmacokinetic/pharmacodynamic relationships during the process of drug development. Without limitation to drug development in a strict sense, this chapter aims at reflecting the main focuses and specifics of microdialysis studies in different organs and tumors, predominantly by giving demonstrative examples. The chapter is arranged according to the organs, essentially descending from cranial to caudal. If data were available from both, animals and humans, in general the latter received more attention.

16.2 Heart

Due to its vital importance and affection by life-threatening diseases the myocardium is a very interesting target for microdialysis studies, but also the pericardium represents a zone of scientific interest. Methodically, the heart poses high challenges to performance of efficient and safe microdialysis studies due to the following characteristics: location in the osseous thorax, constant rapid contractions with pulsing tissue pressure, and relatively thin myocardial walls. Nevertheless, for about two decades several groups have undertaken efforts of performing cardiac microdialysis. In humans, the need and possibility of cardiac surgery for clinical reasons offers an opportunity of applying cardiac microdialysis. The remarkable number of cardiac microdialysis studies in humans shows that researchers have made use of this opportunity.

The main aim of cardiac microdialysis studies has been better understanding of metabolic processes during ischemia occurring in the course of myocardial infarction, but also during cardiac surgery. Microdialysis probes have been inserted into human myocardium particularly at the occasion of bypass surgeries (Poling et al. 2007a, 2007b). But also experiments in pigs allowed measuring interstitial myocardial synthesis of nitric oxide, as well as concentrations of glucose, lactate, glycerol, and pyruvate during artificial ischemia (Mantovani et al. 2006; Martin et al. 2007). Particularly, the lactate/pyruvate ratio measured by microdialysis started rising before any other standard monitoring tool showed abnormal values during ischemia (Bahlmann et al. 2004; Poling et al. 2007). Metabolic changes occur not only during ischemia, but also after establishing revascularization where certain metabolites increase (glucose) while others tend to decrease (lactate, purines) (Poling et al. 2007a, 2007b). In hibernating myocardium, the profile of metabolic parameters behaves in a different manner compared to healthy myocardium (Mazzoni et al. 2008). In rats, it could be illustrated that hypertrophied myocardium is less tolerant to ischemia and reperfusion due to chronically increased hydroxyl radical production (Kalenikova et al. 2004). Remarkably, myocardial metabolite measurements paralleled radiologic patterns obtained from magnetic resonance imaging in patients after myocardial infarction (Poling et al. 2008). These analogies between imaging and metabolic profiles corroborate that microdialysis is a highly sensitive tool for real-time detection of microcellular disorders in vivo (Poling et al. 2008).

The fact that myocardial metabolites show characteristic profiles before, during, and after revascularization suggests the clinical use of microdialysis beyond scientific investigations. Enabling early detection of even minor metabolic changes, microdialysis could serve as a monitoring tool in clinical routine, as already practiced in the brain at specialized neurosurgical centers. A clinical study in 30 patients confirmed that perioperative microdialysis measurements of parameters related to ischemia can be safely performed in a clinical setting, resulting in faster and more reliable detection of ongoing or new ischemia after coronary artery bypass grafting surgery (Bahlmann et al. 2004; Poling et al. 2007). Moreover, it was investigated in pigs how microdialysis samples obtained from the epicardial surface compare to samples from the myocardial wall during acute cardiac ischemia (Abrahamsson et al. 2011). As there were no differences in relative changes for glucose, lactate, and glycerol during baseline, ischemia, and reperfusion in epicardial and myocardial samples it was concluded that cardiac surface microdialysis is a promising and less invasive application of microdialysis for monitoring cardiac ischemia (Abrahamsson et al. 2011).

Notably, microdialysis has been applied for evaluating and comparing different surgical and anesthesiological methods (Poling et al. 2007). For example, the influence of using cold instead of warm cardioplegia was evaluated in a human cardiac microdialysis study (Poling et al. 2006), while the benefits of using either cold blood or cold crystalloid cardioplegia was compared in pigs (Runge et al. 2006). Monitoring human myocardial metabolism by means of microdialysis during coronary surgery revealed lower release of ischemic markers after off-pump surgery with cardioplegia arrest than on-pump (Mantovani et al. 2010). Not only effects of surgery-associated procedures on myocardial hydroxyl radical generation have been monitored during myocardial ischemia, but also those of different drugs, e.g., of fluvastatin in animals (Obata et al. 2009).

Besides, microdialysis was also used for investigating the role of certain proteins in myocardial pathophysiology. Matrix metalloproteinases (MMPs) are believed to contribute to myocardial remodeling. Interstitial MMP activity was continuously monitored in patients undergoing elective cardiac surgery using a fluorogenic MMP substrate, a microdialysis system placed within the myocardium and an in-line fluorescent detection system (Spinale et al. 2008). This allowed characterizing profiles of MMP release during cardiac arrest and cardiopulmonary bypass, demonstrating that dynamic MMP proteolytic activity exists within the human myocardial interstitium. A rabbit model of myocardial ischemia and reperfusion injury demonstrated that vagal nerve stimulation reduced active MMP-9 via increased expression of tissue inhibitor of MMP in cardiomyocytes (Uemura et al. 2007). Interestingly, in these rabbit experiments, the microdialysis probes were used simultaneously both, for measurements of MMP and for topical delivery of acetylcholine to the myocardium via the afferent part of the probe to stimulate the vagal nerve. Myocardial microdialysis experiments in pigs reported that ischemic periods occurring after previous myocardial infarctions contribute significantly to adverse remodeling processes via altered activity of membrane type-I MMPs (Dixon et al. 2010). Thus, subsequent episodes of ischemia and reperfusion activate proteolytic cascades within the infarct region enhancing regional dysfunction.

In contrast, short repetitive periods of ischemia prior to a longer period of ischemia, called ischemic preconditioning, may be beneficial for the myocardium (Waldenstrom et al. 2010). These benefits occur due to altered formation of adenosine triphosphate which was recently confirmed by a myocardial microdialysis study in anaesthetized pigs (Waldenstrom et al. 2010). But, ischemic preconditioning also affects amino acid transamination and glutamate metabolism exerting a beneficial influence on infarct size, as shown in a isolated rat heart model employing microdialysis (Lofgren et al. 2010).

Microdialysis also enables measuring concentrations of endocrine or paracrine compounds in myocardium over time. Concentrations of the bioactive peptide endothelin modulating myocardial contractility are significantly changed in myocardium during and after cardiopulmonary bypass surgery (Multani et al. 2005). Interestingly, ischemia increases the levels of norepinephrine, normetanephrine (Fujii et al. 2004), or bradykinin (Multani et al. 2001) in myocardial interstitium. Nature and relevance of such phenomena remain subject to ongoing research.

The challenging nature of investigating the beating heart has been a stimulus for developing innovative methods in association with microdialysis. Flexibly mounted microdialysis probes have been designed where the tip of the probe moves synchronously with the beating heart to reduce tissue injury (Obata 2001). The combined use of microdialysis probes and selective oxygen electrodes in pigs demonstrated the possibility of simultaneously measuring regional myocardial oxygen tension as well as metabolites during hypoxia (Siaghy et al. 2000). Lately, an ultra-performance liquid chromatography/electrospray-tandem mass spectrometry method has been developed for simultaneous high-throughput detection and quantification of intermediary energy metabolites in microdialysates for cardiac research (Birkler et al. 2010). With a total cycle time of 3 min, this method permits analysis of metabolites from the citric acid cycle, pyruvic acid, lactic acid, adenosine triphosphate, and adenosine along with metabolites of succinic acid or malic acid in much larger sample numbers than with conventional strategies (Birkler et al. 2010). Combining microdialysis sampling and microcoil nuclear magnetic resonance analysis was reported to be useful for tissue targeted myocardial metabonomics because it works with very low sample volumes and has improved mass sensitivity (Price et al. 2005). Microdialysis as a method of delivering neurotransmitters to the myocardium and the use of isolated animal heart models have been mentioned above (Stottrup et al. 2010; Uemura et al. 2007). Finally, microdialysis probes with a high-molecular-mass cutoff allow monitoring levels of high-molecular-weight compounds like myoglobin in the interstitium of the myocardium over time (see also respective chapter in this book) (Kitagawa et al. 2005).

In summary, cardiac research remains one of the most interesting and exciting fields of in vivo microdialysis.

16.3 Lung Tissue

Insertion of microdialysis probes through the chest wall is particularly critical because it may lead to pneumothorax with collapse of lung (Ingvast-Larsson et al. 1992), again limiting microdialysis experiments in human lung to patients scheduled for thoracotomy and thoracic surgery. Usually in humans, microdialysis probes have been implanted under direct vision at the end of lung surgeries (Herkner et al. 2002; Hutschala et al. 2008, 2005; Tomaselli et al. 2003, 2004; Zeitlinger et al. 2007).

As the lung consists of various anatomic structures, the question arises in which compartment microdialysis is measuring (Bergogne-Berezin 1995; Gail and Lenfant 1983). It must be assumed that the probe is not only in touch with interstitial space fluid but also with alveoli, alveolar air, and alveolar lining fluid tangent to the probe. However, exchange of compounds via the membrane cannot take place in air because rapid diffusion occurs only in fluids (Herkner et al. 2002). Therefore, the microdialysis measurements rather represent drug concentrations in the interstitial space of the lung and the alveolar lining fluid of acini. Thus, data obtained from lung microdialysis will considerably diverge from measurements in bronchoalveolar lavage fluid obtained during bronchoscopy which mainly represents surfaces of bigger airways (Braude et al. 1983).

So far, most lung microdialysis studies investigated the pharmacokinetics of pharmaceutics. The first lung microdialysis study was published in 1991 describing pharmacokinetics of theophylline in lung interstitium (Larsson 1991). Apart of theophylline and ambroxol (Larsson 1991; Ren et al. 2009), most other microdialysis studies focused on pharmacokinetics of antimicrobial agents in lung tissue, sometimes in comparison to ELF of humans or animals (Aoki et al. 2008; Dahyot et al. 2006; De La Pena et al. 2001; Eisenberg et al. 1993; Herkner et al. 2002; Hutschala et al. 2008, 2005; Liu et al. 2005, 2002; Marchand et al. 2005, 2008; Tasso et al. 2008; Tomaselli et al. 2003, 2004; Zeitlinger et al. 2007). The high interest in pharmacokinetics of antibiotics in the lung is based on the clinical importance of infections of the lower respiratory tract including pneumonias because impaired penetration of antimicrobial agents into infected lung may lead to failure of antibiotic therapy.

To give examples, lung microdialysis studies revealed that the unbound concentrations of the cephalosporins cefaclor and cefpodoxime in the interstitial space fluid of lung were significantly lower than in plasma (De La Pena et al. 2001; Liu et al. 2005, 2002). This information indicates that dosing of cephalosporins on the basis of plasma concentrations could be insufficient for lung infections due to reduced antimicrobial efficacy (De La Pena et al. 2001; Liu et al. 2005, 2002). In contrast, free imipenem concentrations were virtually identical in blood and lung tissue of both, non-infected as well as bacterially infected rats (Dahyot et al. 2006; Marchand et al. 2005). A microdialysis study comparing ulifloxacin levels in lung interstitium to those in ELF concluded that the accumulation of fluoroquinolones observed in ELF does not occur in the lung tissue (Tasso et al. 2008).

When antibiotic concentrations were measured in the same pneumonic lungs by two separate microdialysis probes within individual patients they varied substantially between both probes (Tomaselli et al. 2003, 2004). This suggested that pneumonic lung is not a homogenous compartment and that pneumonic tissue concentrations are influenced by several other factors (Tomaselli et al. 2003, 2004). Another study investigated the impact of atelectasis on the penetration of levofloxacin in lung tissue of patients undergoing cardiac surgery (Hutschala et al. 2008). As levofloxacin concentrations were substantially lower in atelectatic tissue, it was concluded that clinical dosing of antibiotics should be re-evaluated for diseases associated with formation of atelectases (Hutschala et al. 2008).

Up to now, microdialysis has barely been used to study pathophysiological mechanisms in the lung. Hence, considering the broad spectrum of pulmonary pathologies there is a clear potential that lung microdialysis will be used for such purposes in the future.

16.4 Trachea and Bronchi

Interestingly, microdialysis has also been employed in the trachea. Due to its important role in inflammatory and infectious diseases of the respiratory tract ELF represents a compartment of special interest within the lung. Up to now, microdialysis studies in the big airways and ELF have been performed only in animals.

As bacterial adherence to the bronchial mucosa is an early step of respiratory tract infections, high concentrations of antimicrobial drugs in the tracheal mucosa and ELF might inhibit bacterial adhesion and tissue invasion (Baldwin et al. 1992; Bergogne-Berezin 1995). Therefore, microdialysis probes have been inserted into the trachea of rats via tracheotomy (Eisenberg et al. 1993) or tracheal puncture (Aoki et al. 2008) to measure antibiotic concentrations.

The pathophysiological conditions of the airways are remarkably influenced by local levels of mediators. Therefore, tracheotomy was performed in anesthetized guinea-pigs and microdialysis probes were inserted into the tracheal opening to study local concentrations of eicosanoids (Callaghan et al. 1994). After intravenous injection of arachidonic acid, leukotriene D4, or saline vehicle different eicosanoids were determined in the trachea over 5 h. Significant post-treatment changes confirmed that microdialysis serves for monitoring local eicosanoid production in the trachea.

Mucociliary clearance represents an important defense mechanism, as illustrated by increased rates of pulmonary infections associated with genetic disorders characterized by impaired mucociliary function like primary ciliary dyskinesia or cystic fibrosis (Afzelius 1976; Robinson and Bye 2002). A microdialysis approach was developed to compare mucociliary transport in the airways of animals with and without cystic fibrosis in vivo (Grubb et al. 2004). In this experimental setting a dye is deposited on the airway surface, and a microdialysis probe is placed in a defined distance from the site of dye deposition. The dye is transported toward the probe by ciliary transport and, after reaching the probe, diffuses across the microdialysis membrane. Then, the rate of mucociliary transport can be calculated from the time between dye deposition and recovery. This approach was also used to study the influence of thyme extract, which is broncholytic und secretomotoric. Measuring transport rates of a fluorescence dye in mice showed that the thyme extract could double the rate of ciliary clearance in the trachea (Wienkotter et al. 2007). Similarly, the effects of nigellone and thymoquinone on mucociliary clearance were studied (Wienkotter et al. 2008). The rate of ciliary clearance was highly increased only by nigellone, suggesting its use for treatment of different respiratory diseases. These experiments demonstrate the power of microdialysis to investigate factors and drugs influencing mucociliary transport rates.

Microdialysis was also employed ex vivo in tissue-engineered trachea construction to study the biological effects of an artificial oxygen carrier on epithelial metabolism of the trachea (Tan et al. 2009). Perfusion with the artificial oxygen carrier decreased the tracheal lactate concentration as well as lactate/pyruvate and lactate/glucose ratios, improving epithelial metabolism.

Although the overall number of microdialysis studies in the big airways is small, the illustrated examples show the multifaceted possibilities of using microdialysis to study different questions within one organ.

16.5 Liver

As discussed for the lung, it is probable that the concentrations measured with microdialysis in the liver cannot be assigned exclusively to the interstitial space fluid. The composition of the dialysate might actually be influenced also by bile canaliculi or blood from the portal, venous, or arterial circulation.

With a few exceptions, microdialysis studies in the liver have been performed mainly to evaluate and optimize procedures employed during hepatic surgery and transplantation. Ischemia–reperfusion injury is a major concern in liver surgery (Grubb et al. 2004; Isaksson 2011; Winbladh et al. 2012). Due to its proven ability of monitoring the changing levels of intrahepatic glucose, lactate, glycerol, pyruvate, and of the lactate/pyruvate ratio microdialysis was proposed as routine method for perioperative monitoring of intrahepatic metabolism (Isaksson 2011). Liver resections are less likely to be affected by severe ischemia–reperfusion injury if ischemic preconditioning or intermittent inflow occlusion are applied prior to the surgery. Therefore, hepatic metabolism was studied by means of microdialysis in 32 humans over 5 days after liver resections due to malignant diseases to evaluate the effects of preoperative intermittent inflow occlusion with or without additional ischemic preconditioning (Winbladh et al. 2012).

Analysis of hepatic microdialysates has also been used, partly in combination with other methods, to investigate and improve the outcome of liver transplantation (Hrydziuszko et al. 2010; Nowak et al. 2002; Silva et al. 2008, 2005; Waelgaard et al. 2008). At first, microdialysis served to assess liver function during organ harvest and storage prior to transplantation (Silva et al. 2008, 2006). This procedure contributed to predicting reperfusion injury and outcome after implantation (Silva et al. 2008, 2006). Second, in a group of 20 patients monitored with microdialysis for 1 week after liver transplantation only the two patients with a rejection showed a selective increase in interleukin-8 and complement activation (C5a), starting several days before alanine transferase increased (Waelgaard et al. 2008). The third patient who developed a hepatic artery thrombosis and rejection showed a rapid rise in intrahepatic lactate combined with a complex inflammatory pattern (Waelgaard et al. 2008). Due to this remarkable sensitivity, it was suggested that hepatic microdialysis might be a promising method to control the liver after transplantation. The next years will show if a step toward implementation of microdialysis in surgical routine will be realized.

16.6 Spleen

In a very small number of studies in rodents, microdialysis was used to investigate hormonal reactions in the spleen of rodents. Changes in norepinephrine release in response to immobilization were measured using microdialysis (Shimizu et al. 1996). The results suggested that immobilization enhances the activity of the splenic sympathetic nerve, suppressing the cytotoxicity of splenic natural killer cells.

In another study, microdialysis was used to measure noradrenaline concentrations in the spleen by analyzing dialysates, while similarly studying the influence of different solutions delivered directly to the spleen via the afferent part of the microdialysis probe (Shimizu et al. 1994). The probe perfusion solutions which were locally delivered to the splenic tissue contained different concentrations of potassium, calcium or tetrodotoxin, significantly affecting the splenic secretion of noradrenaline by the sympathetic nerve (Shimizu et al. 1994).

In spite of these interesting study designs, splenic microdialysis has remained a rare and exceptional procedure. The dispensability of the spleen for human survival and the fear of causing abdominal bleeding by probe insertion in humans due to the high local blood perfusion are reasons for the limited number of splenic microdialysis studies.

16.7 Kidney

The spectrum of questions studied using renal microdialysis over the past years is quite broad. In dependence of the scientific question, probes are placed either in cortical or medullar kidney tissue. Like in other organs, investigation of surgery-associated problems has been an important field for microdialysis in the kidney, and the only feasible opportunity of performing renal microdialysis in humans.

Laparoscopic and robotic partial nephrectomy involves temporary clamping of the renal artery, making the kidney susceptible to ischemic damage (Wang et al. 2012). In experimental settings with pigs and in humans with a renal mass or carcinoma who underwent laparoscopic nephrectomy or nephroureterectomy microdialysis was shown to be a tool that provides useful real-time data on the metabolic status of the kidney during ischemia (Weld et al. 2009, 2008). Microdialysis was employed in kidneys of rats to determine whether interstitial isoprostane serves as marker of oxidative injury and renal damage (Wang et al. 2012). Because post clamp levels of isoprostane were higher when clamp times of renal vessels increased, the authors suggested that this compound can serve as a real-time marker of renal ischemia and reperfusion injury. Furthermore, in this study it was demonstrated that pretreatment with allopurinol can exert renoprotective effects.

A feared specific problem associated with kidney transplantation is vascular thrombosis severely affecting morbidity and graft survival (Fonouni 2011; Keller et al. 2009). Renal microdialysis experiments in pigs showed that decreasing glucose levels, increased lactate/pyruvate ratio, and increased glycerol levels are indicating early vascular thromboses after kidney transplantation (Fonouni 2011; Keller et al. 2009). Apart from the probes placed in renal cortex also less invasive extracapsular probes were helpful detecting metabolic changes during renal ischemia (Keller et al. 2008, 2009).

To study pathomechanisms involved in diabetes mellitus, renal microdialysis was used in normal and diabetic rats to monitor several compounds like angiotensin II, 12(S)-hydroxyeicosatetraenoic acid and thromboxane B2 over time and in response to different agents (Abdel-Rahman et al. 2008; Awad et al. 2005). Inducing diabetes mellitus in rats was associated with increased local renal production of angiotensin (Awad et al. 2004). In contrast, nitric oxide levels in the kidney decreased after onset of diabetes which might favor cardiovascular events associated with this disease (Awad et al. 2004). Pointing to therapeutic approaches, the decrease in renal nitric oxide could be reversed by blocking angiotensin I-receptors.

Kidney microdialysis in rats also enhanced our understanding of renal mechanisms involved in the pathogenesis of arterial hypertension (Usa et al. 2007). This first characterization of glucocorticoids in renal interstitium suggested that corticosterone excess in the renal medulla is capable of causing hypertension. Measurements of adenosine triphosphate in renal interstitial fluid of dogs supported the hypothesis that renal adenosine triphosphate affects renal vascular resistance, autoregulation and tubuloglomerular feedback mechanisms (Nishiyama et al. 2006). In a very specific study setting, the effects of renal medullary interstitial infusion of cationic amino acids on hemodynamics in the medulla were examined in anesthetized rats (Kakoki et al. 2004). During intramedullary infusion of L-ornithine, L-lysine, and L-homoarginine nitric oxide concentrations in the inner medulla were significantly decreased while renal medullary blood flow was significantly reduced (Kakoki et al. 2004). In contrast, intramedullary infusion of L-arginine increased nitric oxide and medullary blood flow suggesting that transport of L-arginine is important in the maintenance of blood flow in the renal medulla.

Likewise, in order to investigate the pathophysiology of renal insufficiency induced by different causes microdialysis was experimentally applied to renal cortex and medulla of rats (Fujii et al. 2007; Myers et al. 2006; Shibata et al. 2004). Measuring levels of nitric oxide, prostaglandin E2, thromboxane B2, and norepinephrine during renal failure and in response to different drugs enabled evaluating their role in maintenance of renal function and of cortical or medullary blood flow. Because acute ureteral obstruction may affect kidney function and metabolism an experiment was performed in pigs (Krarup et al. 2007). Interestingly, ureteral obstruction decreased interstitial glucose only in the upper and lower thirds of the kidney, while interstitial urea increased in all regions of the kidney and lactate did not change at all.

Last but not least, extracellular concentrations of the cytostatic drug methotrexate were monitored at steady state in the kidney and other tissues in a rodent model (Ekstrom et al. 1995). While drug levels in muscle and liver linearly correlated with doses, the kidney levels reached a plateau at a dose of 150 mg/kg per 6 h (Ekstrom et al. 1995). This experiment confirmed that, by measuring pharmacokinetics in both targeted and toxicity-prone tissues, microdialysis may be helpful for optimizing dosing of chemotherapy.

16.8 Gut

Early diagnose of intestinal ischemia, of leakage of anastomoses or of increased mucosal permeability is very relevant after surgery of the gastrointestinal tract from the esophagus to the rectum. For example, free intestinal flaps for the reconstruction of the cervical esophagus after carcinoma resection are very difficult to monitor postoperatively. Therefore, most microdialysis studies of the gut had one major objective, i.e., evaluating if this method is capable of detecting intestinal and splanchnic ischemia.

The possibility of detecting intestinal ischemia by means of microdialysis was tested both, in animals and in patients. In this context, the straightest method was placement of the microdialysis probes directly in the intestinal wall (Cibicek et al. 2007; Deeba et al. 2008; Emmertsen et al. 2005; Korth et al. 2003; Oldner et al. 1996; Sommer 2005; Sommer and Larsen 2003; Sorensen 2008). By observing glucose concentrations in microdialysates over time, it was possible to detect cases of critical ischemia of free jejunal flaps in humans (Sorensen 2008). Another reliable parameter seemed to be the lactate/glucose ratio (Sorensen 2008). Hence, microdialysis in the intestinal wall has been considered to be a promising method for monitoring intestinal perfusion and detecting neglected ischemia at much earlier time points than with samples obtained from blood or other compartments (Oldner et al. 1996). Moreover, it was assumed that metabolic monitoring by microdialysis could help avoiding false alarms with very high sensitivity and specificity, and allow to differentiate exactly between ischemic and non-ischemic intestinal segments (Birke-Sorensen and Andersen 2010).

However, there was also a critical report of severe inflammatory reactions in the ileal wall around all microdialysis probes used in experiments (Emmertsen et al. 2005). The authors suggested that microdialysis of the intestinal wall should be confined to experimental research, and that alternative probe locations should be considered for clinical routine (Emmertsen et al. 2005).

One alternative microdialysis probe location with lower risk of inflammation of the intestinal wall might be the intestinal lumen (Solligard et al. 2008, 2007; Tenhunen et al. 1999). The adequacy of splanchnic perfusion of the large bowel was monitored during cardiac surgery using microdialysis probes located in the lumen of the recto-sigmoid junction (Solligard et al. 2007). Interestingly, in this study metabolites of ischemia were detected by the endoluminal microdialysis probes with higher sensitivity than by arterial blood sampling or intramuscular microdialysis (Tenhunen et al. 1999). It was concluded that luminal gut microdialysis can be used in clinical routine to detect not only intestinal, but also cardiac ischemia, as well as pathologic changes in intestinal permeability (Solligard et al. 2008; Tenhunen et al. 1999).

The third probe location evaluated for early detection of visceral ischemia has been the peritoneum (Ellebaek Pedersen et al. 2009; Jansson et al. 2004, 2003; Ungerstedt et al. 2003). To give an example, 3 of 50 patients after surgical resection of rectosigmoid cancer developed late anastomotic leakage which was associated with a significant increase in the lactate/pyruvate ratio in intraperitoneal microdialysates occurring several days prior to development of clinical symptoms (Ellebaek Pedersen et al. 2009). Also in preterm infants after surgery for necrotizing enterocolitis peritoneal microdialysis seemed to be a safe and applicable method for surveillance of inflammatory changes in the peritoneal cavity (Pedersen et al. 2011). Apparently, apart from the lactate/pyruvate ratio, several cytokines such as interleukin-6, interleukin-10, and tumor necrosis factor-alpha were increased in intraperitoneal microdialysates of patients who developed anastomotic leakage before clinical symptoms were evident (Matthiessen et al. 2007). The most appropriate location for intraperitoneal probes after right-sided hemicolectomy seemed to be the juxta-anastomotic region and the small intestinal loop area (Jansson et al. 2004).

Eventually, there was a project using microdialysis in the lumen of the gastrointestinal tract for a completely different issue, namely to measure nitrate and other substances within Barrett's esophagus during acid reflux (Suzuki et al. 2005). Evidently, the interaction between acidic gastric refluxate and nitrite rich saliva activates potentially mutagenic luminal nitrosative chemistry within Barrett's esophagus.

16.9 Pancreas

Studies applying microdialysis in the pancreas, though relatively low in number and limited to animals, have been quite multifaceted. For a pharmacokinetic study, echoendoscopes were inserted into the stomach of Beagle dogs to puncture the pancreatic parenchyma with 19-gauge needles and position specially developed microdialysis probes in the pancreatic tissue (Kitano et al. 2010). In principle, this method seemed eligible to measure the pancreatic concentrations of 5-fluorouracil because neither local bleeding nor fluid accumulation occurred. However, in two of eight dogs the probes broke while being inserted into pancreatic parenchyma. This experience illustrates a possible risk of microdialysis in internal organs, suggesting careful choice and testing of used devices before in vivo application.

An interesting study using microdialysis in drug development was performed in pancreas of pigs. Microdialysis revealed that recombinant trypsin inhibitor injected into the pancreatic duct of pigs has an intrapancreatic half-life of approximately 45 min (Jonsson et al. 1992). Apart from these pharmacokinetic aspects, it was observed that the levels of exocrine trypsinogen were higher inside the pancreas than at its surface under the peritoneum. In a further step, the same research group evaluated three different routes of administering recombinant trypsin inhibitors in pigs with experimental pancreatitis (Jonsson and Ohlsson 1995). Trypsin inhibitor was administered either directly into the pancreatic duct, into the abdominal cavity adjacent to the pancreas or intravenously. The intraglandular concentrations of the trypsin inhibitor were highest after direct injection into the pancreatic duct and substantially lower after intravenous and intraperitoneal administration. Remarkably, the intrapancreatic half-life of the inhibitor after intraductal administration was three to six times longer in pigs with pancreatitis than in healthy pancreas. These findings supported further evaluation of intraductal antiprotease treatment in acute pancreatitis, and, more generally, underlined the importance of studying pharmacokinetics in ill instead of healthy subjects.

In combination with other methods, microdialysis was employed to compare pancreatitis progressing either to mild edematous or to severe necrotic disease in piglets. Only necrotic, but not edematous pancreatitis was characterized by increased vascular permeability to albumin and proteins in the early phase of inflammation (Merilainen et al. 2008).

Also endocrine aspects of the pancreas were studied by means of microdialysis. For example, it was shown that direct oxytocin administration into rat pancreas resulted in significant elevations of insulin and glucagon concentrations (Stock et al. 1990). In a sophisticated approach, microdialysis was used to assess the internal milieu of pancreatic rat islet grafts 1 month after transplantation beneath the renal capsule of nondiabetic or diabetic recipients (Carlsson et al. 2002). Diabetic recipients had a significantly lower local insulin concentration than normoglycemic animals. High lactate/pyruvate ratios suggested that islet grafts have an increased anaerobic glucose metabolism. Finally, microdialysis was used also in isolated dog pancreata ex vivo to study insulin and somatostatin pathways (Nakagawa et al. 1993).

16.10 Gonads

The female gonads have been a focus of extensive investigations employing microdialysis in big animals like cows, monkeys, pigs, and sheep. Numerous studies elucidated ovarian physiology, follicular development and atresia, maintenance or lysis of the corpus luteum, luteal angiogenesis, angiolysis, and local luteal blood flow as well as factors influencing implantation or early pregnancy. For this purpose, microdialysis systems were either surgically implanted in vivo or in vitro into ovary, preovulatory follicles, and different parts of the corpus luteum (Acosta et al. 2000, 2005; Beindorff and Einspanier 2010; Beindorff et al. 2009; Hayashi et al. 2003, 2003, 2003; Miyamoto et al. 2005; Ohtani et al. 2004;

Okuda et al. 2001; Shaw and Britt 2000; Shirasuna et al. 2004a, b, c, 2007; Tanaka et al. 2004; Wuttke et al. 1998). Among the compounds measured at a local level there were progesterone, estradiol, oxytocin, follicle stimulating hormone, human chorionic gonadotropin, relaxin, various prostaglandins, endothelin-1, angiotensin, atrial natriuretic peptide, cortisol, angiopoietins, tumor necrosis factor-alpha, nitric oxide, reactive oxygen species. In this context, intraluteal microdialysis probes were also used to administer different agents directly to the corpus luteum (Jaroszewski and Hansel 2000; Jaroszewski et al. 2003).

In contrast, rather for scientific inactivity than for mercy with laboratory animals, the male gonads have attracted little microdialysis research up to now. There was one project studying metabolites during testicular ischemia in rats (Cheng et al. 2001). Microdialysis probes inserted in both testes documented dynamic and differing profiles of glucose, lactate and pyruvate levels before, during and after ischemia and reperfusion in ipsilateral and contralateral testes.

16.11 Uterus

First, a uterine microdialysis study in rats was dedicated to studying disorders in contractile function of the uterus and the response to nitric oxide (Okawa et al. 2005). On the other hand, endocrine or paracrine signaling has also been studied in the uterus (Licht et al. 2001). Human chorionic gonadotropin is an early embryonic product which might mediate embryo-maternal communication. Local effects of human chorionic gonadotropin on endometrial paracrinology were studied in human females. An intrauterine microdialysis system was employed, allowing both, local administration of exogenous human chorionic gonadotropin, as well as continuous monitoring of tissue response in the uterine cavity. So, it was possible to document that local administration of human chorionic gonadotropin significantly modulated several endometrial factors related to endometrial differentiation, angiogenesis, implantation, and tissue remodeling. Similarly, in 54 women with infertility problems it was shown that intrauterine concentrations of vascular endothelial growth factor are regulated in a cycle-dependent way (Licht et al. 2003).

16.12 Bladder

Microdialysis studies have rarely been conducted in the bladder in vivo. In 2004, a study measured acetylcholine release in the bladder wall and detrusor pressure in rats in vivo after electrical stimulation of the pelvic nerve, investigating mechanisms of voiding dysfunction in diabetes mellitus (Miyamae et al. 2004). Most other bladder microdialysis studies were performed ex vivo in isolated muscle strips obtained from bladder of animals or humans (Iwashita et al. 2004;

Masunaga et al. 2006, 2008; Murakami et al. 2003, 2008; Yono et al. 2005; Yoshida et al. 2006, 2007, 2008a, b). These ex vivo studies studied release of non-neuronal acetylcholine, nitric oxide, or prostaglandin E2 in muscle strips to elucidate functional bladder disorders. In doing so, they compared different underlying conditions, or tested the response to various stimuli or potential therapeutics (Iwashita et al. 2004; Masunaga et al. 2006, 2008; Murakami et al. 2003, 2008; Yono et al. 2005; Yoshida et al. 2006, 2007, 2008a, b).

16.13 Prostate

There are only few examples of microdialysis studies in the prostate. As calcitriol inhibits growth of prostate cancer intraprostatic pharmacokinetics of calcitriol was studied in male rats (Konety et al. 2002). In spite of intravenous calcitriol administration, only a fraction of less than 1 % of the serum concentration was detectable in prostatic tissue within 24 h.

The activity of neurotransmitters and the effects of nitric oxide on prostate function were studied ex vivo in muscle strips isolated from prostates of humans and rabbits employing microdialysis (Yoshida et al. 2003).

16.14 Tumors

Oncologic diseases continue being a leading cause of mortality and a scientific area of major interest. Conventional trials in oncology measure total concentrations of anticancer drugs and their metabolites in plasma correlating them with drug efficacy or toxicity (Kitzen et al. 2006). However, successful clinical outcome of oncologic chemotherapy is linked to adequate drug exposure of tumor cells, and total drug concentrations in the extracellular space of tumors often diverge from those in plasma (Kitzen et al. 2006; Wei et al. 2009). Therefore, plasma profiles of anticancer drugs may be insufficient for estimating clinical outcome of chemotherapy (Brunner and Muller 2002).

Drug concentrations in solid tumors may be substantially reduced due to several diffusion barriers (Brunner and Muller 2002; Wei et al. 2009). The fact that only the unbound fraction of antitumor drugs can passively diffuse or be transported across vascular endothelia to penetrate tumor cells, may, in part, be responsible for inherent mechanisms of resistance and the variability in clinical outcomes (Kitzen et al. 2006). As microdialysis can determine the unbound fraction of drugs in the interstitial fluid of tumors, this technique is utile to assess local pharmacokinetics as well as pharmacodynamics in oncology (Kitzen et al. 2006). Microdialysis is a powerful tool enabling pharmacokinetic/pharmacodynamic modeling and better understanding of exposure–response relationships in oncology. Therefore, use of microdialysis may optimize investigation and dose selection of new or established anticancer drugs (Brunner and Muller 2002; Wei et al. 2009). The following section illustrates interesting aspects and findings of microdialysis studies in tumors.

16.14.1 Pharmacokinetics of Anticancer Drugs in Solid Tumors

First, there are basically pharmacokinetic studies of clinical relevance. For example, in humans mean free concentrations of carboplatin in cutaneous melanoma metastases were reported to reach only about 50–60 % of total serum levels; with mean concentrations in healthy subcutaneous tissue being similar to those in tumor (Blochl-Daum et al. 1996). Likewise, in carboplatin-treated patients concentration–time curves of unbound drug were similar in tumor and normal adipose tissues, essentially following plasma patterns (Konings et al. 2009). In contrast, mean epirubicin exposure in advanced-stage primary breast cancer and in subcutaneous tissue reached only 1 and 11 %, respectively, of plasma values (Hunz et al. 2007).

It is important to perform pharmacokinetic studies with anticancer drugs over a sufficiently long time period or at steady sate. Microdialysis was successfully employed in ambulant tumor patients over several days, which enabled studying tissue pharmacokinetics of anticancer drugs in more detail (Konings et al. 2009). During a 5-day continuous infusion in patients, concentrations of unbound 5-fluorouracil were lower in extracellular fluid of subcutaneous tumor tissue compared to plasma. However, drug exposure of tumor tissue increased during the 5-day infusion period and a circadian rhythmicity of 5-fluorouracil pharmacokinetics was observed in the tumor (Konings et al. 2011). Interestingly, drug concentrations in tumor were higher during the night than during daytime (Konings et al. 2011).

Notably, even within the same type of tumors pharmacokinetic profiles may diverge interindividually to a large extent. In spite of similar plasma concentrations tested neuroblastoma xenografts showed a variable degree of topotecan exposure (Zamboni et al. 1999). Individually different tumor vascularity, capillary permeability, and interstitial pressure have been discussed as factors affecting tumor penetration of antineoplastic drugs (Zamboni et al. 1999). Thus, different drug concentrations must be anticipated in individual tumors in spite of similar systemic exposure.

16.14.2 Pharmacokinetics of Anticancer Drugs in Tumors of the Central Nervous System

Due to the blood brain barrier, the delivery of antitumor drugs to the central nervous system is critical. Therefore, knowing drug disposition early in drug development is particularly important in the brain (Blakeley and Portnow 2010).

Cerebral microdialysis has been used to study penetration of anticancer drugs in the central nervous system and in brain tumors in animal models (Elmeliegy et al. 2011), and innovative approaches have been developed in this context. Murine models were engineered genetically to study diverse aspects of brain tumors. Because exact probe implantation is difficult in small animals, magnetic resonance imaging may be useful to implant microdialysis cannulas at predefined reference points (Elmeliegy et al. 2011). Likewise intra-operative 3D ultrasonography has been used to guide and control probe placement in patients undergoing resection of astrocytomas (Homapour et al. 2010).

After intravenous administration the overall brain exposure to gemcitabine was relatively low in rats (Apparaju et al. 2008). However, drug levels were considerably higher in brain tumors relative to tumor-free regions of the brain suggesting selectively higher cytotoxicity against tumor cells (Apparaju et al. 2008). When methotrexate distribution was measured in two patients with high-grade gliomas the permeability of the blood brain barrier was determined at different microdialysis probe locations, by digitally fusing computed tomography and contrast-enhanced magnetic resonance images (Blakeley et al. 2009). Cerebral drug penetration was significantly greater in contrast-enhancing tumor than in non-enhancing brain tissue (Blakeley et al. 2009). Thus, relating the probe location to areas of blood–brain barrier disruption is useful to increase the significance of microdialysis data from the brain (Blakeley et al. 2009). Concluding, microdialysis is an informative technique for characterizing intratumoral pharmacokinetics of anticancer drugs that do not freely penetrate the blood–brain barrier, providing a rational basis for selecting agents for phase II trials (Blakeley et al. 2009).

16.14.3 Correlating Pharmacokinetics of Anticancer Drugs with Success or Toxicity of Therapy

Beyond merely pharmacokinetic studies microdialysis has been used in oncology also to relate drug exposure of tumors with clinical and biochemical response or toxicity. For example, in 19 tumor patients (melanomas, malignant fibrous histiocytoma, Merkel cell tumor and osteosarcoma) undergoing regional chemotherapy with melphalan by Isolated Limb Infusion, there was a significant association between complete tumor response and drug concentrations in subcutaneous microdialysates (Thompson et al. 2001).

The relationship between pharmacokinetics and toxicity of antitumor drugs is also an important issue in oncology. Microdialysis was employed in rats to describe pharmacokinetic profiles of carboplatin simultaneously in blood, several tissues, and breast tumor (Johansen et al. 2002). Significant disproportionate increases in hepatic carboplatin exposure with increasing intravenous doses suggested a possible explanation for carboplatin-induced hepatic toxicity at higher doses (Johansen et al. 2002).

16.14.4 Coadministration of Drugs Influencing Efficacy or Safety of Antitumor Chemotherapy

Sophisticated study designs with simultaneous measurement of two or more compounds may provide additional insights in interactions, efficacy and safety of coadministered drugs in oncology. Sodium thiosulfate is often coadministered to neutralize cisplatin, reducing its systemic toxicity. Using microdialysis the concentrations of cisplatin and sodium thiosulfate were measured in tumors and plasma of patients with oral cancer in order to compare two administration modes of cisplatin, namely intra-arterial perfusion versus an alternative tumor embolization method (Tegeder et al. 2003). The pharmacokinetic results indicated that cisplatin can be neutralized adequately with both treatment modalities whereas optimal tumor cell killing can be expected only with cisplatin embolization (Tegeder et al. 2003).

Methotrexate tumor delivery was studied in a mouse model to characterize the pharmacokinetic impact of prior administration of the transporter inhibitors probenecid and cyclosporine, alone and in combination (Sani et al. 2010). While cyclosporine significantly enhanced tumor exposure to methotrexate, probenecid did not (Sani et al. 2010). In rats, the administration of angiogenesis inhibitors did not only inhibit tumor growth, but also reduced the concentrations of coadministered anticancer agents in tumor tissue, an important effect to be considered in the design of novel anticancer therapies (Ma et al. 2001).

16.14.5 Tumor Physiology and New Treatment Concepts

The extracellular space of tumors is the bioactive site for the majority of growth factors (Dabrosin 2005). Knowledge of protein activation in this compartment is important for understanding tumor biology. Microdialysis is a recognized method for detection of growth factors in cancer (Dabrosin 2005), and some relevant studies in this area are described herein.

Interleukin-8 has been associated with tumor angiogenesis, metastasis and poor prognosis in breast cancer, and estrogen is crucial in carcinogenesis and tumor progression (Bendrik and Dabrosin 2009). Microdialysis was used in normal human breast tissue of pre- and postmenopausal women, preoperatively in breast cancer, and in experimental breast cancer of mice to study if interleukin-8 secretion is affected by estrogen (Bendrik and Dabrosin 2009). A significant correlation was found between levels of interleukin-8 and estradiol in normal breast and in hormone-dependent breast cancer. Anti-interleukin-8 antibodies inhibited endothelial cell proliferation and decreased tumor angiogenesis (Bendrik and Dabrosin 2009). These results strongly suggested that estradiol has a critical role in the regulation of interleukin-8 and that interleukin-8 may present a novel therapeutic target (Bendrik and Dabrosin 2009). As the mammary tumor microenvironment also contributes to tumor growth and invasion, microdialysis and proteomics have been combined indentifying over 400 proteins in transgenic mouse mammary tumors and mammary controls (Xu et al. 2010). Osteopontin was overexpressed in breast tumor which might stimulate carcinoma cell proliferation and mammary tumor progression (Xu et al. 2010).

Characterization of soluble factors released in the tumor bed of a murine Lewis lung carcinoma model over several days showed that administration of paclitaxel altered the local cytokine network (Zhong et al. 2007). Changing the tumor microenvironment by pretreatment with low-dose paclitaxel improved the antitumor potential of an intratumoral dendritic cell vaccine (Zhong et al. 2007).

Following surgery of high-grade gliomas in humans, microdialysis was used to measure growth factors, cytokines and other proteins (Marcus et al. 2010). Microdialysates from the glioma resection margin demonstrated significantly higher concentrations of interleukin-8 and a higher ratio of matrix metalloproteinase-2 to tissue inhibitor of metalloproteinase 1 than peritumor microdialysates. These characteristics of gliomas suggested an environment favoring tumor invasion and angiogenesis (Marcus et al. 2010). Stereotactic microdialysis was employed to sample extracellular fluid from glioblastomas and from brain adjacent to tumor, before and during conventional radiotherapy (Wibom et al. 2010). Metabolomic patterns of more than 150 metabolites were analyzed. In response to radiotherapy, only some metabolic changes were equal in microdialysates from tumor and normal brain, whereas the overall patterns contained highly discriminating information between tumor and normal brain (Wibom et al. 2010). This points to the possibility of detecting local metabolic marker patterns associated to early treatment response in oncology.

Ascorbic acid is an essential nutrient commonly regarded as an antioxidant. However, real-time microdialysis in mice bearing glioblastoma xenografts showed that parenteral administration of ascorbate induced formation of radicals and hydrogen peroxide selectively in interstitial fluid of tumors but not in blood (Chen et al. 2008). Ascorbate treatment significantly decreased growth rates of glioblastoma, ovarian, and pancreatic tumors (Chen et al. 2008). Thus, ascorbate generated hydrogen-peroxide-dependent cytotoxicity toward cancer cells without adversely affecting normal cells and may be of benefit in cancer treatment (Chen et al. 2008).

16.14.6 Microdialysis as a Tool for Local Delivery of Anticancer Drugs to Tumors

Systemic toxicity of anticancer chemotherapy is a major clinical problem limiting its applicability and efficacy (Martins and de Oliveira 2009). Local drug administration techniques have been sought to increase drug concentrations in tumor tissue while reducing systemic exposure and side-effects (Martins and de Oliveira 2009;

Martins et al. 2012). It was postulated that microdialysis probes inserted in tumor tissue may be a tool for local drug delivery which could revolutionize tumor management (Martins and de Oliveira 2009). Exact mathematical models have been developed to optimize drug delivery by microdialysis probes procedure (Martins et al. 2012). When microdialysis probes are used to administer drugs, they could permit simultaneous monitoring of chemical changes induced by the therapy (Goodman 2011). However, several factors like tumor heterogeneity and size, drug resistance, or metastases must be taken into consideration to avoid set-backs of local cancer chemotherapy. Up to now, local administration of anticancer drugs by means of microdialysis has not yet found its way into clinical routine.

16.14.7 Pitfalls

Like in other applications, there are also pitfalls of microdialysis in oncology. For example, preliminary experiments suggested that docetaxel had a very low recovery (Loos et al. 2007). Different materials had to be tested to eliminate non-specific binding of docetaxel to microdialysis catheters and its effects on the recovery (Loos et al. 2007). This illustrates that, as for other applications, careful preparation and quality controls are also necessary when employing microdialysis in oncology.

16.15 Outlook and Visions

From the current viewpoint, the authors believe that the overall number of microdialysis studies in internal organs and tumors will not grow exponentially, but remain at a quite constant level. Many basically metabolic questions might be answered in a satisfactory manner within the next years. Evidently, organ microdialysis of metabolites has a big potential for evaluating and comparing new surgical and anesthesiological procedures. Regarding the measurement of metabolites, the main question is if microdialysis will succeed in making the big step from being a predominantly scientific device to becoming a tool in clinical routine, e.g., for perioperative detection of organ ischemia. The established implementation of brain microdialysis as routine procedure at highly specialized neurosurgical centers illustrates the clinical capacity and benefit of microdialysis also for other organs. However, further studies and experience would be necessary to prove the benefit of organ microdialysis in clinical routine.

In drug development, microdialysis of organs and tumors is recognized undoubtedly as a valuable tool. Involvement of this technique for measuring pharmacokinetics at the target site instead of plasma and for quantifying pharmacodynamic effects allows proving, optimizing, or dismissing therapeutic concepts during early stages of drug development. This, of course, has a remarkable potential for saving time, labor, and costs in the process of drug development. Apparently, the awareness of the benefits of microdialysis is steadily rising among scientists, pharmaceutical companies, and regulatory authorities. The Federal Drug Administration (FDA) and the European Medicines Agency (EMA) have already focused on tissue concentrations of anti-infective agents rather than plasma concentrations to predict outcome of antimicrobial therapy (EMEA 2000; FDA 1998). It will also be interesting to observe to what extent microdialysis will be employed for developing innovative therapeutics like biologics, e.g., by measuring their influence on cytokines in target organs or tumors.

Whether microdialysis can be established as a routine tool for local drug delivery in the future will depend on the success of proving clinical benefits. Particularly in oncology, local administration of cytostatic drugs to tumors by microdialysis has the potential of avoiding systemic toxic effects.

In conclusion, clinical microdialysis in organs and tumors will remain an exciting and dynamic field with a high potential for improving drug development and innovative research.

References

Abdel-Rahman EM, Abadir PM, Siragy HM (2008) Regulation of renal 12(S)-hydroxyeicosatetraenoic acid in diabetes by angiotensin AT1 and AT2 receptors. Am J Physiol Regul Integr Comp Physiol 295:R1473–R1478

Abrahamsson P, Aberg AM, Johansson G, Winso O, Waldenstrom A, Haney M (2011) Detection of myocardial ischaemia using surface microdialysis on the beating heart. Clin Physiol Funct Imaging 31:175–181

Acosta TJ, Ozawa T, Kobayashi S, Hayashi K, Ohtani M, Kraetzl WD, Sato K, Schams D, Miyamoto A (2000) Periovulatory changes in the local release of vasoactive peptides, prostaglandin f(2alpha), and steroid hormones from bovine mature follicles in vivo. Biol Reprod 63:1253–1261

Acosta TJ, Tetsuka M, Matsui M, Shimizu T, Berisha B, Schams D, Miyamoto A (2005) In vivo evidence that local cortisol production increases in the preovulatory follicle of the cow. J Reprod Dev 51:483–489

Afzelius BA (1976) A human syndrome caused by immotile cilia. Science 193:317–319

Aoki M, Iguchi M, Hayashi H, Suzuki H, Shibasaki S, Kurosawa T, Hayashi M (2008) Proposal of membrane transport mechanism of protein-unbound ulifloxacin into epithelial lining fluid determined by improved microdialysis. Biol Pharm Bull 31:1773–1777

Apparaju SK, Gudelsky GA, Desai PB (2008) Pharmacokinetics of gemcitabine in tumor and non-tumor extracellular fluid of brain: an in vivo assessment in rats employing intracerebral microdialysis. Cancer Chemother Pharmacol 61:223–229

Awad AS, Webb RL, Carey RM, Siragy HM (2005) Increased renal production of angiotensin II and thromboxane B2 in conscious diabetic rats. Am J Hypertens 18:544–548

Awad AS, Webb RL, Carey RM, Siragy HM (2004) Renal nitric oxide production is decreased in diabetic rats and improved by AT1 receptor blockade. J Hypertens 22:1571–1577

Bahlmann L, Misfeld M, Klaus S, Leptien A, Heringlake M, Schmucker P, Sievers HH, Ungerstedt U, Kraatz EG (2004) Myocardial redox state during coronary artery bypass grafting assessed with microdialysis. Intensive Care Med 30:889–894

Baldwin DR, Honeybourne D, Wise R (1992) Pulmonary disposition of antimicrobial agents: methodological considerations. Antimicrob Agents Chemother 36:1171–1175

Beindorff N, Einspanier A (2010) Luteotrophic effects of relaxin, chorionic gonadotrophin and FSH in common marmoset monkeys (Callithrix jacchus). Reproduction 139:923–930

Beindorff N, Honnens A, Penno Y, Paul V, Bollwein H (2009) Effects of human chorionic gonadotropin on luteal blood flow and progesterone secretion in cows and in vitro-microdialyzed corpora lutea. Theriogenology 72:528–534

Bendrik C, Dabrosin C (2009) Estradiol increases IL-8 secretion of normal human breast tissue and breast cancer in vivo. J Immunol 182:371–378

Bergogne-Berezin E (1995) New concepts in the pulmonary disposition of antibiotics. Pulm Pharmacol 8:65–81

Birke-Sorensen H, Andersen NT (2010) Metabolic markers obtained by microdialysis can detect secondary intestinal ischemia: an experimental study of ischemia in porcine intestinal segments. World J Surg 34:923–932

Birkler RI, Stottrup NB, Hermannson S, Nielsen TT, Gregersen N, Botker HE, Andreasen MF, Johannsen M (2010) A UPLC-MS/MS application for profiling of intermediary energy metabolites in microdialysis samples–a method for high-throughput. J Pharm Biomed Anal 53:983–990

Blakeley J, Portnow J (2010) Microdialysis for assessing intratumoral drug disposition in brain cancers: a tool for rational drug development. Expert Opin Drug Metab Toxicol 6:1477–1491

Blakeley JO, Olson J, Grossman SA, He X, Weingart J, Supko JG (2009) Effect of blood brain barrier permeability in recurrent high grade gliomas on the intratumoral pharmacokinetics of methotrexate: a microdialysis study. J Neurooncol 91:51–58

Blochl-Daum B, Muller M, Meisinger V, Eichler HG, Fassolt A, Pehamberger H (1996) Measurement of extracellular fluid carboplatin kinetics in melanoma metastases with microdialysis. Br J Cancer 73:920–924

Braude AC, Hornstein A, Klein M, Vas S, Rebuck AS (1983) Pulmonary disposition of tobramycin. Am Rev Respir Dis 127:563–565

Brunner M, Muller M (2002) Microdialysis:an in vivo approach for measuring drug delivery in oncology. Eur J Clin Pharmacol 58:227–234

Callaghan DH, Yergey JA, Rousseau P, Masson P (1994) Respiratory tract eicosanoid measurement using microdialysis sampling and GC/MS detection. Pulm Pharmacol 7:35–41

Carlsson PO, Kiuru A, Nordin A, Olsson R, Lin JM, Bergsten P, Hillered L, Andersson A, Jansson L (2002) Microdialysis measurements demonstrate a shift to nonoxidative glucose metabolism in rat pancreatic islets transplanted beneath the renal capsule. Surgery 132:487–494

Chen Q, Espey MG, Sun AY, Pooput C, Kirk KL, Krishna MC, Khosh DB, Drisko J, Levine M (2008) Pharmacologic doses of ascorbate act as a prooxidant and decrease growth of aggressive tumor xenografts in mice. Proc Natl Acad Sci U S A 105:11105–11109

Cheng CL, Yang CR, Yang DY, Kuo JS, Huang YF, Cheng FC (2001) Simultaneous monitoring of extracellular glucose, pyruvate, and lactate in rat testies during ischemia: a microdialysis study. Urol Res 29:272–277

Cibicek N, Zivna H, Zadak Z, Kulir J, Cermakova E, Palicka V (2007) Colon submucosal microdialysis: a novel in vivo approach in barrier function assessment—a pilot study in rats. Physiol Res 56:611–617

Dabrosin C (2005) Microdialysis—an in vivo technique for studies of growth factors in breast cancer. Front Biosci 10:1329–1335

Dahyot C, Marchand S, Pessini GL, Pariat C, Debaene B, Couet W, Mimoz O (2006) Microdialysis study of imipenem distribution in skeletal muscle and lung extracellular fluids of Acinetobacter baumannii-infected rats. Antimicrob Agents Chemother 50:2265–2267

De La Pena A, Dalla Costa T, Talton JD, Rehak E, Gross J, Thyroff-Friesinger U, Webb AI, Muller M, Derendorf H (2001) Penetration of cefaclor into the interstitial space fluid of skeletal muscle and lung tissue in rats. Pharm Res 18:1310–1314

Deeba S, Corcoles EP, Hanna GB, Pareskevas P, Aziz O, Boutelle MG, Darzi A (2008) Use of rapid sampling microdialysis for intraoperative monitoring of bowel ischemia. Dis Colon Rectum 51:1408–1413

Dixon JA, Gaillard WF 2nd, Rivers WT, Koval CN, Stroud RE, Mukherjee R, Spinale FG (2010) Heterogeneity in MT1-MMP activity with ischemia-reperfusion and previous myocardial infarction: relation to regional myocardial function. Am J Physiol Heart Circ Physiol 299:H1947–H1958

Eisenberg EJ, Conzentino P, Eickhoff WM, Cundy KC (1993) Pharmacokinetic measurement of drugs in lung epithelial lining fluid by microdialysis: aminoglycoside antibiotics in rat bronchi. J Pharmacol Toxicol Methods 29:93–98

Ekstrom PO, Anderson A, Warren DJ, Giercksky KE, Slordal L (1995) Pharmacokinetics of different doses of methotrexate at steady state by in situ microdialysis in a rat model. Cancer Chemother Pharmacol 36:283–289

Ellebaek Pedersen M, Qvist N, Bisgaard C, Kelly U, Bernhard A, Moller Pedersen S (2009) Peritoneal microdialysis. Early diagnosis of anastomotic leakage after low anterior resection for rectosigmoid cancer. Scand J Surg 98:148–154

Elmeliegy MA, Carcaboso AM, Chow LML, Zhang ZM, Calabrese C, Throm SL, Wang F, Baker SJ, Stewart CF (2011) Magnetic resonance imaging-guided microdialysis cannula implantation in a spontaneous high-grade glioma murine model. J Pharm Sci

EMEA (2000) Points to consider on pharmacokinetics and pharmacodynamics in the development of antibacterial medicinal products. (www.emea.eu.int/pdfs/human/ewp/265599en.pdf)

Emmertsen KJ, Wara P, Soerensen FB, Stolle LB (2005) Intestinal microdialysis–applicability, reproducibility and local tissue response in a pig model. Scand J Surg 94:246–251

Guidance for industry. Developing antimicrobial drugs—general considerations for clinical trials.(www.fda.gov/cder/guidance/2580dft.pdf)

Fonouni H, Tahmasbi Rad M, Golriz M, Faridar A, Esmaeilzadeh M, Jarahian P, Hafezi M, Jafarieh S, Macher-Goeppinger S, Longerich T, Orakcioglu B, Sakowitz O, Schmidt J, Mehrabi A (2011) Using microdialysis for early detection of vascular thrombosis after kidney transplantation in an experimental porcine model. Nephrol Dial Transplant 27:541–547

Fujii T, Sugiura T, Ohkita M, Kobuchi S, Takaoka M, Matsumura Y (2007) Selective antagonism of the postsynaptic alpha(1)-adrenoceptor is protective against ischemic acute renal failure in rats. Eur J Pharmacol 574:185–191

Fujii T, Yamazaki T, Akiyama T, Sano S, Mori H (2004) Extraneuronal enzymatic degradation of myocardial interstitial norepinephrine in the ischemic region. Cardiovasc Res 64:125–131

Gail DB, Lenfant CJ (1983) Cells of the lung: biology and clinical implications. Am Rev Respir Dis 127:366–387

Goodman JC (2011) Clinical microdialysis in neuro-oncology: principles and applications. Chin J Cancer 30:173–181

Grubb BR, Jones JH, Boucher RC (2004) Mucociliary transport determined by in vivo microdialysis in the airways of normal and CF mice. Am J Physiol Lung Cell Mol Physiol 286:L588–L595

Hayashi K, Acosta TJ, Berisha B, Kobayashi S, Ohtani M, Schams D, Miyamoto A (2003a) Changes in prostaglandin secretion by the regressing bovine corpus luteum. Prostaglandins Other Lipid Mediat 70:339–349

Hayashi K, Miyamoto A, Konari A, Ohtani M, Fukui Y (2003b) Effect of local interaction of reactive oxygen species with prostaglandin F(2alpha) on the release of progesterone in ovine corpora lutea in vivo. Theriogenology 59:1335–1344

Hayashi KG, Acosta TJ, Tetsuka M, Berisha B, Matsui M, Schams D, Ohtani M, Miyamoto A (2003c) Involvement of angiopoietin-tie system in bovine follicular development and atresia: messenger RNA expression in theca interna and effect on steroid secretion. Biol Reprod 69:2078–2084

Herkner H, Muller MR, Kreischitz N, Mayer BX, Frossard M, Joukhadar C, Klein N, Lackner E, Muller M (2002) Closed-chest microdialysis to measure antibiotic penetration into human lung tissue. Am J Respir Crit Care Med 165:273–276

Homapour B, Bowen JE, Want EJ, O'Neill K, Apostolopoulos V, Nandi D, Van Dellen JR, Roncaroli F (2010) Intra-operative, real-time, three-dimensional ultrasound assisted positioning of catheters in the microdialysis of glial tumours. J Clin Neurosci 17:506–510

Hrydziuszko O, Silva MA, Perera MT, Richards DA, Murphy N, Mirza D, Viant MR (2010) Application of metabolomics to investigate the process of human orthotopic liver transplantation: a proof-of-principle study. OMICS 14:143–150

Hunz M, Jetter A, Warm M, Pantke E, Tuscher M, Hempel G, Jaehde U, Untch M, Kurbacher C, Fuhr U (2007) Plasma and tissue pharmacokinetics of epirubicin and paclitaxel in patients receiving neoadjuvant chemotherapy for locally advanced primary breast cancer. Clin Pharmacol Ther 81:659–668

Hutschala D, Kinstner C, Skhirtladze K, Mayer-Helm BX, Zeitlinger M, Wisser W, Muller M, Tschernko E (2008) The impact of perioperative atelectasis on antibiotic penetration into lung tissue: an in vivo microdialysis study. Intensive Care Med 34:1827–1834

Hutschala D, Skhirtladze K, Zuckermann A, Wisser W, Jaksch P, Mayer-Helm BX, Burgmann H, Wolner E, Muller M, Tschernko EM (2005) In vivo measurement of levofloxacin penetration into lung tissue after cardiac surgery. Antimicrob Agents Chemother 49:5107–5111

Ingvast-Larsson C, Appelgren LE, Nyman G (1992) Distribution studies of theophylline: microdialysis in rat and horse and whole body autoradiography in rat. J Vet Pharmacol Ther 15:386–394

Isaksson B, D'Souza AM, Jersenius U, Ungerstedt J, Lundell L, Permert J, Bjornstedt M, Nowak G (2011) Continuous assessment of intrahepatic metabolism by microdialysis during and after portal triad clamping. J Surg Res 169:214–219

Iwashita H, Yoshida M, Nishi T, Otani M, Ueda S (2004) In vivo transfer of a neuronal nitric oxide synthase expression vector into the rat bladder by electroporation. BJU Int 93:1098–1103

Jansson K, Strand I, Redler B, Magnuson A, Ungerstedt U, Norgren L (2004) Results of intraperitoneal microdialysis depend on the location of the catheter. Scand J Clin Lab Invest 64:63–70

Jansson K, Ungerstedt J, Jonsson T, Redler B, Andersson M, Ungerstedt U, Norgren L (2003) Human intraperitoneal microdialysis: increased lactate/pyruvate ratio suggests early visceral ischaemia. A pilot study. Scand J Gastroenterol 38:1007–1011

Jaroszewski JJ, Hansel W (2000) Intraluteal administration of a nitric oxide synthase blocker stimulates progesterone and oxytocin secretion and prolongs the life span of the bovine corpus luteum. Proc Soc Exp Biol Med 224:50–55

Jaroszewski JJ, Skarzynski DJ, Hansel W (2003) Nitric oxide as a local mediator of prostaglandin F2alpha-induced regression in bovine corpus luteum: an in vivo study. Exp Biol Med (Maywood) 228:1057–1062

Johansen MJ, Thapar N, Newman RA, Madden T (2002) Use of microdialysis to study platinum anticancer agent pharmacokinetics in preclinical models. J Exp Ther Oncol 2:163–173

Jonsson P, Borgstrom A, Ohlsson K (1992) Measurements of exocrine proteins in the pig pancreas using microdialysis. Gastroenterol Jpn 27:529–535

Jonsson P, Ohlsson K (1995) Intrapancreatic turnover of recombinant human pancreatic secretory trypsin inhibitor in experimental porcine pancreatitis. Scand J Clin Lab Invest 55:223–227

Kakoki M, Kim HS, Arendshorst WJ, Mattson DL (2004) L-Arginine uptake affects nitric oxide production and blood flow in the renal medulla. Am J Physiol Regul Integr Comp Physiol 287:R1478–R1485

Kalenikova EI, Gorodetskaya EA, Murashev AN, Ruuge EK, Medvedev OS (2004) Role of reactive oxygen species in the sensitivity of rat hypertrophied myocardium to ischemia. Biochemistry (Mosc) 69:311–316

Keller AK, Jorgensen TM, Olsen LH, Stolle LB (2008) Early detection of renal ischemia by in situ microdialysis: an experimental study. J Urol 179:371–375

Keller AK, Jorgensen TM, Ravlo K, Nielsen TK, Olsen LH, Stolle LB (2009) Microdialysis for detection of renal ischemia after experimental renal transplantation. J Urol 182:1854–1859

Kitagawa H, Yamazaki T, Akiyama T, Sugimachi M, Sunagawa K, Mori H (2005) Microdialysis separately monitors myocardial interstitial myoglobin during ischemia and reperfusion. Am J Physiol Heart Circ Physiol 289:H924–H930

Kitano M, Sakamoto H, Das K, Komaki T, Kudo M (2010) EUS-guided in vivo microdialysis of the pancreas: a novel technique with potential diagnostic and therapeutic application. Gastrointest Endosc 71:176–179

Kitzen JJ, Verweij J, Wiemer EA, Loos WJ (2006) The relevance of microdialysis for clinical oncology. Curr Clin Pharmacol 1:255–263

Konety BR, Somogyi G, Atan A, Muindi J, Chancellor MB, Getzenberg RH (2002) Evaluation of intraprostatic metabolism of 1,25-dihydroxyvitamin D(3) (calcitriol) using a microdialysis technique. Urology 59:947–952

Konings IR, Engels FK, Sleijfer S, Verweij J, Wiemer EA, Loos WJ (2009) Application of prolonged microdialysis sampling in carboplatin-treated cancer patients. Cancer Chemother Pharmacol 64:509–516

Konings IR, Sleijfer S, Mathijssen RH, de Bruijn P, Ghobadi Moghaddam-Helmantel IM, van Dam LM, Wiemer EA, Verweij J, Loos WJ (2011) Increasing tumoral 5-fluorouracil concentrations during a 5-day continuous infusion: a microdialysis study. Cancer Chemother Pharmacol 67:1055–1062

Korth U, Krieter H, Denz C, Janke C, Ellinger K, Bertsch T, Henn C, Klein J (2003) Intestinal ischaemia during cardiac arrest and resuscitation: comparative analysis of extracellular metabolites by microdialysis. Resuscitation 58:209–217

Krarup PM, Stolle LB, Rawashdeh YF, Skott O, Djurhuus JC, Froekiaer J (2007) Regional changes in renal cortical glucose, lactate and urea during acute unilateral ureteral obstruction. Scand J Urol Nephrol 41:47–53

Larsson CI (1991) The use of an "internal standard" for control of the recovery in microdialysis. Life Sci 49:PL73–PL78

Licht P, Russu V, Lehmeyer S, Wildt L (2001) Molecular aspects of direct LH/hCG effects on human endometrium–lessons from intrauterine microdialysis in the human female in vivo. Reprod Biol 1:10–19

Licht P, Russu V, Lehmeyer S, Wissentheit T, Siebzehnrubl E, Wildt L (2003) Cycle dependency of intrauterine vascular endothelial growth factor levels is correlated with decidualization and corpus luteum function. Fertil Steril 80:1228–1233

Liu P, Fuhrherr R, Webb AI, Obermann B, Derendorf H (2005) Tissue penetration of cefpodoxime into the skeletal muscle and lung in rats. Eur J Pharm Sci 25:439–444

Liu P, Muller M, Grant M, Webb AI, Obermann B, Derendorf H (2002) Interstitial tissue concentrations of cefpodoxime. J Antimicrob Chemother 50(Suppl):19–22

Lofgren B, Povlsen JA, Rasmussen LE, Stottrup NB, Solskov L, Krarup PM, Kristiansen SB, Botker HE, Nielsen TT (2010) Amino acid transamination is crucial for ischaemic cardioprotection in normal and preconditioned isolated rat hearts–focus on L-glutamate. Exp Physiol 95:140–152

Loos WJ, Zamboni WC, Engels FK, de Bruijn P, Lam MH, de Wit R, Verweij J, Wiemer EA (2007) Pitfalls of the application of microdialysis in clinical oncology: controversial findings with docetaxel. J Pharm Biomed Anal 45:288–294

Ma J, Pulfer S, Li S, Chu J, Reed K, Gallo JM (2001) Pharmacodynamic-mediated reduction of temozolomide tumor concentrations by the angiogenesis inhibitor TNP-470. Cancer Res 61:5491–5498

Mantovani V, Kennergren C, Bugge M, Sala A, Lonnroth P, Berglin E (2010) Myocardial metabolism assessed by microdialysis: a prospective randomized study in on- and off-pump coronary bypass surgery. Int J Cardiol 143:302–308

Mantovani V, Kennergren C, Goiny M, Ungerstedt U, Lonnroth P, Sala A, Berglin E (2006) Microdialysis for myocardial metabolic surveillance: developing a clinical technique. Clin Physiol Funct Imaging 26:224–231

Marchand S, Dahyot C, Lamarche I, Mimoz O, Couet W (2005) Microdialysis study of imipenem distribution in skeletal muscle and lung extracellular fluids of noninfected rats. Antimicrob Agents Chemother 49:2356–2361

Marchand S, Frasca D, Dahyot-Fizelier C, Breheret C, Mimoz O, Couet W (2008) Lung microdialysis study of levofloxacin in rats following intravenous infusion at steady state. Antimicrob Agents Chemother 52:3074–3077

Marcus HJ, Carpenter KL, Price SJ, Hutchinson PJ (2010) In vivo assessment of high-grade glioma biochemistry using microdialysis: a study of energy-related molecules, growth factors and cytokines. J Neurooncol 97:11–23

Martin C, Schulz R, Post H, Boengler K, Kelm M, Kleinbongard P, Gres P, Skyschally A, Konietzka I, Heusch G (2007) Microdialysis-based analysis of interstitial NO in situ: NO synthase-independent NO formation during myocardial ischemia. Cardiovasc Res 74:46–55

Martins FC, de Oliveira CF (2009) Chemotherapy and the future: microdialysis as a local administration technique. Eur J Gynaecol Oncol 30:5–8

Martins FC, Santos JL, Oliveira CF (2012) Microdialysis: improving local chemotherapy in cancer using a mathematical model. Front Biosci (Elite Ed) 4:401–409

Masunaga K, Yoshida M, Inadome A, Iwashita H, Miyamae K, Ueda S (2006) Prostaglandin E2 release from isolated bladder strips in rats with spinal cord injury. Int J Urol 13:271–276

Masunaga K, Yoshida M, Inadome A, Murakami S, Sugiyama Y, Satoji Y, Maeda Y, Ueda S (2008) Pharmacological effects of solifenacin on human isolated urinary bladder. Pharmacology 82:43–52

Matthiessen P, Strand I, Jansson K, Tornquist C, Andersson M, Rutegard J, Norgren L (2007) Is early detection of anastomotic leakage possible by intraperitoneal microdialysis and intraperitoneal cytokines after anterior resection of the rectum for cancer? Dis Colon Rectum 50:1918–1927

Mazzoni M, Bortone F, Ceriani R, Caprioli G, Passaretti B, Arena V, Parodi O (2008) Metabolism of hibernating myocardium assessed by microdialysis during surgical revascularization: report of a case. J Card Surg 23:374–376

Merilainen S, Makela J, Anttila V, Koivukangas V, Kaakinen H, Niemela E, Ohtonen P, Risteli J, Karttunen T, Soini Y, Juvonen T (2008) Acute edematous and necrotic pancreatitis in a porcine model. Scand J Gastroenterol 43:1259–1268

Miyamae K, Yoshida M, Inadome A, Murakami S, Otani M, Iwashita H, Masunaga K, Ueda S (2004) Acetylcholine release from urinary bladder smooth muscles of non-insulin-dependent diabetic rats. Urol Int 73:74–80

Miyamoto A, Shirasuna K, Wijayagunawardane MP, Watanabe S, Hayashi M, Yamamoto D, Matsui M, Acosta TJ (2005) Blood flow: a key regulatory component of corpus luteum function in the cow. Domest Anim Endocrinol 29:329–339

Multani MM, Ikonomidis JS, Kim PY, Miller EA, Payne KJ, Mukherjee R, Dorman BH, Spinale FG (2005) Dynamic and differential changes in myocardial and plasma endothelin in patients undergoing cardiopulmonary bypass. J Thorac Cardiovasc Surg 129:584–590

Multani MM, Krombach RS, Goldberg AT, King MK, Hendrick JW, Sample JA, Baicu SC, Joffs C, deGasparo M, Spinale FG (2001) Myocardial bradykinin following acute angiotensin-converting enzyme inhibition, AT1 receptor blockade, or combined inhibition in congestive heart failure. J Cardiovasc Pharmacol Ther 6:369–376

Murakami S, Yoshida M, Iwashita H, Otani M, Miyamae K, Masunaga K, Miyamoto Y, Inadome A, Ueda S (2003) Pharmacological effects of KRP-197 on the human isolated urinary bladder. Urol Int 71:290–298

Murakami S, Yoshida M, Masunaga K, Maeda Y, Ueda S (2008) Change in acetylcholine release from rat bladder with partial outlet obstruction. BJU Int 101:633–639

Myers SI, Wang L, Liu F, Bartula LL (2006) Oxygen-radical regulation of renal blood flow following suprarenal aortic clamping. J Vasc Surg 43:577–586

Nakagawa A, Samols E, Stagner JI (1993) Exocrine interstitial insulin and somatostatin in the perfused dog pancreas. Am J Physiol 264:G728–G734

Nishiyama A, Jackson KE, Majid DS, Rahman M, Navar LG (2006) Renal interstitial fluid ATP responses to arterial pressure and tubuloglomerular feedback activation during calcium channel blockade. Am J Physiol Heart Circ Physiol 290:H772–H777

Nowak G, Ungerstedt J, Wernerman J, Ungerstedt U, Ericzon BG (2002) Clinical experience in continuous graft monitoring with microdialysis early after liver transplantation. Br J Surg 89:1169–1175

Obata T (2001) In vivo microdialysis technique on rat myocardium. Nihon Yakurigaku Zasshi 118:379–382

Obata T, Yonemoti H, Aomine M (2009) The protective effect of fluvastatin on hydroxyl radical generation by inhibiting low-density lipoprotein (LDL) oxidation in the rat myocardium. Microvasc Res 77:163–165

Ohtani M, Takase S, Wijayagunawardane MP, Tetsuka M, Miyamoto A (2004) Local interaction of prostaglandin F 2alpha with endothelin-1 and tumor necrosis factor-alpha on the release of progesterone and oxytocin in ovine corpora lutea in vivo: a possible implication for a luteolytic cascade. Reproduction 127:117–124

Okawa T, Asano K, Takahashi H, Sato A, Vedernikov YP, Saade GR, Gafield RE (2005) Nitric oxide donor-induced inhibition of pregnant rat uterine spontaneous contractile activity and release of nitric oxide from uterus measured by microdialysis. J Endocrinol Invest 28:998–1002

Okuda K, Uenoyama Y, Berisha B, Lange IG, Taniguchi H, Kobayashi S, Miyamoto A, Schams D (2001) Estradiol-17beta is produced in bovine corpus luteum. Biol Reprod 65:1634–1639

Oldner A, Goiny M, Ungerstedt U, Sollevi A (1996) Splanchnic homeostasis during endotoxin challenge in the pig as assessed by microdialysis and tonometry. Shock 6:188–193

Pedersen ME, Dahl M, Qvist N (2011) Intraperitoneal microdialysis in the postoperative surveillance after surgery for necrotizing enterocolitis: a preliminary report. J Pediatr Surg 46:352–356

Poling J, Leptien A, Klaus S, Rees W, Kraatz E, Wiebe K, Bahlmann L (2007a) Analysis of the myocardial metabolism by microdialysis during open beating heart surgery. Scand Cardiovasc J 41:114–119

Poling J, Rees W, Klaus S, Bahlmann L, Hubner N, Heringlake M, Mantovani V, Warnecke H (2008) Functional recovery of chronic ischemic myocardium after surgical revascularization correlates with magnitude of oxidative metabolism. Cardiology 110:174–181

Poling J, Rees W, Klaus S, Bahlmann L, Hubner N, Mantovani V, Warnecke H (2007b) Myocardial metabolic monitoring with the microdialysis technique during and after open heart surgery. Acta Anaesthesiol Scand 51:341–346

Poling J, Rees W, Mantovani V, Klaus S, Bahlmann L, Ziaukas V, Hubner N, Warnecke H (2006) Evaluation of myocardial metabolism with microdialysis during bypass surgery with cold blood- or calafiore cardioplegia. Eur J Cardiothorac Surg 30:597–603

Price KE, Vandaveer SS, Lunte CE, Larive CK (2005) Tissue targeted metabonomics: metabolic profiling by microdialysis sampling and microcoil NMR. J Pharm Biomed Anal 38:904–909

Ren YC, Wang L, He HB, Tang X (2009) Pulmonary selectivity and local pharmacokinetics of ambroxol hydrochloride dry powder inhalation in rat. J Pharm Sci 98:1797–1803

Robinson M, Bye PT (2002) Mucociliary clearance in cystic fibrosis. Pediatr Pulmonol 33:293–306

Runge M, Hughes P, Peter Gotze J, Petersen RH, Steinbruchel DA (2006) Evaluation of myocardial metabolism with microdialysis after protection with cold blood- or cold crystalloid cardioplegia. A porcine model. Scand Cardiovasc J 40:186–193

Sani SN, Henry K, Bohlke M, Kim J, Stricker-Krongrad A, Maher TJ (2010) The effects of drug transporter inhibitors on the pharmacokinetics and tissue distribution of methotrexate in normal and tumor-bearing mice: a microdialysis study. Cancer Chemother Pharmacol 66:159–169

Shaw DW, Britt JH (2000) In vivo oxytocin release from microdialyzed bovine corpora lutea during spontaneous and prostaglandin-induced regression. Biol Reprod 62:726–730

Shibata Y, Takaoka M, Maekawa D, Kuwahara C, Matsumura Y (2004) Involvement of nitric oxide in the suppressive effect of 17beta-estradiol on endothelin-1 overproduction in ischemic acute renal failure. J Cardiovasc Pharmacol 44(Suppl 1):S459–S461

Shimizu N, Hori T, Nakane H (1994) An interleukin-1 beta-induced noradrenaline release in the spleen is mediated by brain corticotropin-releasing factor: an in vivo microdialysis study in conscious rats. Brain Behav Immun 8:14–23

Shimizu N, Kaizuka Y, Hori T, Nakane H (1996) Immobilization increases norepinephrine release and reduces NK cytotoxicity in spleen of conscious rat. Am J Physiol 271:R537–R544

Shirasuna K, Asaoka H, Acosta TJ, Wijayagunawardane MP, Matsui M, Ohtani M, Miyamoto A (2004a) Endothelin-1 within the corpus luteum during spontaneous luteolysis in the cow: local interaction with prostaglandin F2alpha and angiotensin II. J Cardiovasc Pharmacol 44(Suppl 1):S252–S255

Shirasuna K, Asaoka H, Acosta TJ, Wijayagunawardane MP, Ohtani M, Hayashi KG, Matsui M, Miyamoto A (2004b) Real-time dynamics of prostaglandin F2alpha release from uterus and corpus luteum during spontaneous luteolysis in the cow. Reproduction 128:189–195

Shirasuna K, Asaoka H, Acosta TJ, Wijayagunawardane MP, Ohtani M, Hayashi M, Matsui M, Miyamoto A (2004c) Real-time relationships in intraluteal release among prostaglandin F2alpha, endothelin-1, and angiotensin II during spontaneous luteolysis in the cow. Biol Reprod 71:1706–1711

Shirasuna K, Shimizu T, Hayashi KG, Nagai K, Matsui M, Miyamoto A (2007) Positive association, in local release, of luteal oxytocin with endothelin 1 and prostaglandin F2alpha during spontaneous luteolysis in the cow: a possible intermediatory role for luteolytic cascade within the corpus luteum. Biol Reprod 76:965–970

Siaghy EM, Devaux Y, Sfaksi N, Carteaux JP, Ungureanu-Longrois D, Zannad F, Villemot JP, Burlet C, Mertes PM (2000) Consequences of inspired oxygen fraction manipulation on myocardial oxygen pressure, adenosine and lactate concentrations: a combined myocardial microdialysis and sensitive oxygen electrode study in pigs. J Mol Cell Cardiol 32:493–504

Silva MA, Mirza DF, Murphy N, Richards DA, Reynolds GM, Wigmore SJ, Neil DA (2008) Intrahepatic complement activation, sinusoidal endothelial injury, and lactic acidosis are associated with initial poor function of the liver after transplantation. Transplantation 85:718–725

Silva MA, Murphy N, Richards DA, Wigmore SJ, Bramhall SR, Buckels JA, Adams DH, Mirza DF (2006) Interstitial lactic acidosis in the graft during organ harvest, cold storage, and reperfusion of human liver allografts predicts subsequent ischemia reperfusion injury. Transplantation 82:227–233

Silva MA, Richards DA, Bramhall SR, Adams DH, Mirza DF, Murphy N (2005) A study of the metabolites of ischemia-reperfusion injury and selected amino acids in the liver using microdialysis during transplantation. Transplantation 79:828–835

Solligard E, Juel IS, Spigset O, Romundstad P, Gronbech JE, Aadahl P (2008) Gut luminal lactate measured by microdialysis mirrors permeability of the intestinal mucosa after ischemia. Shock 29:245–251

Solligard E, Wahba A, Skogvoll E, Stenseth R, Gronbech JE, Aadahl P (2007) Rectal lactate levels in endoluminal microdialysate during routine coronary surgery. Anaesthesia 62:250–258

Sommer T (2005) Microdialysis of the bowel: the possibility of monitoring intestinal ischemia. Expert Rev Med Devices 2:277–286

Sommer T, Larsen JF (2003) Detection of intestinal ischemia using a microdialysis technique in an animal model. World J Surg 27:416–420

Sorensen HB (2008) Free jejunal flaps can be monitored by use of microdialysis. J Reconstr Microsurg 24:443–448

Spinale FG, Koval CN, Deschamps AM, Stroud RE, Ikonomidis JS (2008) Dynamic changes in matrix metalloprotienase activity within the human myocardial interstitium during myocardial arrest and reperfusion. Circulation 118:S16–S23

Stock S, Fastbom J, Bjorkstrand E, Ungerstedt U, Uvnas-Moberg K (1990) Effects of oxytocin on in vivo release of insulin and glucagon studied by microdialysis in the rat pancreas and autoradiographic evidence for [3H] oxytocin binding sites within the islets of Langerhans. Regul Pept 30:1–13

Stottrup NB, Lofgren B, Birkler RD, Nielsen JM, Wang L, Caldarone CA, Kristiansen SB, Contractor H, Johannsen M, Botker HE, Nielsen TT (2010) Inhibition of the malate-aspartate shuttle by pre-ischaemic aminooxyacetate loading of the heart induces cardioprotection. Cardiovasc Res 88:257–266

Suzuki H, Iijima K, Scobie G, Fyfe V, McColl KE (2005) Nitrate and nitrosative chemistry within Barrett's oesophagus during acid reflux. Gut 54:1527–1535

Tan Q, El-Badry AM, Contaldo C, Steiner R, Hillinger S, Welti M, Hilbe M, Spahn DR, Jaussi R, Higuera G, van Blitterswijk CA, Luo Q, Weder W (2009) The effect of perfluorocarbon-based artificial oxygen carriers on tissue-engineered trachea. Tissue Eng Part A 15:2471–2480

Tanaka J, Acosta TJ, Berisha B, Tetsuka M, Matsui M, Kobayashi S, Schams D, Miyamoto A (2004) Relative changes in mRNA expression of angiopoietins and receptors tie in bovine corpus luteum during estrous cycle and prostaglandin F2alpha-induced luteolysis: a possible mechanism for the initiation of luteal regression. J Reprod Dev 50:619–626

Tasso L, Bettoni CC, Oliveira LK, Dalla Costa T (2008) Evaluation of gatifloxacin penetration into skeletal muscle and lung by microdialysis in rats. Int J Pharm 358:96–101

Tegeder I, Brautigam L, Seegel M, Al-Dam A, Turowski B, Geisslinger G, Kovacs AF (2003) Cisplatin tumor concentrations after intra-arterial cisplatin infusion or embolization in patients with oral cancer. Clin Pharmacol Ther 73:417–426

Tenhunen JJ, Kosunen H, Alhava E, Tuomisto L, Takala JA (1999) Intestinal luminal microdialysis: a new approach to assess gut mucosal ischemia. Anesthesiology 91:1807–1815

Thompson JF, Siebert GA, Anissimov YG, Smithers BM, Doubrovsky A, Anderson CD, Roberts MS (2001) Microdialysis and response during regional chemotherapy by isolated limb infusion of melphalan for limb malignancies. Br J Cancer 85:157–165

Tomaselli F, Dittrich P, Maier A, Woltsche M, Matzi V, Pinter J, Nuhsbaumer S, Pinter H, Smolle J, Smolle-Juttner FM (2003) Penetration of piperacillin and tazobactam into pneumonic human lung tissue measured by in vivo microdialysis. Br J Clin Pharmacol 55:620–624

Tomaselli F, Maier A, Matzi V, Smolle-Juttner FM, Dittrich P (2004) Penetration of meropenem into pneumonic human lung tissue as measured by in vivo microdialysis. Antimicrob Agents Chemother 48:2228–2232

Uemura K, Li M, Tsutsumi T, Yamazaki T, Kawada T, Kamiya A, Inagaki M, Sunagawa K, Sugimachi M (2007) Efferent vagal nerve stimulation induces tissue inhibitor of metalloproteinase-1 in myocardial ischemia-reperfusion injury in rabbit. Am J Physiol Heart Circ Physiol 293:H2254–H2261

Ungerstedt J, Nowak G, Ericzon BG, Ungerstedt U (2003) Intraperitoneal microdialysis (IPM): a new technique for monitoring intestinal ischemia studied in a porcine model. Shock 20:91–96

Usa K, Singh RJ, Netzel BC, Liu Y, Raff H, Liang M (2007) Renal interstitial corticosterone and 11-dehydrocorticosterone in conscious rats. Am J Physiol Renal Physiol 293:F186–F192

Waelgaard L, Thorgersen EB, Line PD, Foss A, Mollnes TE, Tonnessen TI (2008) Microdialysis monitoring of liver grafts by metabolic parameters, cytokine production, and complement activation. Transplantation 86:1096–1103

Waldenstrom A, Haney M, Biber B, Kavianipour M, Moritz T, Stranden P, Wikstrom G, Ronquist G (2010) Ischaemic preconditioning is related to decreasing levels of extracellular adenosine that may be metabolically useful in the at-risk myocardium: an experimental study in the pig. Acta Physiol (Oxf) 199:1–9

Wang Z, Colli JL, Keel C, Bailey K, Grossman L, Majid D, Lee BR (2012) Isoprostane: quantitation of renal ischemia and reperfusion injury after renal artery clamping in an animal model. J Endourol 26:21–25

Wei YH, Xu LZ, Shen Q, Li FZ (2009) Microdialysis a technique for pharmacokinetic-pharmacodynamic studies of oncological drugs. Curr Pharm Biotechnol 10:631–640

Weld KJ, Evearitt K, Dixon P, Cespedes RD (2009) Effects of ischemia on human renal interstitial fluid metabolites. J Urol 181:878–883

Weld KJ, Montiglio C, Bush AC, Harroff HH, Cespedes RD (2008) Real-time analysis of renal interstitial metabolites during induced renal ischemia*. J Endourol 22:571–574

Wibom C, Surowiec I, Moren L, Bergstrom P, Johansson M, Antti H, Bergenheim AT (2010) Metabolomic patterns in glioblastoma and changes during radiotherapy: a clinical microdialysis study. J Proteome Res 9:2909–2919

Wienkotter N, Begrow F, Kinzinger U, Schierstedt D, Verspohl EJ (2007) The effect of thyme extract on beta2-receptors and mucociliary clearance. Planta Med 73:629–635

Wienkotter N, Hopner D, Schutte U, Bauer K, Begrow F, El-Dakhakhny M, Verspohl EJ (2008) The effect of nigellone and thymoquinone on inhibiting trachea contraction and mucociliary clearance. Planta Med 74:105–108

Winbladh A, Bjornsson B, Trulsson L, Offenbartl K, Gullstrand P, Sandstrom P (2012) Ischemic preconditioning prior to intermittent pringle maneuver in liver resections. J Hepatobiliary Pancreat Sci 19:159–170

Wuttke W, Spiess S, Knoke I, Pitzel L, Leonhardt S, Jarry H (1998) Synergistic effects of prostaglandin F2alpha and tumor necrosis factor to induce luteolysis in the pig. Biol Reprod 58:1310–1315

Xu BJ, Yan W, Jovanovic B, Shaw AK, An QA, Eng J, Chytil A, Link AJ, Moses HL (2010) Microdialysis combined with proteomics for protein identification in breast tumor microenvironment in vivo. Cancer Microenviron 4:61–71

Yono M, Latifpour J, Yoshida M, Ueda S (2005) Age-related alterations in the biochemical and functional properties of the bladder in type 2 diabetic GK rats. J Recept Signal Transduct Res 25:147–157

Yoshida J, Aikawa K, Yoshimura Y, Shishido K, Yanagida T, Yamaguchi O (2007) The effects of ovariectomy and estrogen replacement on acetylcholine release from nerve fibres and passive stretch-induced acetylcholine release in female rat bladder. Neurourol Urodyn 26:1050–1055

Yoshida M, Inadome A, Maeda Y, Satoji Y, Masunaga K, Sugiyama Y, Murakami S (2006) Non-neuronal cholinergic system in human bladder urothelium. Urology 67:425–430

Yoshida M, Inadome A, Murakami S (2003) Pharmacological analysis of neurotransmitters contributing to lower urinary tract function. Nihon Yakurigaku Zasshi 121:307–316

Yoshida M, Iwashita H, Otani M, Masunaga K, Inadome A (2008a) Delivery of DNA into bladder via electroporation. Methods Mol Biol 423:249–257

Yoshida M, Masunaga K, Satoji Y, Maeda Y, Nagata T, Inadome A (2008b) Basic and clinical aspects of non-neuronal acetylcholine: expression of non-neuronal acetylcholine in urothelium and its clinical significance. J Pharmacol Sci 106:193–198

Zamboni WC, Houghton PJ, Hulstein JL, Kirstein M, Walsh J, Cheshire PJ, Hanna SK, Danks MK, Stewart CF (1999) Relationship between tumor extracellular fluid exposure to topotecan and tumor response in human neuroblastoma xenograft and cell lines. Cancer Chemother Pharmacol 43:269–276

Zeitlinger MA, Traunmuller F, Abrahim A, Muller MR, Erdogan Z, Muller M, Joukhadar C (2007) A pilot study testing whether concentrations of levofloxacin in interstitial space fluid of soft tissues may serve as a surrogate for predicting its pharmacokinetics in lung. Int J Antimicrob Agents 29:44–50

Zhong H, Han B, Tourkova IL, Lokshin A, Rosenbloom A, Shurin MR, Shurin GV (2007) Low-dose paclitaxel prior to intratumoral dendritic cell vaccine modulates intratumoral cytokine network and lung cancer growth. Clin Cancer Res 13:5455–5462

Biography

Prof. Markus Müller, MD is Head of the Department of Clinical Pharmacology at the Medical University of Vienna in Austria. He is specialized in Internal Medicine and Clinical Pharmacology and is an international expert in clinical research. He has published over 80 articles on the use and methodology of microdialysis in clinical drug development.

M. Müller (ed.), *Microdialysis in Drug Development*, AAPS Advances in the Pharmaceutical Sciences Series, DOI: 10.1007/978-1-4614-4815-0,

Index

M. Müller (ed.), *Microdialysis in Drug Development*, AAPS Advances in the Pharmaceutical Sciences Series, DOI: 10.1007/978-1-4614-4815-0,

MIX
Papier aus verantwortungsvollen Quellen
Paper from responsible sources
FSC® C105338

If you have any concerns about our products,
you can contact us on
ProductSafety@springernature.com

In case Publisher is established outside the EU,
the EU authorized representative is:
Springer Nature Customer Service Center GmbH
Europaplatz 3, 69115 Heidelberg, Germany

Printed by Libri Plureos GmbH
in Hamburg, Germany